Optimization and Optimal Control

Series on Computers and Operations Research

Series Editor: P. M. Pardalos *(University of Florida)*

Published

Vol. 1 Optimization and Optimal Control
eds. P. M. Pardalos, I. Tseveendorj and R. Enkhbat

Series on Computers and Operations Research Vol. 1

Optimization and Optimal Control

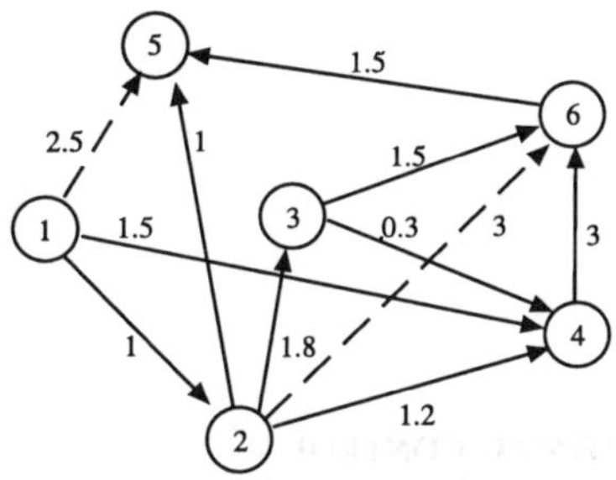

Editors

Panos M. Pardalos
Department of Industrial and Systems Engineering, University of Florida, USA

Ider Tseveendorj
Department of Mathematics, University of Orleans, France

Rentsen Enkhbat
School of Economic Studies, National University of Mongolia, Ulaanbaatar, Mongolia

World Scientific
New Jersey • London • Singapore • Hong Kong

Published by

World Scientific Publishing Co. Pte. Ltd.
5 Toh Tuck Link, Singapore 596224
USA office: Suite 202, 1060 Main Street, River Edge, NJ 07661
UK office: 57 Shelton Street, Covent Garden, London WC2H 9HE

British Library Cataloguing-in-Publication Data
A catalogue record for this book is available from the British Library.

OPTIMIZATION AND OPTIMAL CONTROL
Series on Computers and Operations Research — Vol. 1

ISBN 981-238-597-5

Printed in Singapore by World Scientific Printers (S) Pte Ltd

In memory of a dear teacher and friend
Oleg V. Vasiliev

Preface

Numerous mathematical models in many diverse fields of science and engineering lead to the formulation of optimization or optimal control problems. In optimization we often solve problems in finite dimensional spaces. On the other hand, in optimal control problems we often deal with infinite dimensional spaces where the solution is described by a curve instead of a finite vector. Optimization and optimal control has flourished in the last half century. In this book we cover a small subset of recent important topics in optimization and optimal control with emphasis on recent theoretical developments and applications.

An international conference on *Optimization and Optimal Control* was held during August 13-17, 2002 at Ulaanbaatar, Mongolia. The conference was endorsed by France Télécom R&D, the Mongolian Mathematical Society, the Mongolian Academy of Sciences, the Mongolian State University, DAAD and the Center for Applied Optimization at the University of Florida. The conference provided a forum for researchers working on different aspects of optimization and optimal control to present their recent discoveries, and to initiate scientific collaborations. This volume contains a selective collection of carefully refereed papers based on talks presented at this conference. The two themes of *optimization* and *optimal control* pervade this book.

The target audience of the book includes graduate students in engineering and operations research, academic researchers, as well as practitioners, who use optimization and optimal control methods for their specific needs and applications.

We would like to take this opportunity to thank the sponsors, the authors of the papers, the anonymous referees, and World Scientific, for making the publication of this book possible.

P.M. Pardalos, I. Tseveendorj, and R. Enkhbat
June 2003

Contents

Optimization and Optimal Control

Optimization and Optimal Control, pp. 1 – 28
P.M. Pardalos, I. Tseveendorj, and R. Enkhbat, Editors

Extragradient approach to the solution of two person non-zero sum games

Anatoly Antipin (antipin@ccas.ru)
Computing Center of Russian Academy of Sciences,
Vavilov str., 40, 119991 Moscow, Russia

Abstract.

The two person non-zero sum game is considered in the following statements: in classical form, under the availability of inequality-constraints and coupled constraints. A extragradient method for computing the Nash equilibria for these statements is suggested and its convergence is investigated.

Keywords: Non-zero sum game, extragradient method, coupled constraints, convergence.

1 Introduction

Let $\Omega = X_1 \times X_2$ be a rectangle where $X_1 \in R_1^{n_1}, X_2 \in R_2^{n_2}$ are convex closed sets from finite-dimensional Euclidean spaces, generally speaking, various dimensionality. Let functions $f_1(x_1, x_2) + \varphi_1(x_1), f_2(x_1, x_2) + \varphi_2(x_2)$ be determined on the product space $R_1^{n_1} \times R_2^{n_2}$. Consider an extreme mapping $Y(x) = y_1(x_2) \times y_2(x_1)$, which maps any point $x = (x_1, x_2) \in \Omega$ to some convex closed subset from Ω

$$y_1(x_2) \in \text{Argmin}\{f_1(z_1, x_2) + \varphi_1(z_1) \mid z_1 \in X_1\},$$

$$y_2(x_1) \in \text{Argmin}\{f_2(x_1, z_2) + \varphi_2(z_2) \mid z_2 \in X_2\}. \tag{1}$$

The subset $Y(x)$ represents by itself the direct product of optimal solution sets for problem (1). If functions $f_1(z_1, x_2) + \varphi(z_1)$, $f_2(x_1, z_2) + \varphi(z_2)$ are continuous and convex in these own variables, i.e. the first function is convex in z_1, the second one is convex in z_2 for any x_1 and x_2, $X_i, i = 1, 2$ are convex compact sets, then there exists a fixed point $x^* = (x_1^*, x_2^*)$ of this mapping [10]. This point satisfies the system of extreme inclusions

$$x_1^* \in \text{Argmin}\{f_1(z_1, x_2^*) + \varphi_1(z_1) \mid z_1 \in X_1\},$$
$$x_2^* \in \text{Argmin}\{f_2(x_1^*, z_2) + \varphi_2(z_2) \mid z_2 \in X_2\}. \qquad (2)$$

The problem of computing a fixed point (2) was formulated by Nash in 1950 as a two person game. It is a particular case of a n-person game. In this game the set $X_i, i = 1, 2$ is interpreted as the strategy set of $i-$th player, where $z_i \in X_i$ is a individual strategy of the player, $f_1(z_1, x_2) + \varphi_1(z_1)$ and $f_2(x_1, z_2) + \varphi_2(z_2)$ is a payoff function of $i-$th player. A characteristic sense of a fixed point x^* of (2) is that any player is not interested to disturb the equilibrium state as anybody of them cannot reduce the value of its payoff function in the one-sided order. Any solution of game (2) we shall call also as equilibrium solution or equilibrium. This solution, in particular, means a condition of compromise with a summarized prize $f_1(x_1^*, x_2^*)+\varphi_1(x_1^*)+f_2(x_1^*, x_2^*)+\varphi_2(x_2^*)$. After publication [14] (Nash, 1950) the efforts of researchers were undertaken to develop solution methods for games. As a result the methods of solving for two person zero-sum game were found out. Formally this game can be reduced to a saddle point problem and, therefore, saddle point methods provide some tools to solve a zero-sum game. We mark some papers in this topic. First of all it is the iterative method for an evaluation of value of matrix game. The method was offered by Brown 1951 [9], the convergence of it was proved by Robinson 1951 [16]. Bakushinsky and Goncharsky 1994 [8] extended the approach to convex-concave games and proved convergence of it for two person zero-sum games. The number of the approaches uses idea of an transformation of game to other type of problems. For example, Hansen and Scarf 1974 [11] approximated a game by means of some mapping to calculate a fixed point of it with the help of simplicial partitions of a set. Lemke and Hawson 1964 [12] reduced the game to complementarity problem and applied the pivoting-approach to solve it. However from a point of view of game mathematical modelling the game methods look more attractive on a comparison with approaches of a reduction them to other types of problems. We pick out another paper, Mills 1960 [13], where a bimatrix game is reduced to non-convex quadratic programming problem. In the present article the gradient descent idea for the solution of non-zero sum game is used. Earlier, this gradient idea but in other form was considered and justified for potential game by Rosen 1965 [17]. The outcomes obtained there follow as a particular case from outcomes of the present paper. It is known, Nikaido and Isoda 1955 [15] that any game always can be presented in equivalent form of computing a fixed point of extreme inclusion induced by normalized function of two vectorial variables of the same dimensionality. If the initial problem was a zero-sum game, then the normalized function is antisymmetric. In this paper the sizeable extension of class of games with an antisymmetric normalized function (i.e. zero-sum games) is introduced up to a game class with positive semidefinite (skew-symmetric) normalized function. The introdiced class of game problems includes itself all zero-sum games and this class can be considered as analog of convex programming in the class of nonlinear programming problems. The theory of solution methods for bilinear games is developed and presented in Antipin 2002 [7].

2 Discussion of the problem

Problem (2) represents a system of extreme inclusions, which it always is possible to scalarize and to present in the form of problem for computing a fixed point of extreme map. To that end we introduce a normalized function of the kind

$$\Phi(v,w)+\varphi(w)=f_1(z_1,x_2)+\varphi_1(z_1)+f_2(x_1,z_2)+\varphi_2(z_2),$$

where $w=(z_1,z_2), v=(x_1,x_2),\ v,w\in\Omega=X_1\times X_2$. In terms of new variables the problem (2) can be written in the shape

$$v^*\in \text{Argmin}\{\Phi(v^*,w)+\varphi(w)\mid \quad w\in\Omega\} \tag{3}$$

or that is the same

$$\Phi(v^*,v^*)+\varphi(v^*)\le\Phi(v^*,w)+\varphi(w)\ \ \forall w\in\Omega. \tag{4}$$

It is difficult to be convinced of equivalence of problems (2) and (3). However, we present (4) as

$$\begin{gathered}f_1(x_1^*,x_2^*)+\varphi_1(x_1^*)+f_2(x_1^*,x_2^*)+\varphi_2(x_2^*)\\ \le f_1(z_1,x_2^*)+\varphi_1(z_1)+f_2(x_1^*,z_2)+\varphi_2(z_2)\ \ z_1\in X_1,z_2\in X_2.\end{gathered}$$

By virtue of the separability of function $\Phi(v,w)$ and modularity of the set Ω the last inequality can be splitted into a system of inequalities

$$f_1(x_1^*,x_2^*)+\varphi_1(x_1^*)\le f_2(z_1,x_2^*)+\varphi_1(z_1)\ \ \forall z_1\in X_1,$$

$$f_2(x_1^*,x_2^*)+\varphi_2(x_2^*)\le f_2(x_1^*,z_2)+\varphi_2(z_2)\ \ \forall z_2\in X_2,$$

i.e any fixed point (4) is the solution (2). The inverse proposition is true too. If game (2) satisfies the condition $f_1(x_1,x_2)+\varphi_1(x_1)+f_2(x_1,x_2)+\varphi_2(x_2)=0\ \forall x_1\in X_1,x_2\in X_2$, then it is called a zero sum game. It follows from this condition immediately that $f_1(x_1,x_2)=-f_2(x_1,x_2)=f(x_1,x_2),\varphi_1(x_1)=0,\varphi_2(x_1)=0\ \ \forall x_1\in X_1,x_2\in X_2$, i.e.

$$x_1^*\in\text{Argmin}\{f(z_1,x_2^*)\mid\ z_1\in X_1\}$$

$$x_2^*\in\text{Argmin}\{-f(x_1^*,z_2)\mid\ z_2\in X_2\}. \tag{5}$$

It is clear that the problem can be rewritten in the form of a system of inequalities

$$f(x_1^*,z_2)\le f(x_1^*,x_2^*)\le f(z_1,x_2^*)\ \ \forall z_1\in X_1,\ z_2\in X_2. \tag{6}$$

In this case, the pair of vectors x_1^*, x_2^* is the saddle point of $f(x_1, x_2)$ on set $X_1 \times X_2$. It is useful to mark that problem (3) also can be interpreted as two person game, where the strategies of the first player are described by means of variable $v \in \Omega$, and second one are in variable $w \in \Omega$. The choice of strategy of the first player consists in presentation of specific vector $v \in \Omega$, the response of the second player is in presentation of set $Y(v) \subset \Omega$. In this situation it is required to chose $v = v^*$ such that $v^* \in Y(v^*)$. We have proved that if $v^* = (x_1^*, x_2^*)$ is a fixed point in (3), then the pair of vectors (x_1^*, x_2^*) is the saddle point of function $f(x_1, x_2)$ in (5). Antipin 2001 [6] exhibits that the saddle point property is the key property to prove the convergence of various methods to fixed points of extreme maps. Therefore the question arises on whether can the pair v^*, v^* be a saddle point some function connected with this problem, if v^* is a solution of (3) ? To answer this question we introduce a condition, which is enough to describe the situation as a whole. A function $\Phi(v, w)$ is called positive semi-definite or skew-symmetric on $\Omega \times \Omega$ if it obeys the inequality, Antipin 1995 [1]

$$\Phi(w, w) - \Phi(w, v) - \Phi(v, w) + \Phi(v, v) \geq 0 \quad \forall v, w \in \Omega \times \Omega. \tag{7}$$

This condition can be considered, on the one hand, as generalization of the antisymmetric property, i.e. when the condition $\Phi(v, w) = -\Phi(w, v)$ is held, and on the other hand, as generalization of concept of the positive semi-definiteness of matrices. Indeed, if a function has a bilinear structure $\Phi(v, w) = \langle \Phi v, w \rangle$, where Φ is a square matrix, then condition (7) takes the form of positive semi-definiteness for a matrix, i.e. $\langle \Phi(v - w), v - w \rangle \geq 0 \quad \forall (v - w) \in R^n$. It is easy to check that if the function $\Phi(v, w)$ is positive semi-definite, then $\Phi(v, w) + \varphi(w)$ is positive semi-definite as well. Positive semi-definite condition (7) allows to mark out the class of equilibrium problems which may be considered as generalization of classes of convex and saddle point programming problems. The equilibrium problems subjected to the condition of positive semi-definiteness have the important properties, namely, if v^* is a solution of equilibrium problem (3), then the pair v^*, v^* is the saddle point of shift function $\Psi(v, w) = \Phi(v, w) + \varphi(w) - \Phi(v, v) - \varphi(v)$ (it is identically equal to zero on the diagonal of square $\Omega \times \Omega$). Indeed, if v^* is a solution, then at $v = v^*$ from (4) and (7) we have

$$\begin{aligned} &\Phi(w, w) + \varphi(w) - \Phi(w, v^*) - \varphi(v^*) \\ &\geq \Phi(v^*, w) + \varphi(w) - \Phi(v^*, v^*) - \varphi(v^*) \geq 0 \end{aligned}$$

for all $w \in \Omega$. Hence

$$\Psi(v, v^*) \leq \Psi(v^*, v^*) \leq \Psi(v^*, w) \quad \forall v, w \in \Omega \times \Omega. \tag{8}$$

Thus, the v^*, v^* is the saddle point for the shift function $\Psi(v, w)$. This function is convex in w and, generally speaking, is not concave in v. Below we shall show that the saddle point condition of function $\Psi(v, w)$ is a crucial property in the substantiation of convergence for many iterative methods to a equilibrium solution of (3). The (v^*, v^*) may be a saddle point for another functions, too, for example, for $\Psi_1(v, w) =$

$\langle \nabla_2\Phi(v,v), w-v\rangle$, where $\nabla_2\Phi(v,w)$ is partial gradient $\Phi(v,w)$ in w for any v. The last function is very convenient to use it for the substantiation of convergence for gradient-type methods. Therefore we prove this property. In this paper the function $\Phi(v,w)+\varphi(w)$ is supposed to be convex in w for any v, i.e. satisfies the system of inequalities

$$\langle \nabla f(x), y-x\rangle \leq f(y)-f(x) \leq \langle \nabla f(y), y-x\rangle \tag{9}$$

for all x y from a set. We use this system of inequalities for (7) for the case $\Phi(v,w)+\varphi(w)$

$$\Phi(w,w)+\varphi(w)-\Phi(w,v)-\varphi(v)-\Phi(v,w)-\varphi(w)+\Phi(v,v)+\varphi(v)\geq 0 \tag{10}$$

for all $v,w \in \Omega\times\Omega$, then we get

$$\langle \nabla_2\Phi(w,w)+\nabla\varphi(w)-\nabla_2\Phi(v,v)+\nabla\varphi(v), w-v\rangle \geq 0 \;\; \forall w,v\in\Omega\times\Omega, \tag{11}$$

i.e. gradient-restriction $(\nabla_2\Phi(v,w)+\nabla\varphi(w))|_{v=w}$ is the monotone operator on the diagonal of square $\Omega\times\Omega$. We shall write out a necessary condition for problem (3) in the form of a variational inequality

$$\langle \nabla_2\Phi(v^*,v^*)+\nabla\varphi(v^*), w-v^*\rangle \geq 0 \;\; \forall w\in\Omega. \tag{12}$$

Let us set $v=v^*$ in (11) and compare with (12), then

$$\langle \nabla_2\Phi(w,w)+\nabla\varphi(w), w-v^*\rangle \geq 0 \;\; \forall w\in\Omega. \tag{13}$$

Under the notation of function $\Psi_1(v,w)=\langle\nabla_2\Phi(v,v), w-v\rangle$ both inequalities can be rewritten in the form of a saddle point condition

$$\Psi_1(v,v^*)\leq \Psi_1(v^*,v^*)\leq \Psi_1(v^*,w) \;\; \forall v,w\in\Omega\times\Omega.$$

Hereinafter, we shall assume that the objective function $\Phi(v,w)+\varphi(w)$ of (3) has two main properties: it is convex in w for any v, and $\Phi(v,w)$ is positive semi-definite. Both properties guarantee the existence of saddle point property for a point v^*,v^* that, in turn, provides convergence of majority iterative processes to equilibrium solutions. The above reasoning is correct provided that the function $\Phi(v,w)$ is positive semi-definite. In case it is not, it is true that any function always can be reduced to the positive semi-definite kind. The latter circumstance is very important, as it has universal character. We show that any function $\Phi(v,w)$ can be splitted on two components: symmetric and antisymmetric. We select two linear subspaces in the linear space of the real-valued functions $\Phi(v,w)$, which are determined by the following conditions

$$\Phi(v,w)-\Phi(w,v)=0 \;\; \forall w\in\Omega, \;\; \forall v\in\Omega, \tag{14}$$

$$\Phi(v,w)+\Phi(w,v)=0 \;\; \forall w\in\Omega, \;\; \forall v\in\Omega. \tag{15}$$

The functions of the first subspace are called symmetric; those of the second class, anti-symmetric. Recall that a pair of points with coordinates w, v and v, w is situated symmetrically concerning the diagonal of the square $\Omega \times \Omega$, i.e., with respect to the linear manifold $v = w$. This allows us to introduce the concept of a transposed function, Antipin 1998 [3]. If we assign the value of $\Phi(w, v)$ calculated at the point w, v to every point with coordinates v, w, that is $v, w \to \Phi(w, v)$, then we obtain the transposed function $\Phi^T(v, w) = \Phi(w, v)$. In terms of this function conditions (14) and (15) look like

$$\Phi(v, w) = \Phi^T(v, w), \quad \Phi(v, w) = -\Phi^T(v, w).$$

Using the obvious relations $\Phi(v, w) = (\Phi^T(v, w))^T$, $(\Phi_1(v, w) + \Phi_2(v, w))^T = \Phi_1^T(v, w) + \Phi_2^T(v, w)$, we can readily verify that any real function $\Phi(v, w)$ can be represented as the sum

$$\Phi(v, w) = S(v, w) + K(v, w), \tag{16}$$

where $S(v, w)$ and $K(v, w)$ are symmetric and antisymmetric functions, respectively. This expansion is unique

$$S(v, w) = \frac{1}{2}\left(\Phi(v, w) + \Phi^T(v, w)\right),$$

$$K(v, w) = \frac{1}{2}\left(\Phi(v, w) - \Phi^T(v, w)\right). \tag{17}$$

Antipin 1998 [3] had shown that the gradient-restriction $\nabla_2 S(v, w)|_{v=w}$ coincides with the gradient $(1/2)\nabla S(v, v)$ of the restriction for symmetric function $S(v, v) = S(v, w)|_{v=w}$. The latter means that alongside with the function $\Phi(v, w) + \varphi(w)$ it always is possible to introduce other function $K(v, w) + S(w, w) + \varphi(w) = K(v, w) + \varphi_1(w)$ such that gradient-restrictions of both functions are the same, and $K(v, w)$ is a positive semi-definite function. Thus, if it is assumed that $\Phi(v, w) + \varphi(w)$ is not positive semi-definite function, then there always exists an other positive semi-definite function $K(v, w) + \varphi_1(w)$ such that the gradient-restrictions for the both functions are the same and, certainly, a solution sets of both equilibrium problem generated both functions are the same as well. Therefore, we can say that the positive semi-definiteness condition is not restraining very much, although under transition from one function to other there will be difficulties with fulfilment of the second condition: convexity condition of $K(v, w) + \varphi_1(w)$ in w for any v.

3 Extragradient game methods

In game (2) each of the participants has to decide a minimization problem with convex functions in its own variable under fixed values of parameters, which are simultaneously variables for each contenders. There is the question, when methods developed for solving convex problems can be transferred for solving a system of such problems. First of all we note that the game (2) has several equivalent formulations. For example, in the form of variational inequalities

$$\langle \nabla_1 f_1(x_1^*, x_2^*) + \nabla\varphi_1(x_1^*), z_1 - x_1^* \rangle \geq 0 \ \forall z_1 \in X_1,$$
$$\langle \nabla_2 f_2(x_1^*, x_2^*) + \nabla\varphi_2(x_2^*), z_2 - x_2^* \rangle \geq 0 \ \forall z_2 \in X_2, \tag{18}$$

or in the kind of operator equations

$$x_1^* = \pi_{X_1}(x_1^* - \alpha(\nabla_1 f_1(x_1^*, x_2^*) + \nabla\varphi_1(x_1^*))),$$
$$x_2^* = \pi_{X_2}(x_2^* - \alpha(\nabla_2 f_2(x_1^*, x_2^*) + \nabla\varphi_2(x_2^*))). \tag{19}$$

where $\nabla_1 f_1(x_1, x_2), \nabla_2 f_2(x_1, x_2)$ are partial derivatives of functions $f_1(x_1, x_2)$ and $f_2(x_1, x_2)$ in first and second variables accordingly, $\nabla\varphi_1(x_1)$, $\nabla\varphi_2(x_2)$ are gradients of appropriate functions, $\pi_{X_i}(...), i = 1, 2$ are projection operators of vectors onto a set X_i, $\alpha > 0$ is a parameter like steplength. Conditions (18), (19) represent necessary, and in the convex case, sufficient conditions of a minimum in (2). Recall that the solution set of positive semi-sefinite equilibrium (game) problems introduced above is a convex closed set, Antipin 2002 [7]. For the solution of system (19) the extragradient method is used. It includes two half-steps Antipin 1998 [3], the first half-step is

$$\bar{x}_1^n = \pi_{X_1}(x_1^n - \alpha(\nabla_1 f_1(x_1^n, x_2^n) + \nabla\varphi_1(x_1^n))),$$
$$\bar{x}_2^n = \pi_{X_2}(x_2^n - \alpha(\nabla_2 f_2(x_1^n, x_2^n) + \nabla\varphi_2(x_2^n))), \tag{20}$$

and second half-step is

$$x_1^{n+1} = \pi_{X_1}(x_1^n - \alpha(\nabla_1 f_1(\bar{x}_1^n, \bar{x}_2^n) + \nabla\varphi_1(\bar{x}_1^n))),$$
$$x_2^{n+1} = \pi_{X_2}(x_2^n - \alpha(\nabla_2 f_2(\bar{x}_1^n, \bar{x}_2^n) + \nabla\varphi_2(\bar{x}_2^n))). \tag{21}$$

A steplength $\alpha > 0$ in this process is determined from some condition, which will be mentioned below. The first half-step is treated as a calculation of a prediction point where direction of the future motion can be determined and then the second half-step implements the motion in the computed direction. Note that the ordinary gradient-type method

$$x_1^{n+1} = \pi_{X_1}(x_1^n - \alpha(\nabla_1 f_1(x_1^n, x_2^n) + \nabla\varphi_1(x_1^n))),$$
$$x_2^{n+1} = \pi_{X_2}(x_2^n - \alpha(\nabla_2 f_2(x_1^n, x_2^n) + \nabla\varphi_2(x_2^n)))$$

does not converge to a solution of (19). But, in the particular case, if the operator

$$\nabla_2 \Phi(v, w)|_{v=w} = (\nabla_1 f_1(x_1, x_2), \nabla_2 f_2(x_1, x_2))$$

is potential (i.e. Jacobian of $\nabla_2 \Phi(v, v)$ is the symmetric matrix), then the ordinary gradient method is converging, Rosen 1965[17]. In this case the equilibrium problem is equivalent to optimization problem, Antipin 2001 [6].

From (20), (21) we have estimates

$$|\bar{x}_1^n - x_1^{n+1}| \le \alpha|\nabla_1 f_1(x_1^n, x_2^n) + \nabla\varphi_1(x_1^n) - \nabla_1 f_1(\bar{x}_1^n, \bar{x}_2^n) - \nabla\varphi_1(\bar{x}_1^n)|,$$

$$|\bar{x}_2^n - x_2^{n+1}| \le \alpha|\nabla_2 f_2(x_1^n, x_2^n) + \nabla\varphi_2(x_2^n) - \nabla_2 f_2(\bar{x}_1^n, \bar{x}_2^n) - \nabla\varphi_2(\bar{x}_2^n)|. \quad (22)$$

We present the process (20), (21) in the form of variational inequalities. Equation (20) rewrites accordingly with the definition of the projection operator as

$$\langle \bar{x}_1^n - x_1^n + \alpha(\nabla_1 f_1(x_1^n, x_2^n) + \nabla\varphi_1(x_1^n)), z_1 - \bar{x}_1^n\rangle \ge 0 \ \ \forall z_1 \in X_1,$$

$$\langle \bar{x}_2^n - x_2^n + \alpha(\nabla_2 f_2(x_1^n, x_2^n) + \nabla\varphi_2(x_2^n)), z_2 - \bar{x}_2^n\rangle \ge 0 \ \ \forall z_2 \in X_2. \quad (23)$$

Equation (21) is written as

$$\langle x_1^{n+1} - x_1^n + \alpha(\nabla_1 f_1(\bar{x}_1^n, \bar{x}_2^n) + \nabla\varphi_1(\bar{x}_1^n)), z_1 - x_1^{n+1}\rangle \ge 0 \ \ \forall z_1 \in X_1,$$

$$\langle x_2^{n+1} - x_2^n + \alpha(\nabla_2 f_2(\bar{x}_1^n, \bar{x}_2^n) + \nabla\varphi_2(\bar{x}_2^n)), z_2 - x_2^{n+1}\rangle \ge 0 \ \ \forall z_2 \in X_2. \quad (24)$$

When discussing questions of convergence of the game method (20), (21) it is important to underline that two players are a whole system, which in due course evolves to equilibrium state and character of this evolution is determined, in main, by system properties, i.e. system properties of the players, as a whole unit. These system properties we formulate in terms of the normalized function.

$$\Phi(v, w) + \varphi(w) = f_1(z_1, x_2) + \varphi_1(z_1) + f_2(x_1, z_2) + \varphi_2(z_2), \quad (25)$$

where $w = (z_1, z_2), v = (x_1, x_2), \ v, w \in \Omega = X_1 \times X_2$. They include: the positive semi-definite property of $\Phi(v, w)$, the Lipschitz condition of gradient-restriction of this function and convexity property of $\Phi(v, w)$ in w for any v. Certainly, all these properties, in turn, are determined by properties of the functions $f_1(z_1, x_2)$, $\varphi_1(z_1)$, $f_2(x_1, z_2)$, $\varphi_2(z_2)$. For example, if these functions are convex in z_1 and z_2 for any values x_1 and x_2, then $\Phi(v, w) + \varphi(w)$ is convex in w for any v.

Theorem 1 *Suppose that a solution set of game (2) is non-empty, the normalized function of this game $\Phi(v, w) + \varphi(w)$ is positive semi-definite and convex in w for any v, its gradient-restriction $\nabla_2\Phi(v, w)|_{v=w} + \nabla\varphi(w)$ satisfies the Lipschitz condition with constant L, $\Omega \subseteq R^n$ is convex closed set. Then, the sequence x_1^n, x_2^n generated by the method (20),(21) with steplength α chosen from the condition $0 < \alpha < 1/(\sqrt{2}L)$ converges to a game solution, i.e. $x_1^n, x_2^n \to x_1^*, x_2^*$ as $n \to \infty$ monotonically in the norm.*

PROOF. By putting $z_1 = x_1^*, z_2 = x_2^*$ in (24), then

$$\langle x_1^{n+1} - x_1^n + \alpha(\nabla_1 f_1(\bar{x}_1^n, \bar{x}_2^n) + \nabla\varphi_1(\bar{x}_1^n)), x_1^* - x_1^{n+1}\rangle \ge 0,$$

$$\langle x_2^{n+1} - x_2^n + \alpha(\nabla_2 f_2(\bar{x}_1^n, \bar{x}_2^n) + \nabla\varphi_2(\bar{x}_2^n)), x_2^* - x_2^{n+1}\rangle \ge 0. \quad (26)$$

By setting $z_1 = x_1^{n+1}, z_2 = x_2^{n+1}$ in (23)

$$\langle \bar{x}_1^n - x_1^n + \alpha(\nabla_1 f_1(x_1^n, x_2^n) + \nabla\varphi_1(x_1^n)), x_1^{n+1} - \bar{x}_1^n\rangle \geq 0,$$

$$\langle \bar{x}_2^n - x_2^n + \alpha(\nabla_2 f_2(x_1^n, x_2^n) + \nabla\varphi_2(\bar{x}_2^n)), x_2^{n+1} - \bar{x}_2^n\rangle \geq 0.$$

Then

$$\langle \bar{x}_1^n - x_1^n, x_1^{n+1} - \bar{x}_1^n\rangle + \alpha\langle \nabla_1 f_1(\bar{x}_1^n, \bar{x}_2^n) + \nabla\varphi_1(\bar{x}_1^n), x_1^{n+1} - \bar{x}_1^n\rangle$$
$$+\alpha\langle \nabla_1 f_1(x_1^n, x_2^n) + \nabla\varphi_1(x_1^n) - \nabla_1 f_1(\bar{x}_1^n, \bar{x}_2^n) - \nabla\varphi_1(\bar{x}_1^n), x_1^{n+1} - \bar{x}_1^n\rangle \geq 0$$

or taking into account of (22) we have:

$$\langle \bar{x}_1^n - x_1^n, x_1^{n+1} - \bar{x}_1^n\rangle + \alpha\langle \nabla_1 f_1(\bar{x}_1^n, \bar{x}_2^n) + \nabla\varphi_1(\bar{x}_1^n), x_1^{n+1} - \bar{x}_1^n\rangle$$
$$+\alpha^2|\nabla_1 f_1(x_1^n, x_2^n) + \nabla\varphi_1(x_1^n) - \nabla_1 f_1(\bar{x}_1^n, \bar{x}_2^n) - \nabla\varphi_1(\bar{x}_1^n)|^2 \geq 0. \tag{27}$$

We get an analogous inequality with respect to variables z_2

$$\langle \bar{x}_2^n - x_2^n, x_2^{n+1} - \bar{x}_2^n\rangle + \alpha\langle \nabla_2 f_2(\bar{x}_1^n, \bar{x}_2^n) + \nabla\varphi_2(\bar{x}_2^n), x_2^{n+1} - \bar{x}_2^n\rangle$$
$$+\alpha^2|\nabla_2 f_2(x_1^n, x_2^n) + \nabla\varphi_2(x_2^n) - \nabla_2 f_2(\bar{x}_1^n, \bar{x}_2^n) - \nabla\varphi_2(\bar{x}_2^n)|^2 \geq 0. \tag{28}$$

We add the pair of inequalities (26) and (27),(28) to obtain

$$\langle x_1^{n+1} - x_1^n, x_1^* - x_1^{n+1}\rangle + \langle \bar{x}_1^n - x_1^n, x_1^{n+1} - \bar{x}_1^n\rangle$$
$$+\langle x_2^{n+1} - x_2^n, x_2^* - x_2^{n+1}\rangle + \langle \bar{x}_2^n - x_2^n, x_2^{n+1} - \bar{x}_2^n\rangle$$
$$+\alpha\langle \nabla_1 f_1(\bar{x}_1^n, \bar{x}_2^n) + \nabla\varphi_1(\bar{x}_1^n), x_1^* - \bar{x}_1^n\rangle$$
$$+\alpha\langle \nabla_2 f_2(\bar{x}_1^n, \bar{x}_2^n) + \nabla\varphi_2(\bar{x}_2^n), x_2^* - \bar{x}_2^n\rangle$$
$$+\alpha^2|\nabla_1 f_1(x_1^n, x_2^n) + \nabla\varphi_1(x_1^n) - \nabla_1 f_1(\bar{x}_1^n, \bar{x}_2^n) - \nabla\varphi_1(\bar{x}_1^n)|^2$$
$$+\alpha^2|\nabla_2 f_2(x_1^n, x_2^n) + \nabla\varphi_2(x_2^n) - \nabla_2 f_2(\bar{x}_1^n, \bar{x}_2^n) - \nabla\varphi_2(\bar{x}_2^n)|^2 \geq 0. \tag{29}$$

Represent (29) in the vector form :

$$(x_1^{n+1} - x_1^n, x_2^{n+1} - x_2^n)\begin{pmatrix} x_1^* - x_1^{n+1} \\ x_2^* - x_2^{n+1}\end{pmatrix}$$

$$+(\bar{x}_1^n - x_1^n, \bar{x}_2^n - x_2^n)\begin{pmatrix} x_1^{n+1} - \bar{x}_1^n \\ x_2^{n+1} - \bar{x}_2^n\end{pmatrix}$$

$$+(\nabla_1 f_1(\bar{x}_1^n, \bar{x}_2^n) + \nabla\varphi_1(\bar{x}_1^n), \nabla_2 f_2(\bar{x}_1^n, \bar{x}_2^n) + \nabla\varphi_2(\bar{x}_2^n))\begin{pmatrix} x_1^* - \bar{x}_1^n \\ x_2^* - \bar{x}_2^n\end{pmatrix}$$
$$+\alpha^2|\nabla_1 f_1(x_1^n, x_2^n) + \nabla\varphi_1(x_1^n) - \nabla_1 f_1(\bar{x}_1^n, \bar{x}_2^n) - \nabla\varphi_1(\bar{x}_1^n)|^2$$
$$+\alpha^2|\nabla_2 f_2(x_1^n, x_2^n) + \nabla\varphi_2(x_2^n) - \nabla_2 f_2(\bar{x}_1^n, \bar{x}_2^n) - \nabla\varphi_2(\bar{x}_2^n)|^2 \geq 0.$$

Recall the notations for vectors introduced previous to $v = \begin{pmatrix} x_1 \\ x_2 \end{pmatrix}$, $v^* = \begin{pmatrix} x_1^* \\ x_2^* \end{pmatrix}$. If we differentiate the function $\Phi(v,w) + \varphi(w)$ in w for any v and take the restriction of gradient for this function onto the diagonal of square $\Omega \times \Omega$, (i.e. at $v = w$) then we get from (25)

$$\nabla_2\Phi(v,w)|_{v=w} + \nabla\varphi(w) = \begin{pmatrix} \nabla_1 f_1(z_1, x_2) \\ \nabla_2 f_2(x_1, z_2) \end{pmatrix}_{z_1=x_1, z_2=x_2} + \begin{pmatrix} \nabla\varphi_1(z_1) \\ \nabla\varphi_2(z_2) \end{pmatrix}$$

In particular, at $z_1 = \bar{x}_1^n, z_2 = \bar{x}_2^n$ we have

$$\nabla_2\Phi(\bar{v}^n, \bar{v}^n) + \nabla\varphi(\bar{v}^n) = \begin{pmatrix} \nabla_1 f_1(\bar{x}_1^n, \bar{x}_2^n) + \nabla\varphi_1(\bar{x}_1^n) \\ \nabla_2 f_2(\bar{x}_1^n, \bar{x}_2^n) + \nabla\varphi_2(\bar{x}_2^n) \end{pmatrix} \tag{30}$$

In view of the introduced notations we rewrite the last inequality in the form

$$\langle v^{n+1} - v^n, v^* - v^{n+1}\rangle + \langle \bar{v}^n - v^n, v^{n+1} - \bar{v}^n\rangle$$
$$+\alpha\langle \nabla_2\Phi(\bar{v}^n, \bar{v}^n) + \nabla\varphi(\bar{v}^n), v^* - \bar{v}^n\rangle$$
$$+\alpha^2|\nabla_2\Phi(v^n, v^n) + \nabla\varphi(v^n) - \nabla_2\Phi(\bar{v}^n, \bar{v}^n) - \nabla\varphi(\bar{v}^n)|^2 \geq 0. \tag{31}$$

If the operator $\nabla_2\Phi(v,v), \nabla\varphi(v)$ satisfies the Lipschitz condition, then the estimate is correct

$$|\nabla_2\Phi(v^n, v^n) + \nabla\varphi(v^n) - \nabla_2\Phi(\bar{v}^n, \bar{v}^n) - \nabla\varphi(\bar{v}^n)| \leq L|v^n - \bar{v}^n|$$

On the other hand, third third addend from (31) is nonpositive by virtue of estimation (13) at $w = \bar{v}^n$. Using the above inequality (31) takes the form

$$\langle v^{n+1} - v^n, v^* - v^{n+1}\rangle + \langle \bar{v}^n - v^n, v^{n+1} - \bar{v}^n\rangle + (\alpha L)^2|v^n - \bar{v}^n|^2 \geq 0. \tag{32}$$

By means of the identity

$$|v_1 - v_3|^2 = |v_1 - v_2|^2 + 2\langle v_1 - v_2, v_2 - v_3\rangle + |v_2 - v_3|^2, \tag{33}$$

we split the first two scalar products in (32)

$$|v^{n+1} - v^*|^2 + |v^{n+1} - \bar{v}^n|^2 + |\bar{v}^n - v^n|^2$$
$$+2(\alpha L)^2|\bar{v}^n - v^n)|^2 \leq |v^n - v^*|^2.$$

Then, we yield

$$|v^{n+1} - v^*|^2 + |v^{n+1} - \bar{v}^n|^2 + (1 - 2(\alpha L)^2)|\bar{v}^n - v^n|^2 \leq |v^n - v^*|^2. \tag{34}$$

Let us sum up the inequality obtained from $n = 0$ to $n = N$

$$|v^{N+1} - v^*|^2 + \sum_{k=0}^{k=N} |v^{k+1} - \bar{v}^k|^2 + d\sum_{k=0}^{k=N} |\bar{v}^k - v^k|^2 \leq |v^0 - v^*|^2,$$

where $d = 1 - 2(\alpha L)^2 > 0$ under the theorem conditions. From this inequality the boundedness of trajectory follows

$$|v^{N+1} - v^*|^2 \le |v^0 - v^*|^2,$$

and the series are convergent $\sum_{k=0}^{\infty} |v^{k+1} - \bar{v}^k|^2 < \infty$, $\sum_{k=0}^{\infty} |\bar{v}^k - v^k|^2 < \infty$. Consequently, the values tend to zero $|v^{n+1} - \bar{v}^n|^2 \to 0$, $|\bar{v}^n - v^n|^2 \to 0$, $n \to \infty$. We also have that the sequence v^n is limited. That is, there exists a point v' such that $v^{n_i} \to v'$ as $n_i \to \infty$, and $|v^{n_i+1} - v^{n_i}|^2 \to 0, |\bar{v}^{n_i} - v^{n_i}|^2 \to 0$. We put $n = n_i$ in (23) or (24) and passing to the limit as $n_i \to \infty$, we get

$$\langle \nabla_1 f_1(x_1', x_2') + \nabla\varphi_1(x_1'), z_1 - x_1' \rangle \ge 0 \ \forall z_1 \in X_1,$$

$$\langle \nabla_2 f_2(x_1', x_2') + \nabla\varphi_2(x_2'), z_2 - x_2' \rangle \ge 0 \ \forall z_2 \in X_2.$$

The obtained conditions are similar to (18) therefore $x_1', x_2' = x_1^*, x_2^*$, i.e. any limiting point of sequence x_1^n, x_2^n is a solution of problem. The monotonicity condition for decrease of value $|v^n - v^*|$ guarantees the uniqueness of the limit point, i.e. convergence provides $v^n = (x_1^n, x_2^n) \to v^* = (x_1^*, x_2^*)$ as $n \to \infty$. The theorem is proved.

4 Extragradient game methods using Lagrange function

We consider now convex two person non-zero sum game, where each player of the game has, in addition, thefunctional constraints

$$x_1^* \in \text{Argmin}\{f_1(z_1, x_2^*) + \varphi_1(z_1) \ | \ g_1(z_1) \le 0, \ z_1 \in X_1\},$$

$$x_2^* \in \text{Argmin}\{f_2(x_1^*, z_2) + \varphi_2(z_2) \ | \ g_2(z_2) \le 0, \ z_2 \in X_2\}. \tag{35}$$

Each of the participants of the game decides a convex programming problem in its own variable at fixed value parameters. We introduce the Lagrange functions for each of the players. These functions depend on parameters $v = (x_1, x_2)$

$$\mathcal{L}_1(z_1, x_2, p_1) = f_1(z_1, x_2) + \varphi_1(z_1) + \langle p_1, g_1(z_1) \rangle \ \ \forall z_1 \in X_1, \ p_1 \ge 0,$$

$$\mathcal{L}_2(x_1, z_2, p_2) = f_2(x_1, z_2) + \varphi_2(z_2) + \langle p_2, g_2(z_2) \rangle \ \ \forall z_2 \in X_2, \ p_2 \ge 0. \tag{36}$$

We assume that under equilibrium conditions, i.e. at $x_1 = x_1^*, x_2 = x_2^*$ points x_1^*, p_1^* and x_2^*, p_2^* are saddle points for the Lagrange functions $\mathcal{L}_1(z_1, x_2^*, p_1)$ and $\mathcal{L}_2(x_1^*, z_2, p_2)$. The latter means that the inequality system is satisfied

$$\mathcal{L}_1(x_1^*, x_2^*, p_1) \le \mathcal{L}_1(x_1^*, x_2^*, p_1^*) \le \mathcal{L}_1(z_1, x_2^*, p_1^*) \ \ \forall z_1 \ge 0, \ \forall p_1 \ge 0, \tag{37}$$

$$\mathcal{L}_2(x_1^*, x_2^*, p_2) \le \mathcal{L}_2(x_1^*, x_2^*, p_2^*) \le \mathcal{L}_2(x_1^*, z_2, p_2^*) \ \ \forall z_2 \ge 0, \ \forall p_2 \ge 0. \tag{38}$$

We rewrite inequalities (37),(38) in the form of the system of problems

$$x_1^* \in \mathrm{Argmin}\{f_1(z_1, x_2^*) + \varphi_1(z_1) + \langle p_1^*, g_1(z_1)\rangle \mid \forall\ z_1 \in X_1\},$$

$$p_1^* \in \mathrm{Argmax}\{\langle p_1, g_1(x_1^*)\rangle \quad \mid \quad p_1 \geq 0\}, \tag{39}$$

$$x_2^* \in \mathrm{Argmin}\{f_2(x_1^*, z_2) + \varphi_2(z_2) + \langle p_2^*, g_2(z_2)\rangle \mid \forall z_2 \in X_2\},$$

$$p_2^* \in \mathrm{Argmax}\{\langle p_2, g_2(x_2^*)\rangle \quad \mid \quad p_2 \geq 0\} \tag{40}$$

or in the form of variational inequalities

$$\langle \nabla_1 f_1(x_1^*, x_2^*) + \nabla\varphi_1(x_1^*) + \nabla g_1^T(x_1^*)p_1^*, z_1 - x_1^*\rangle \geq 0\ \forall z_1 \in X_1,$$

$$-\langle g_1(x_1^*), p_1 - p_1^*\rangle \geq 0 \quad \forall p_1 \geq 0,$$

$$\langle \nabla_2 f_2(x_1^*, x_2^*) + \nabla\varphi_2(x_2^*) + \nabla g_2^T(x_2^*)p_2^*, z_2 - x_2^*\rangle \geq 0\ \forall z_2 \in X_2,$$

$$-\langle g_2(x_2^*), p_2 - p_2^*\rangle \geq 0 \quad \forall p_2 \geq 0, \tag{41}$$

where $\nabla g_1^T(x_1), \nabla g_2^T(x_2)$ are $m_1 \times n$ and $m_2 \times n$ matrices, and $\nabla g_{1,i}(x_1)$, $\nabla g_{2,j}(x_2)$, $i = 1, 2, ..., m_1, j = 1, 2, ..., m_2$ are vector-lines. We differentiate the Lagrange functions (36) in $w = (z_1, z_2)$

$$\nabla_1 \mathcal{L}_1(z_1, x_2, p_1) = \nabla_1 f_1(z_1, x_2) + \nabla\varphi_1(z_1) + \nabla g_1^T(z_1)p_1,$$

$$\nabla_2 \mathcal{L}_2(x_1, z_2, p_2) = \nabla_2 f_2(x_1, z_2) + \nabla\varphi_2(z_2) + \nabla g_2^T(z_2)p_2, \tag{42}$$

and present the system of variational inequalities (41) in the equivalent form of operator equations

$$x_1^* = \pi_{X_1}(x_1^* - \alpha(\nabla_1 \mathcal{L}_1(x_1^*, x_2^*, p_1^*)),$$

$$p_1^* = \pi_+(p_1^* + \alpha g_1(x_1^*)),$$

$$x_2^* = \pi_{X_2}(x_2^* - \alpha(\nabla_2 \mathcal{L}_2(x_1^*, x_2^*, p_2^*)),$$

$$p_2^* = \pi_+(p_2^* + \alpha g_2(x_2^*)), \tag{43}$$

where $\pi_+(...)$ is the projection operator of some vector into the positive orthant R_+^n, $\alpha > 0$ is a parameter like steplength. For the solution of system (43) we use the extragradient method with respect to primal and dual variables (Antipin 1997 [2], Antipin 2000 [4], Antipin 2001 [5]). The method includes two half steps: the first half-step

$$\bar{p}_1^n = \pi_+(p_1^n + \alpha g_1(x_1^n)),$$

$$\bar{x}_1^n = \pi_{X_1}(x_1^n - \alpha(\nabla_1 \mathcal{L}_1(x_1^n, x_2^n, \bar{p}_1^n)),$$

$$\bar{p}_2^n = \pi_+(p_2^n + \alpha g_2(x_2^n)),$$

$$\bar{x}_2^n = \pi_{X_2}(x_2^n - \alpha(\nabla_2 \mathcal{L}_2(x_1^n, x_2^n, \bar{p}_2^n)), \tag{44}$$

and second half-step

$$
\begin{aligned}
p_1^{n+1} &= \pi_+(p_1^n + \alpha g_1(\bar{x}_1^n)),\\
x_1^{n+1} &= \pi_{X_1}(x_1^n - \alpha(\nabla_1 \mathcal{L}_1(\bar{x}_1^n, \bar{x}_2^n, \bar{p}_1^n)),\\
p_2^{n+1} &= \pi_+(p_2^n + \alpha g_2(\bar{x}_2^n)),\\
x_2^{n+1} &= \pi_{X_2}(x_2^n - \alpha(\nabla_2 \mathcal{L}_2(\bar{x}_1^n, \bar{x}_2^n, \bar{p}_2^n)).
\end{aligned} \tag{45}
$$

The steplength α in (44),(45) is determined from an interval $0 < \alpha < \alpha_0$, where right-hand side of this segment will be determined later. For the justification of correctness of selecting out parameter α we receive evaluations of deviations for vectors $\bar{x}_1^n, \bar{x}_2^n$ and x_1^{n+1}, x_2^{n+1} and $\bar{p}_1^n, \bar{p}_2^n$ and p_1^{n+1}, p_2^{n+1} from (44), (45)

$$
\begin{aligned}
&|\bar{p}_1^n - p_1^{n+1}| \le \alpha |g_1(x_1^n) - g_1(\bar{x}_1^n)|, \quad |\bar{p}_2^n - p_2^{n+1}| \le \alpha |g_2(x_2^n) - g_2(\bar{x}_2^n)|,\\
&|\bar{x}_1^n - x_1^{n+1}| \le \alpha |\nabla_1 \mathcal{L}_1(x_1^n, x_2^n, \bar{p}_1^n) - \nabla_1 \mathcal{L}_1(\bar{x}_1^n, \bar{x}_2^n, \bar{p}_1^n)|,\\
&|\bar{x}_2^n - x_2^{n+1}| \le \alpha |\nabla_2 \mathcal{L}_2(x_1^n, x_2^n, \bar{p}_2^n) - \nabla_2 \mathcal{L}_2(\bar{x}_1^n, \bar{x}_2^n, \bar{p}_2^n)|.
\end{aligned} \tag{46}
$$

We rewrite (44),(45) in the form of variational inequalities. According with the definition of the projection operator we present equations from (44) in the form

$$
\begin{aligned}
&\langle \bar{p}_1^n - p_1^n - \alpha g_1(x_1^n), p_1 - \bar{p}_1^n \rangle \ge 0 \quad \forall p_1 \ge 0,\\
&\langle \bar{p}_2^n - p_2^n - \alpha g_2(x_2^n), p_2 - \bar{p}_2^n \rangle \ge 0 \quad \forall p_2 \ge 0
\end{aligned} \tag{47}
$$

and

$$
\begin{aligned}
&\langle \bar{x}_1^n - x_1^n + \alpha \nabla_1 \mathcal{L}_1(x_1^n, x_2^n, \bar{p}_1^n), z_1 - \bar{x}_1^n \rangle \ge 0 \quad \forall z_1 \in X_1,\\
&\langle \bar{x}_2^n - x_2^n + \alpha \nabla_2 \mathcal{L}_2(x_1^n, x_2^n, \bar{p}_2^n), z_2 - \bar{x}_2^n \rangle \ge 0 \quad \forall z_2 \in X_2.
\end{aligned} \tag{48}
$$

We rewrite equations (45) as

$$
\begin{aligned}
&\langle p_1^{n+1} - p_1^n - \alpha g_1(\bar{x}_1^n), p_1 - p_1^{n+1} \rangle \ge 0 \quad \forall p_1 \ge 0,\\
&\langle p_2^{n+1} - p_2^n - \alpha g_2(\bar{x}_2^n), p_2 - p_2^{n+1} \rangle \ge 0 \quad \forall p_2 \ge 0
\end{aligned} \tag{49}
$$

and

$$
\begin{aligned}
&\langle x_1^{n+1} - x_1^n + \alpha \nabla_1 \mathcal{L}_1(\bar{x}_1^n, \bar{x}_2^n, \bar{p}_1^n), z_1 - x_1^{n+1} \rangle \ge 0 \quad \forall z_1 \in X_1,\\
&\langle x_2^{n+1} - x_2^n + \alpha \nabla_2 \mathcal{L}_2(\bar{x}_1^n, \bar{x}_2^n, \bar{p}_2^n), z_2 - x_2^{n+1} \rangle \ge 0 \quad \forall z_2 \in X_2.
\end{aligned} \tag{50}
$$

In theorem 1 we used inequality (13), which is the corollary of monotonicity property for gradient-restriction $\nabla_2 \Phi(v, v) + \nabla \varphi(v)$. In the considered situation we generalize this inequality for the case when the functional constraints of equilibrium problem are taking into account by means of Lagrange functions. To this end we put $z_1 = \bar{x}_1^n, z_2 = \bar{x}_2^n$ in the first and third inequalities (41)

$$
\langle \nabla_1 f_1(x_1^*, x_2^*) + \nabla \varphi_1(x_1^*) + \nabla g_1^T(x_1^*) p_1^*, \bar{x}_1^n - x_1^* \rangle \ge 0,
$$

$$\langle \nabla_2 f_2(x_1^*, x_2^*) + \nabla \varphi_2(x_2^*) + \nabla g_2^T(x_2^*) p_2^*, \bar{x}_2^n - x_2^* \rangle \geq 0.$$

Then

$$\langle \nabla_1 f_1(x_1^*, x_2^*) + \nabla \varphi_1(x_1^*), \bar{x}_1^n - x_1^* \rangle + \langle p_1^*, \nabla g_1(x_1^*)(\bar{x}_1^n - x_1^*) \rangle \geq 0,$$
$$\langle \nabla_2 f_2(x_1^*, x_2^*) + \nabla \varphi_2(x_2^*), \bar{x}_2^n - x_2^* \rangle + \langle p_2^*, \nabla g_2(x_2^*)(\bar{x}_2^n - x_2^*) \rangle \geq 0.$$

Using the convexity of vector functions $g_1(x_1), g_2(x_2)$, we add both inequalities to obtain

$$\langle \nabla_1 f_1(x_1^*, x_2^*) + \nabla \varphi_1(x_1^*), \bar{x}_1^n - x_1^* \rangle + \langle \nabla_2 f_2(x_1^*, x_2^*) + \nabla \varphi_2(x_2^*), \bar{x}_2^n - x_2^* \rangle$$
$$+\langle p_1^*, g_1(\bar{x}_1^n) - g_1(x_1^*) \rangle + \langle p_2^*, g_2(\bar{x}_2^n) - g_2(x_2^*) \rangle \geq 0. \quad (51)$$

We use the obtained estimate in proving the following theorem.

Theorem 2 *Suppose that a solution set of game (35) is non-empty, the Lagrange function of each player has got a saddle point, the normalized function of this game $\Phi(v, w) + \varphi(w)$ is positive semi-definite and convex in w for any v, its gradient-restriction $\nabla_2 \Phi(v, w)|_{v=w} + \nabla \varphi(w)$ satisfies the Lipschitz condition with constant L_1, vector and matrix functions $g(w), \nabla g^T(w)$ satisfy the Lipschitz conditions with constants L_3 L_2, sequence $\bar{p}^n \leq C$ is limited for all n, $\Omega \subseteq R^n$ is convex closed set. Then, the sequence $x_1^n, x_2^n, p_1^n, p_2^n$, generated by method (44),(45) with steplength α chosen from condition $0 < \alpha < 1/(\sqrt{2(L_1^2 + (CL_2)^2) + L_3^2}$ converges to a game solution, i.e. .?. $x_1^n, x_2^n \to x_1^*, x_2^*$, $p_1^n, p_2^n \to p_1^*, p_2^*$ as $n \to \infty$. monotonically in the norm.*

Proof. We put $z_1 = x_1^*, z_2 = x_2^*$ in (50), then

$$\langle x_1^{n+1} - x_1^n + \alpha \nabla_1 \mathcal{L}_1(\bar{x}_1^n, \bar{x}_2^n, \bar{p}_1^n), x_1^* - x_1^{n+1} \rangle \geq 0,$$
$$\langle x_2^{n+1} - x_2^n + \alpha \nabla_2 \mathcal{L}_2(\bar{x}_1^n, \bar{x}_2^n, \bar{p}_2^n), x_2^* - x_2^{n+1} \rangle \geq 0. \quad (52)$$

We set $z_1 = x_1^{n+1}, z_2 = x_2^{n+1}$ in (48)

$$\langle \bar{x}_1^n - x_1^n + \alpha \nabla_1 \mathcal{L}_1(x_1^n, x_2^n, \bar{p}_1^n), x_1^{n+1} - \bar{x}_1^n \rangle \geq 0,$$
$$\langle \bar{x}_2^n - x_2^n + \alpha \nabla_2 \mathcal{L}_2(x_1^n, x_2^n, \bar{p}_2^n), x_2^{n+1} - \bar{x}_2^n \rangle \geq 0. \quad (53)$$

Hence

$$\langle \bar{x}_1^n - x_1^n, x_1^{n+1} - \bar{x}_1^n \rangle + \alpha \langle \nabla_1 \mathcal{L}_1(\bar{x}_1^n, \bar{x}_2^n, \bar{p}_1^n), x_1^{n+1} - \bar{x}_1^n \rangle$$
$$+\alpha \langle \nabla_1 \mathcal{L}_1(x_1^n, x_2^n, \bar{p}_1^n) - \nabla_1 \mathcal{L}_1(\bar{x}_1^n, \bar{x}_2^n, \bar{p}_1^n), x_1^{n+1} - \bar{x}_1^n \rangle \geq 0,$$

taking into account (46), we have

$$\langle \bar{x}_1^n - x_1^n, x_1^{n+1} - \bar{x}_1^n \rangle + \alpha \langle \nabla_1 \mathcal{L}_1(\bar{x}_1^n, \bar{x}_2^n, \bar{p}_1^n), x_1^{n+1} - \bar{x}_1^n \rangle$$
$$+\alpha^2 |\nabla_1 \mathcal{L}_1(x_1^n, x_2^n, \bar{p}_1^n) - \nabla_1 \mathcal{L}_1(\bar{x}_1^n, \bar{x}_2^n, \bar{p}_1^n)|^2 \geq 0. \quad (54)$$

We obtain the similar inequality with respect to variable z_2

$$\langle \bar{x}_2^n - x_2^n, x_2^{n+1} - \bar{x}_2^n \rangle + \alpha \langle \nabla_2 \mathcal{L}_2(\bar{x}_1^n, \bar{x}_2^n, \bar{p}_2^n), x_2^{n+1} - \bar{x}_2^n \rangle$$
$$+\alpha^2 |\nabla_2 \mathcal{L}_2(x_1^n, x_2^n, \bar{p}_2^n) - \nabla_2 \mathcal{L}_2(\bar{x}_1^n, \bar{x}_2^n, \bar{p}_2^n)|^2 \geq 0. \quad (55)$$

Add the systems of pairs of inequalities (52) and (54), (55) to obtain

$$\langle x_1^{n+1} - x_1^n, x_1^* - x_1^{n+1} \rangle + \langle \bar{x}_1^n - x_1^n, x_1^{n+1} - \bar{x}_1^n \rangle$$
$$+\langle x_2^{n+1} - x_2^n, x_2^* - x_2^{n+1} \rangle + \langle \bar{x}_2^n - x_2^n, x_2^{n+1} - \bar{x}_2^n \rangle$$
$$+\alpha \langle \nabla_1 \mathcal{L}_1(\bar{x}_1^n, \bar{x}_2^n, \bar{p}_1^n), x_1^* - \bar{x}_1^n \rangle + \alpha \langle \nabla_2 \mathcal{L}_2(\bar{x}_1^n, \bar{x}_2^n, \bar{p}_2^n), x_2^* - \bar{x}_2^n \rangle$$
$$+\alpha^2 |\nabla_1 \mathcal{L}_1(x_1^n, x_2^n, \bar{p}_1^n) - \nabla_1 \mathcal{L}_1(\bar{x}_1^n, \bar{x}_2^n, \bar{p}_1^n)|^2$$
$$+\alpha^2 |\nabla_2 \mathcal{L}_2(x_1^n, x_2^n, \bar{p}_2^n) - \nabla_2 \mathcal{L}_2(\bar{x}_1^n, \bar{x}_2^n, \bar{p}_2^n)|^2 \geq 0.$$

In view of convexity of $g_1(x_1), g_2(x_2)$ we estimate the fifth and sixth term in the obtained inequality separately:

$$\langle \nabla_1 \mathcal{L}_1(\bar{x}_1^n, \bar{x}_2^n, \bar{p}_1^n), x_1^* - \bar{x}_1^n \rangle + \langle \nabla_2 \mathcal{L}_2(\bar{x}_1^n, \bar{x}_2^n, \bar{p}_2^n), x_2^* - \bar{x}_2^n \rangle$$
$$= \langle \nabla_1 f_1(\bar{x}_1^n, \bar{x}_2^n) + \nabla \varphi_1(\bar{x}_1^n), x_1^* - \bar{x}_1^n \rangle + \langle \nabla g_1^T(\bar{x}_1^n) \bar{p}_1^n, x_1^* - \bar{x}_1^n) \rangle$$
$$+\langle \nabla_2 f_2(\bar{x}_1^n, \bar{x}_2^n) + \nabla \varphi_2(\bar{x}_2^n), x_2^* - \bar{x}_2^n \rangle + \langle \nabla g_2^T(\bar{x}_2^n) \bar{p}_2^n, x_2^* - \bar{x}_2^n \rangle$$
$$\leq \langle \nabla_1 f_1(\bar{x}_1^n, \bar{x}_2^n) + \nabla \varphi_1(\bar{x}_1^n), x_1^* - \bar{x}_1^n \rangle + \langle \bar{p}_1^n, g_1(x_1^*) - g_1(\bar{x}_1^n) \rangle$$
$$+\langle \nabla_2 f_2(\bar{x}_1^n, \bar{x}_2^n) + \nabla \varphi_2(\bar{x}_2^n), x_2^* - \bar{x}_2^n \rangle + \langle \bar{p}_2^n, g_2(x_2^*) - g_2(\bar{x}_2^n) \rangle.$$

Then

$$\langle x_1^{n+1} - x_1^n, x_1^* - x_1^{n+1} \rangle + \langle \bar{x}_1^n - x_1^n, x_1^{n+1} - \bar{x}_1^n \rangle$$
$$+\langle x_2^{n+1} - x_2^n, x_2^* - x_2^{n+1} \rangle + \langle \bar{x}_2^n - x_2^n, x_2^{n+1} - \bar{x}_2^n \rangle$$
$$+\alpha (\langle \nabla_1 f_1(\bar{x}_1^n, \bar{x}_2^n) + \nabla \varphi_1(\bar{x}_1^n), x_1^* - \bar{x}_1^n \rangle + \langle \bar{p}_1^n, g_1(x_1^*) - g_1(\bar{x}_1^n) \rangle)$$
$$+\alpha (\langle \nabla_2 f_2(\bar{x}_1^n, \bar{x}_2^n) + \nabla \varphi_2(\bar{x}_2^n), x_2^* - \bar{x}_2^n \rangle + \langle \bar{p}_2^n, g_2(x_2^*) - g_2(\bar{x}_2^n) \rangle$$
$$+\alpha^2 |\nabla_1 \mathcal{L}_1(x_1^n, x_2^n, \bar{p}_1^n) - \nabla_1 \mathcal{L}_1(\bar{x}_1^n, \bar{x}_2^n, \bar{p}_1^n)|^2$$
$$+\alpha^2 |\nabla_2 \mathcal{L}_2(x_1^n, x_2^n, \bar{p}_2^n) - \nabla_2 \mathcal{L}_2(\bar{x}_1^n, \bar{x}_2^n, \bar{p}_2^n)|^2 \geq 0. \quad (56)$$

Using notations (30), we add inequality (51) and (56), then get

$$\langle x_1^{n+1} - x_1^n, x_1^* - x_1^{n+1} \rangle + \langle \bar{x}_1^n - x_1^n, x_1^{n+1} - \bar{x}_1^n \rangle$$
$$+\langle x_2^{n+1} - x_2^n, x_2^* - x_2^{n+1} \rangle + \langle \bar{x}_2^n - x_2^n, x_2^{n+1} - \bar{x}_2^n \rangle$$
$$+\langle \nabla_2 \Phi(\bar{v}_1^n, \bar{v}_2^n) + \nabla \varphi(\bar{v}_1^n) - \nabla_2 \Phi(v_1^*, v_2^*) - \nabla \varphi(v_1^*), v_1^* - \bar{v}_1^n \rangle$$
$$+\langle \bar{p}_1^n - p_1^*, g_1(x_1^*) - g_1(\bar{x}_1^n) \rangle + \langle \bar{p}_2^n - p_2^*, g_2(x_2^*) - g_2(\bar{x}_2^n) \rangle$$
$$+\alpha^2 |\nabla_1 \mathcal{L}_1(x_1^n, x_2^n, \bar{p}_1^n) - \nabla_1 \mathcal{L}_1(\bar{x}_1^n, \bar{x}_2^n, \bar{p}_1^n)|^2$$

$$+\alpha^2|\nabla_2\mathcal{L}_2(x_1^n,x_2^n,\bar{p}_2^n)-\nabla_2\mathcal{L}_2(\bar{x}_1^n,\bar{x}_2^n,\bar{p}_2^n)|^2\geq 0. \tag{57}$$

To the obtained inequality we return back later and now consider inequalities (47) and (49). Let us put $p_1=p_1^*,\ p_2=p_2^*$ in (49)

$$\langle p_1^{n+1}-p_1^n,p_1^*-p_1^{n+1}\rangle-\alpha\langle g_1(\bar{x}_1^n),p_1^*-p_1^{n+1}\rangle\geq 0,$$

$$\langle p_2^{n+1}-p_2^n,p_2^*-p_2^{n+1}\rangle-\alpha\langle g_2(\bar{x}_2^n),p_2^*-p_2^{n+1}\rangle\geq 0 \tag{58}$$

and $p_1=p_1^{n+1},\ p_2=p_2^{n+1}$ in (47), then

$$\langle \bar{p}_1^n-p_1^n,p_1^{n+1}-\bar{p}_1^n\rangle+\alpha\langle g_1(\bar{x}_1^n)-g_1(x_1^n),p_1^{n+1}-\bar{p}_1^n\rangle$$

$$-\alpha\langle g_1(\bar{x}_1^n),p_1^{n+1}-\bar{p}_1^n\rangle\geq 0,$$

$$\langle \bar{p}_2^n-p_2^n,p_2^{n+1}-\bar{p}_2^n\rangle+\alpha\langle g_2(\bar{x}_2^n)-g_2(x_2^n),p_2^{n+1}-\bar{p}_2^n\rangle$$

$$-\alpha\langle g_2(\bar{x}_2^n),p_2^{n+1}-\bar{p}_2^n\rangle\geq 0. \tag{59}$$

We combine both inequalities (58) and (59)

$$\langle p_1^{n+1}-p_1^n,p_1^*-p_1^{n+1}\rangle+\langle p_2^{n+1}-p_2^n,p_2^*-p_2^{n+1}\rangle$$

$$+\langle \bar{p}_1^n-p_1^n,p_1^{n+1}-\bar{p}_1^n\rangle+\langle \bar{p}_2^n-p_2^n,p_2^{n+1}-\bar{p}_2^n\rangle$$

$$+\alpha\langle g_1(\bar{x}_1^n)-g_1(x_1^n),p_1^{n+1}-\bar{p}_1^n\rangle+\alpha\langle g_2(\bar{x}_2^n)-g_2(x_2^n),p_2^{n+1}-\bar{p}_2^n\rangle$$

$$-\alpha\langle g_1(\bar{x}_1^n),p_1^*-\bar{p}_1^n\rangle-\alpha\langle g_2(\bar{x}_2^n),p_2^*-\bar{p}_2^n\rangle\geq 0.$$

Using (46), rewrite the last inequality in the form

$$\langle p_1^{n+1}-p_1^n,p_1^*-p_1^{n+1}\rangle+\langle p_2^{n+1}-p_2^n,p_2^*-p_2^{n+1}\rangle$$

$$+\langle \bar{p}_1^n-p_1^n,p_1^{n+1}-\bar{p}_1^n\rangle+\langle \bar{p}_2^n-p_2^n,p_2^{n+1}-\bar{p}_2^n\rangle$$

$$+\alpha^2|g_1(\bar{x}_1^n)-g_1(x_1^n)|^2+\alpha^2|g_2(\bar{x}_2^n)-g_2(x_2^n)|^2$$

$$-\alpha\langle g_1(\bar{x}_1^n),p_1^*-\bar{p}_1^n\rangle-\alpha\langle g_2(\bar{x}_2^n),p_2^*-\bar{p}_2^n\rangle\geq 0. \tag{60}$$

We put $p_1=\bar{p}_1^n,p_2=\bar{p}_2^n$ in the second and the fourth inequalities (41) and add both to (60). Then

$$\langle p_1^{n+1}-p_1^n,p_1^*-p_1^{n+1}\rangle+\langle p_2^{n+1}-p_2^n,p_2^*-p_2^{n+1}\rangle$$

$$+\langle \bar{p}_1^n-p_1^n,p_1^{n+1}-\bar{p}_1^n\rangle+\langle \bar{p}_2^n-p_2^n,p_2^{n+1}-\bar{p}_2^n\rangle$$

$$+\alpha^2|g_1(\bar{x}_1^n)-g_1(x_1^n)|^2+\alpha^2|g_2(\bar{x}_2^n)-g_2(x_2^n)|^2$$

$$+\alpha\langle g_1(x_1^*)-g_1(\bar{x}_1^n),p_1^*-\bar{p}_1^n\rangle+\alpha\langle g_2(x_2^*)-g_2(\bar{x}_2^n),p_2^*-\bar{p}_2^n\rangle\geq 0. \tag{61}$$

At last, we combine inequalities (57) and (61) and take into account the monotonicity condition (11) of the operator $\nabla_2\Phi(v,v)+\nabla\varphi(v)$, to obtain

$$\langle x_1^{n+1}-x_1^n,x_1^*-x_1^{n+1}\rangle+\langle \bar{x}_1^n-x_1^n,x_1^{n+1}-\bar{x}_1^n\rangle$$

$$+\langle x_2^{n+1} - x_2^n, x_2^* - x_2^{n+1}\rangle + \langle \bar{x}_2^n - x_2^n, x_2^{n+1} - \bar{x}_2^n\rangle$$
$$+\langle p_1^{n+1} - p_1^n, p_1^* - p_1^{n+1}\rangle + \langle p_2^{n+1} - p_2^n, p_2^* - p_2^{n+1}\rangle$$
$$+\langle \bar{p}_1^n - p_1^n, p_1^{n+1} - \bar{p}_1^n\rangle + \langle \bar{p}_2^n - p_2^n, p_2^{n+1} - \bar{p}_2^n\rangle$$
$$+\alpha^2|\nabla_1\mathcal{L}_1(x_1^n, x_2^n, \bar{p}_1^n) - \nabla_1\mathcal{L}_1(\bar{x}_1^n, \bar{x}_2^n, \bar{p}_1^n)|^2$$
$$+\alpha^2|\nabla_2\mathcal{L}_2(x_1^n, x_2^n, \bar{p}_2^n) - \nabla_2\mathcal{L}_2(\bar{x}_1^n, \bar{x}_2^n, \bar{p}_2^n)|^2$$
$$+\alpha^2|g_1(\bar{x}_1^n) - g_1(x_1^n)|^2 + \alpha^2|g_2(\bar{x}_2^n) - g_2(x_2^n)|^2 \geq 0. \tag{62}$$

Further, following the logic to the theorem 1, we scalarized the obtained inequality and pass to the variables $v = (x_1, x_2)^T, p = (p_1, p_2)^T$. To this end we consider the three last parts in (62) in more details. We expand the convolution (25) with the help of scalarization of two Lagrange functions (36)

$$\mathcal{L}(v, w, p) = \Phi(v, w) + \varphi(w) + \langle p, g(w)\rangle = \mathcal{L}_1(z_1, x_2, p_1) + \mathcal{L}_2(x_1, z_2, p_2)$$
$$= f_1(z_1, x_2) + \varphi_1(z_1) + \langle p_1, g_1(z_1)\rangle + f_2(x_1, z_2) + \varphi_2(z_2) + \langle p_2, g_2(z_2)\rangle.$$

Differentiating the function $\mathcal{L}(v, w, p)$ in w, we consider gradient-restriction of this function $\nabla_2\mathcal{L}(v, w, p)|_{v=w}$ on the diagonal of square $\Omega \times \Omega$

$$\nabla_2\mathcal{L}(v, w, p)|_{v=w} = \nabla_2\Phi(v, w)|_{v=w} + \nabla\varphi(w) + \nabla g^T(w)p$$
$$= \begin{pmatrix} \nabla_1\mathcal{L}_1(z_1, x_2, p_1) \\ \nabla_2\mathcal{L}_2(x_1, z_2, p_2) \end{pmatrix}_{z_1=x_1, z_2=x_2}$$

In particular, at $z_1 = \bar{x}_1^n, z_2 = \bar{x}_2^n, p_1 = \bar{p}_1^n, p_2 = \bar{p}_2^n$ we have

$$\nabla_2\Phi(\bar{v}^n, \bar{v}^n) + \nabla\varphi(\bar{v}^n) + \nabla g^T(\bar{v}^n)\bar{p}^n = \begin{pmatrix} \nabla_1\mathcal{L}_1(\bar{x}_1^n, \bar{x}_2^n, \bar{p}_1^n) \\ \nabla_2\mathcal{L}_2(\bar{x}_1^n, \bar{x}_2^n, \bar{p}_2^n) \end{pmatrix}$$

Using the conditions

$$|\nabla_2\Phi(v + h, v + h) + \nabla\varphi(v + h) - \nabla_2\Phi(\bar{v}, \bar{v}) - \nabla\varphi(\bar{v})| \leq L_1|h|,$$
$$|\nabla g^T(v + h) - \nabla g^T(\bar{v})| \leq L_2|h|, \ |g(v + h) - g(v)| \leq L_3|h|,$$
$$|\bar{p}^n| \leq C, \ n \to \infty.$$

estimate the three last terms from (62)

$$|\nabla_1\mathcal{L}_1(x_1^n, x_2^n, \bar{p}_1^n) - \nabla_1\mathcal{L}_1(\bar{x}_1^n, \bar{x}_2^n, \bar{p}_1^n)|^2$$
$$+|\nabla_2\mathcal{L}_2(x_1^n, x_2^n, \bar{p}_2^n) - \nabla_2\mathcal{L}_2(\bar{x}_1^n, \bar{x}_2^n, \bar{p}_2^n)|^2$$
$$= |\nabla_2\Phi(v^n, v^n) + \nabla\varphi(v^n) - \nabla_2\Phi(\bar{v}^n, \bar{v}^n) - \nabla\varphi(\bar{v}^n)$$
$$+(\nabla g^T(v^n) - \nabla g^T(\bar{v}^n))\bar{p}^n|^2$$

$$\leq L_1^2|v^n - \bar{v}^n|^2 + (CL_2)^2|v^n - \bar{v}^n|^2 = (L_1^2 + (CL_2)^2)|v^n - \bar{v}^n|^2$$

and

$$|g(v^n) - g(\bar{v}^n)|^2 = |g_1(\bar{x}_1^n) - g_1(x_1^n)|^2$$
$$+|g_2(\bar{x}_2^n) - g_2(x_2^n)|^2 = L_3^2|v^n - \bar{v}^n|^2.$$

In view of what we discussed above we present (62) as

$$\langle v^{n+1} - v^n, v^* - v^{n+1}\rangle + \langle \bar{v}^n - v^n, v^{n+1} - \bar{v}^n\rangle$$

$$+\langle p^{n+1} - p^n, p^* - p^{n+1}\rangle + \langle \bar{p}^n - p^n, p^{n+1} - \bar{p}^n\rangle$$
$$+\alpha^2(L_1^2 + (CL_2)^2)|v^n - \bar{v}^n| + \alpha^2 L_3^2|v^n - \bar{v}^n|^2 \geq 0.$$

Applying identity (33), we reduce the obtained inequality to the form

$$|v^{n+1} - v^*|^2 + |v^{n+1} - \bar{v}^n|^2 + d|\bar{v}^n - v^n|^2$$

$$+|p^{n+1} - p^*|^2 + |p^{n+1} - \bar{p}^n|^2 + |\bar{p}^n - p^n|^2 \leq |v^n - v^*|^2 + |p^n - p^*|^2, \tag{63}$$

where $d = 1 - 2\alpha^2(L_1^2 + (CL_2)^2) + L_3^2) \geq 0$ is under conditions of the theorem, as $0 < \alpha < 1/(\sqrt{2(L_1^2 + (CL_2)^2) + L_3^2})$. The inequality (63) is an exact analog of (34), therefore if we continue the reasoning of the theorem 1, then we easily establish the monotone convergence of sequence $x_1^n, x_2^n \to x_1^*, x_2^*$, $p_1^n, p_2^n \to p_1^*, p_2^*$ as $n \to \infty$ to game solution of the problem. The theorem is proved.

5 Game problems with coupled constraints

Let the objective functions $f_1(x_1, x_2) + \varphi_1(x_1)$, $f_2(x_1, x_2) + \varphi_2(x_2)$ and the functional vectorial constraint $g(x_1, x_2) \leq 0$, where $g(x_1, x_2) = (g_1(x_1, x_2), ..., g_m(x_1, x_2))$ be defined on the product space $R_1^{n_1} \times R_2^{n_2}$. It is supposed that functions $f_1(z_1, x_2) + \varphi(z_1)$, $f_2(x_1, z_2) + \varphi(z_2)$ are continuous and convex. The constraint functions $g(x_1, x_2)$ are convex in variables x_1 and x_2 separately and, generally speaking, are not convex in these variables jointly, for example, it can be a saddle-type function $g(x_1, x_2) = \langle x_1, x_2\rangle$. We consider the following extreme map.
Let $G = \{x_1, x_2 \mid G(x_1, x_2) \leq 0, x_1 \in X_1, x_2 \in X_2\}$ be an admissible set. Having taken a certain point $x = (x_1, x_2) \in G$ in this set we take two cross-sections of a kind $G_1(x) = \{z_1 \in X_1 \mid (z_1, x_2) \in G\}$ and $G_2(x) = \{z_2 \in X_2 \mid (x_1, z_2) \in G\}$. Assuming that the objective functions are convex in their own variables for any values of other variables, we consider a pair of optimization problems, each of them is defined on its cross-section

$$y_1(x_2) \in \text{Argmin}\{f_1(z_1, x_2) + \varphi_1(z_1) \mid g(z_1, x_2) \leq 0, \quad z_1 \in X_1\},$$

$$y_2(x_1) \in \text{Argmin}\{f_2(x_1, z_2) + \varphi_2(z_2) \mid g(x_1, z_2) \leq 0, \quad z_2 \in X_2\}. \tag{64}$$

The system of problems (64) determines the operator $Y(x) = y_1(x_2) \times y_2(x_1)$, which maps any point $x = (x_1, x_2) \in G$ into some convex closed subset from $\pi_1 G \times \pi_2 G$, where $\pi_1 G$ is a projection of set G onto the space of variable x_1 and $\pi_2 G$ is a projection of set G onto the space of variable x_2. The subset $Y(x)$ represents a direct product of optimal sets of problems (64). If the functions $f_1(z_1, x_2) + \varphi(z_1)$, $f_2(x_1, z_2) + \varphi(z_2)$ are continuous and convex in their own variables, and $X_i, i = 1, 2$ are convex compact sets, then the map has a fixed point $x^* = (x_1^*, x_2^*)$, (Aubin, Frankowska 1990 [10]). This point satisfies the following system of extreme inclusions

$$x_1^* \in \text{Argmin}\{f_1(z_1, x_2^*) + \varphi_1(z_1) \mid g(z_1, x_2^*) \leq 0, \quad z_1 \in X_1\},$$
$$x_2^* \in \text{Argmin}\{f_2(x_1^*, z_2) + \varphi_2(z_2) \mid g(x_1^*, z_2) \leq 0, \quad z_2 \in X_2\}. \tag{65}$$

The problem (65) represents by itself a two-person game with coupled constraints dependent on the parameters x_1, x_2. For certain values of parameters each of the game participants tries to solve the convex programming problem in its own variables. Both problems of this system have, actually, the same constraint but with respect to different variables: the first problem in variable x_1, the second one in variable x_2. In this game situation both players are hardly coupled both objective functions and constraints. The level of independence of players in this situation is not high, therefore it is natural to consider the problem in scalarized form. To this end we introduce two normalized functions of the form

$$\Phi(v, w) + \varphi(w) = f_1(z_1, x_2) + \varphi_1(z_1) + f_2(x_1, z_2) + \varphi_2(z_2),$$
$$G(v, w) = g(z_1, x_2) + g(x_1, z_2) \tag{66}$$

where $w = (z_1, z_2), v = (x_1, x_2)$, $v, w \in \Omega = X_1 \times X_2$. In terms of the new variables problem(65) can be presented in the form

$$v^* \in \text{Argmin}\{\Phi(v^*, w) + \varphi(w) \mid \quad G(v^*, w) \leq 0, w \in \Omega\} \tag{67}$$

or that is the same

$$\Phi(v^*, v^*) + \varphi(v^*) \leq \Phi(v^*, w) + \varphi(w), \quad G(v^*, w) \leq 0 \;\; \forall w \in \Omega. \tag{68}$$

It is hard to be convinced of the equivalence of problems (65) and (67) or (68). We present (68) as

$$f_1(x_1^*, x_2^*) + \varphi_1(x_1^*) + f_2(x_1^*, x_2^*) + \varphi_2(x_2^*)$$
$$\leq f_1(z_1, x_2^*) + \varphi_1(z_1) + f_2(x_1^*, z_2) + \varphi_2(z_2),$$
$$g(z_1, x_2^*) + g(x_1^*, z_2) \leq 0, \quad z_1 \in X_1, z_2 \in X_2.$$

The objective function and the functional constraint in this problem have the separable structure. This circumstance is used to split the problem and to reduce it to a two-person game. To this end we introduce the Lagrange function

$$\mathcal{L}(x_1, x_2, z_1, z_2, \lambda) = f_1(z_1, x_2) + \varphi_1(z_1) + f_2(x_1, z_2) + \varphi_2(z_2) + \langle \lambda, g(z_1, x_2) + g(x_1, z_2) \rangle, \tag{69}$$

where $x_1 \in X_1, x_2 \in X_2, z_1 \in X_1, z_2 \in X_2, \lambda \geq 0$. Assuming that the optimization problem from (68) is subject to constraint qualification, it is possible to assert that x_1^*, x_2^*, λ^* is a saddle point for the Lagrange function in the state of equilibrium, that is

$$\mathcal{L}(x_1^*, x_2^*, x_1^*, x_2^*, \lambda) \leq \mathcal{L}(x_1^*, x_2^*, x_1^*, x_2^*, \lambda^*) \leq \mathcal{L}(x_1^*, x_2^*, z_1, z_2, \lambda^*) \tag{70}$$

for all $z_1 \in X_1,\ z_2 \in X_2,\ \lambda \geq 0$. The right-hand side of inequality for this system represents by itself the optimization problem for a separable function on square $X_1 \times X_2$. By virtue of separability of the objective function in z_1 and z_2 and the block structure of constraints the problem is splitted into two independent subtasks, each of them is determined in its own space, namely:

$$f_1(x_1^*, x_2^*) + \varphi_1(x_1^*) + \langle \lambda^*, g(x_1^*, x_2^*)\rangle \leq f_1(z_1, x_2^*) + \varphi_1(z_1) + \langle \lambda^*, g(z_1, x_2^*)\rangle, \quad z_1 \in X_1,$$

$$f_2(x_1^*, x_2^*) + \varphi_2(x_2^*) + \langle \lambda^*, g(x_1^*, x_2^*) \leq f_2(x_1^*, z_2) + \varphi_2(z_2) + \langle \lambda^*, g(x_1^*, z_2)\rangle, \quad z_2 \in X_2.$$

It is clear that these problems can be rewritten as

$$f_1(x_1^*, x_2^*) + \varphi_1(x_1^*) \leq f_1(z_1, x_2^*) + \varphi_1(z_1) + \langle \lambda^*, g(z_1, x_2^*) - g(x_1^*, x_2^*)\rangle, \quad z_1 \in X_1,$$

$$f_2(x_1^*, x_2^*) + \varphi_2(x_2^*) \leq f_2(x_1^*, z_2) + \varphi_2(z_2) + \langle \lambda^*, g(x_1^*, z_2) - g(x_1^*, x_2^*)\rangle, \quad z_2 \in X_2.$$

Or

$$x_1^* \in \text{Argmin}\{f_1(z_1, x_2^*) + \varphi_1(z_1) \mid \langle \lambda^*, g(z_1, x_2^*) - g(x_1^*, x_2^*)\rangle \leq 0, \quad z_1 \in X_1\},$$

$$x_2^* \in \text{Argmin}\{f_2(x_1^*, z_2) + \varphi_2(z_2) \mid \langle \lambda^*, g(x_1^*, z_2) - g(x_1^*, x_2^*)\rangle \leq 0, \quad z_2 \in X_2\}.$$

We carry through some transformations of the constraints for above problems. From the left-hand side inequality (70), we get $\langle \lambda - \lambda^*, g(x_1^*, x_2^*) + g(x_1^*, x_2^*)\rangle \geq 0\ \forall \lambda \geq 0$. Assuming, at first $\lambda = 0$, and then $\lambda = 2\lambda^*$ in this inequality, we obtain $\langle \lambda^*, g(x_1^*, x_2^*)\rangle = 0$ and $\lambda^* \geq 0$. In view of that the last two problems get the form

$$x_1^* \in \text{Argmin}\{f_1(z_1, x_2^*) + \varphi_1(z_1) \mid \langle \lambda^*, g(z_1, x_2^*)\rangle \leq 0,\ z_1 \in X_1\},$$

$$x_2^* \in \text{Argmin}\{f_2(x_1^*, z_2) + \varphi_2(z_2) \mid \langle \lambda^*, g(x_1^*, z_2)\rangle \leq 0,\ z_2 \in X_2\}. \tag{71}$$

The obtained statements mean that x_1^* and x_2^* are optimal points for functions $f_1(z_1, x_2^*) + \varphi_1(z_1)$ and $f_2(x_1^*, z_2) + \varphi_2(z_2)$ on sets $\{z_1 | \langle \lambda^*, g(z_1, x_2^*)\rangle \leq 0,\ z_1 \in X_1\}$ and $\{z_2 |\ \langle \lambda^*, g(x_1^*, z_2)\rangle \leq 0,\ z_2 \in X_2\}$ in equilibrium state. It is not easy to see that these points are optimal solutions of the same functions on sets of a kind $\{z_1 | g(z_1, x_2^*) \leq 0,\ z_1 \in X_1\}$ and $\{z_2 | g(x_1^*, z_2) \leq 0,\ z_2 \in X_2\}$. If not, under some regularity condition for these sets it is easily to receive inconsistency supposing that there exists another point $x_1^{'}$, for example for the first condition, such that $f_1(x_1^{'}, x_2^*) + \varphi_1(x_1^{'}) < f_1(x_1^*, x_2^*) + \varphi_1(x_1^*)$ and $g(x_1^{'}, x_2^*) \leq 0$. Multiplying the last inequality to non-negative vector λ^*, we get at once an inconsistency with above assertion. Thus, from (71) we have

$$x_1^* \in \text{Argmin}\{f_1(z_1, x_2^*) + \varphi_1(z_1) \mid g(z_1, x_2^*) \leq 0,\ z_1 \in X_1\},$$

$$x_2^* \in \text{Argmin}\{f_2(x_1^*, z_2) + \varphi_2(z_2) \mid g(x_1^*, z_2) \leq 0,\ z_2 \in X_2\}. \tag{72}$$

The equivalence of (65) and (67) is established.

6 Extragradient coupled constraints game methods

To solve the game (72) it is enough to solve the equilibrium problem (67). With this purpose we present the Lagrange function for (69) in new variables

$$\mathcal{L}(v, w, \lambda) = \Phi(v, w) + \varphi(w) + \langle \lambda, G(v, w) \rangle, \tag{73}$$

where $v \in \Omega, w \in \Omega, \lambda \geq 0$. The system of inequalities (70) in these variables looks like

$$\mathcal{L}(v^*, v^*, p) \leq \mathcal{L}(v^*, v^*, p^*) \leq \mathcal{L}(v^*, w, p^*) \ \ \forall w \in \Omega, p \geq 0. \tag{74}$$

We rewrite this system of inequalities in the equivalent form

$$v^* \in \text{Argmin}\{\Phi(v^*, w) + \varphi(w) + \langle \lambda^*, G(v^*, w) \rangle \mid \ w \in \Omega\},$$

$$\lambda_1^* \in \text{Argmax}\{\langle \lambda, G(v^*, v^*) \rangle \ \ \mid \ \lambda \geq 0\}, \tag{75}$$

The obtained system of problems generates in turn, necessary (sufficient in the convex case) conditions of a minimum both in the form of variational inequality and in the form of operator equation. Of course, both forms are equivalent, the first of them looks like

$$\langle \nabla_2 \Phi(v^*, v^*) + \nabla \varphi(v^*) + \nabla_2 G^T(v^*, v^*) \lambda^*, w - v^* \rangle \geq 0 \ \forall w \in \Omega,$$

$$-\langle G(v^*, v^*), \lambda - \lambda^* \rangle \geq 0 \ \forall \lambda \geq 0, \tag{76}$$

where $\nabla_2 \Phi(v, w), \varphi(w)$ are a partial gradient in w for any v and a gradient for functions $\Phi(v, w), \varphi(w)$ respectively. $\nabla_2 G^T(v, w)$ is $m_1 \times n$ matrix, where $\nabla_2 g_i(v, w)$, $i = 1, 2, ..., m$ are vector-lines. The second form of necessary conditions for (74) takes the form of operator equation

$$v^* = \pi_\Omega(v^* - \alpha(\nabla_2 \Phi(v^*, v^*) + \nabla \varphi(v^*) + \nabla_2 G^T(v^*, v^*)),$$

$$\lambda^* = \pi_+(\lambda^* + (\alpha/2) G(v^*, v^*)), \tag{77}$$

where $\pi_\Omega(...), \pi_+(...)$ are projection operators of some vector into Ω and the positive orthant R^n_+ respectively, $\alpha > 0$ is a parameter like steplength. Before we start to discussing the solution methods of the system of equations (77) we consider properties of the function $G(v, w)$ in more details . First of all we note the properties of symmetry for this function. Indeed, from $G(v, w) = g(z_1, x_2) + g(x_1, z_2) = g(x_1, z_2) + g(z_1, x_2) = G(w, v)$ it follows

$$G(v, w) = G(w, v) \ \ \forall v \in \Omega, w \in \Omega. \tag{78}$$

Differentiating identity (78) in w, we obtain

$$\nabla_2 G^T(v, w) = \nabla_1 G^T(w, v) \ \ \forall v \in \Omega, w \in \Omega, \tag{79}$$

where $\nabla_1 G^T(.,.), \nabla_2 G^T(.,.)$ are partial gradients (derivatives) in first and second variables. We prove a key property of the symmetrical function $G(v,w)$, namely: "private" in w gradient-restriction of the function $G(v,w)$ on the diagonal of square $\Omega\times\Omega$ is equal to a half of gradient of the restricted function $G(v,w)$ onto this diagonal

$$2\nabla_2 G^T(v,w)|_{v=w} = \nabla G^T(v,v) \;\; \forall v \in \Omega. \tag{80}$$

Indeed, by the definition of the differentiability for the function $G(v,w)$ we get (Antipin 1998 [3]):

$$G(v+h, w+k) = G(v,w) + \nabla_1 G^T(v,w)h + \nabla_2 G^T(v,w)k + \omega(v,w,h,k), \tag{81}$$

where $\omega(v,w,h,k)/(|h|^2+|k|^2)^{1/2} \to 0$ as $|h|^2+|k|^2 \to 0$. Let $w=v$ and $h=k$ be, then using (79), we get

$$G(v+h, v+h) = G(v,v) + 2\nabla_2 G^T(v,w)|_{v=w}h + \omega(v,h), \tag{82}$$

where $\omega(v,h)/|h| \to 0$ as $|h| \to 0$. Since (82) is a particular case of (81) it means that gradient-restriction $2\nabla_2 G^T(v,w)|_{v=w}$ is the gradient $\nabla G^T(v,v)$ of $G(v,v)$, i.e. (80) is true. If the function $G(v,w)$ is convex in w for any v, then the operator $\nabla_2 G^T(v,w)$ is monotone in w for any v but the gradient-restriction $\nabla_2 G^T(v,w)|_{v=w}$, generally speaking, is not monotone. To make sure the monotonicity of gradient-restriction (is not mandatory for symmetrical functions) we introduce a class of positive-semidefinite functions. To this end, we note in $g_0(v,w) = \Phi(v,w), G(v,w) = (g_1(v,w),...,G_m(v,w))$ and for each functions $g_i(v,w), i=0,1,...,m$ we extend (7) (Antipin 1995 [1]). A function $g_i(v,w)$ is called positive-semidefinite onto $\Omega\times\Omega$, if it obeys the inequality

$$g_i(w,w) - g_i(w,v) - g_i(v,w) + g_i(v,v) \geq 0 \;\; \forall w,v \in \Omega. \tag{83}$$

The condition of positive-semidefiniteness of (83) is sufficient to guarantee the monotonicity of gradient-restriction $\nabla_2 g_i(v,w)|_{v=w}$, if the function $g_i(v,w)$ is convex in w for any v. Indeed, using the system of inequalities (9) from (83) we get the monotonicity of gradient-restriction

$$\langle \nabla_2 g_i(w,w) - \nabla_2 g_i(v,v), w-v\rangle \geq 0 \quad \forall v,w \in \Omega. \tag{84}$$

Using the above inequalities, we transform separately third term in the first inequality (76). Taking into account the key property of symmetric functions (80) and convexity of vectorial function $G(v,v)$ componently, we have

$$\langle \nabla_2 G^T(v^*,v^*)\lambda^*, w-v^*\rangle = \frac{1}{2}\langle \lambda^*, \nabla G^T(v^*,v^*)(w-v^*)\rangle \leq \frac{1}{2}\langle \lambda^*, G(w,w)-G(v^*,v^*)\rangle \geq 0.$$

In view of an obtained evaluation we rewrite the first inequality from (76) in the form

$$\langle \nabla_2 \Phi(v^*,v^*) + \nabla\varphi(v^*), w-v^*\rangle + (1/2)\langle \lambda^*, G(w,w) - G(v^*,v^*)\rangle \geq 0 \;\; \forall w \in \Omega. \tag{85}$$

If the operator $\nabla_2\Phi(v,v)+\varphi(v)$ is monotone, then by virtue of (84) we get from (85)

$$\langle\nabla_2\Phi(w,w)+\nabla\varphi(w),w-v^*\rangle+(1/2)\langle\lambda^*,G(w,w)-G(v^*)\rangle\geq 0\ \forall w\in\Omega. \quad (86)$$

These estimates are underlying the convergence analysis of gradient-type methods to the equilibrium solutions (Antipin 2001 [5]). To solve the system of equations (77) we use the extragradient approach

$$\bar\lambda^n=\pi_+(\lambda^n+(\alpha/2)G(v^n,v^n)),$$

$$\bar v^n=\pi_\Omega(v^n-\alpha(\nabla_2\Phi(v^n,v^n)+\nabla\varphi(v^n)+\nabla_2G^T(v^n,v^n)\bar\lambda^n)) \quad (87)$$

$$\bar\lambda^{n+1}=\pi_+(\lambda^n+(\alpha/2)G(\bar v^n,\bar v^n)),$$

$$v^{n+1}=\pi_\Omega(v^n-\alpha(\nabla_2\Phi(\bar v^n,\bar v^n)+\nabla\varphi(\bar v^n)+\nabla_2G^T(\bar v^n,\bar v^n)\bar\lambda^n)), \quad (88)$$

The steplength α in (87),(88) is determined from the interval

$$0<\varepsilon\leq\alpha<1/\sqrt{2(L_1+L_2C)^2+(1/2)L_3^2},\quad \varepsilon>0, \quad (89)$$

where constants L_1,L_2,L_3,C are determined below. For the justification of correctness of selecting the parameter α we obtain evaluations of deviations for vectors $\bar v^n$ and v^{n+1}, $\bar\lambda^n$ and λ^{n+1} in (87), (88)

$$|\bar\lambda^n-\lambda^{n+1}|\leq(\alpha/2)|G(v^n,v^n)-G(\bar v^n,\bar v^n)|\leq(\alpha/2)L_3|v^n-\bar v^n|, \quad (90)$$

$$|\bar v^n-v^{n+1}|\leq\alpha(|\nabla_2\Phi(v^n,v^n)+\nabla\varphi(v^n)-\nabla_2\Phi(\bar v^n,\bar v^n)-\nabla\varphi(\bar v^n)|$$
$$+|\nabla_2G^T(v^n,v^n)-\nabla_2G^T(\bar v^n,\bar v^n)||\bar\lambda^n|)|\bar v^n-v^n|\leq\alpha(L_1+L_2|\bar\lambda^n|)|\bar v^n-v^n|$$
$$\leq\alpha(L_1+L_2C)|\bar v^n-v^n|, \quad (91)$$

where

$$(|\nabla_2\Phi(v^n,v^n)+\nabla\varphi(v^n)-\nabla_2\Phi(\bar v^n,\bar v^n)-\nabla\varphi(\bar v^n)|\leq L_1|\bar v^n-v^n|,$$

$$|\nabla_2G^T(v^n,v^n)-\nabla_2G^T(\bar v^n,\bar v^n)|\leq L_2|\bar v^n-v^n|,\quad |\bar\lambda^n|\leq,C\ \forall n\to\infty.$$

We rewrite (87),(88) in the form of variational inequalities.

$$\langle\bar\lambda^n-\lambda^n-(\alpha/2)G(v^n,v^n),\lambda-\bar\lambda^n\rangle\geq 0,\ \forall\lambda\geq 0,$$

$$\langle\bar v^n-v^n+\alpha(\nabla_2\Phi(v^n,v^n)+\nabla\varphi(v^n)+\nabla_2G^T(v^n,v^n)\bar\lambda^n)),v-\bar v^n\rangle\geq 0,\quad\forall v\in\Omega, \quad (92)$$

and

$$\langle\lambda^{n+1}-\lambda^n-(\alpha/2)G(\bar v^n,\bar v^n),\lambda-\lambda^{n+1}\rangle\geq 0,\quad\forall\lambda\geq 0,$$

$$\langle v^{n+1}-v^n+\alpha(\nabla_2\Phi(\bar v^n,\bar v^n)+\nabla\varphi(\bar v^n)+\nabla_2G^T(\bar v^n,\bar v^n)\bar\lambda^n)),v-v^{n+1}\rangle\geq 0,\quad\forall v\in\Omega, \quad (93)$$

We show that the process (87)-(89) converges monotonically under the norm to one of equilibrium solutions.

Theorem 3 *Suppose that a solution set of game problem (72) is non-empty, the function $\Phi(v,w)$, $G(v,w)$ are positive semidefinite and convex in w for any v, the Lipschitz conditions hold in (90), (91), dual sequence $|\bar{\lambda}^n| \le C$ is bounded for all n, $\Omega \subseteq R^n$ is convex closed set. Then, the sequence v^n, λ^n, generated by method (87)-(89) converges monotonically under the norm to one of the equilibrium solutions, i.e. $v^n, \lambda^n \to v^*, \lambda^* \in \Omega^* \times R^N_+$ as $n \to \infty$.*

PROOF. By putting $w = v^*$ in (93), we get

$$\langle v^{n+1} - v^n, v^* - v^{n+1}\rangle + \alpha\langle \nabla_2\Phi(\bar{v}^n, \bar{v}^n) + \nabla\varphi(\bar{v}^n), v^* - v^{n+1}\rangle$$
$$+\alpha\langle \nabla_2 G^T(\bar{v}^n, \bar{v}^n)\bar{\lambda}^n, v^* - v^{n+1}\rangle \ge 0. \tag{94}$$

Take $w = v^{n+1}$ in (92)

$$\langle \bar{v}^n - v^n + \alpha(\nabla_2\Phi(v^n, v^n) + \nabla\varphi(v^n) + \nabla_2 G^T(v^n, v^n)\bar{\lambda}^n), v^{n+1} - \bar{v}^n\rangle \ge 0.$$

Hence

$$\langle \bar{v}^n - v^n, v^{n+1} - \bar{v}^n\rangle + \alpha\langle \nabla_2\Phi(\bar{v}^n, \bar{v}^n) + \nabla\varphi(\bar{v}^n), v^{n+1} - \bar{v}^n\rangle$$
$$-\alpha\langle \nabla_2\Phi(\bar{v}^n, \bar{v}^n) + \nabla\varphi(\bar{v}^n) - \nabla_2\Phi(v^n, v^n) - \nabla\varphi(v^n), v^{n+1} - \bar{v}^n\rangle$$
$$+\alpha\langle \nabla_2 G^T(\bar{v}^n, \bar{v}^n)\bar{\lambda}^n, v^{n+1} - \bar{v}^n\rangle - \alpha\langle (\nabla_2 G^T(\bar{v}^n, \bar{v}^n) - \nabla_2 G^T(v^n, v^n))\bar{\lambda}^n, v^{n+1} - \bar{v}^n\rangle \ge 0,$$

or taking into account (91)

$$\langle \bar{v}^n - v^n, v^{n+1} - \bar{v}^n\rangle + \alpha\langle \nabla_2\Phi(\bar{v}^n, \bar{v}^n) + \nabla\varphi(\bar{v}^n), v^{n+1} - \bar{v}^n\rangle$$
$$+\alpha\langle \nabla_2 G^T(\bar{v}^n, \bar{v}^n)\bar{\lambda}^n, v^{n+1} - \bar{v}^n\rangle + \alpha^2(L_1 + L_2 C)^2|\bar{v}^n - v^n| \ge 0. \tag{95}$$

We add inequalities (94) and (95)

$$\langle v^{n+1} - v^n, v^* - v^{n+1}\rangle + \langle \bar{v}^n - v^n, v^{n+1} - \bar{v}^n\rangle + \alpha\langle \nabla_2\Phi(\bar{v}^n, \bar{v}^n) + \nabla\varphi(\bar{v}^n), v^* - \bar{v}^n\rangle$$
$$+\alpha\langle \nabla_2 G^T(\bar{v}^n, \bar{v}^n)\bar{\lambda}^n, v^* - \bar{v}^n\rangle + \alpha^2(L_1 + L_2 C)^2|\bar{v}^n - v^n|^2 \ge 0. \tag{96}$$

Using (80) and the convexity of function $G(v,v)$ by virtue of (84), we transform apart the fourth term from (96)

$$\langle \bar{\lambda}^n, \nabla_2 G(\bar{v}^n, \bar{v}^n)(v^* - \bar{v}^n)\rangle = (1/2)\langle \bar{\lambda}^n, \nabla G(\bar{v}^n, \bar{v}^n)(v^* - \bar{v}^n)\rangle$$
$$\le (1/2)\langle \bar{\lambda}^n, G(v^*, v^*) - G(\bar{v}^n, \bar{v}^n)\rangle.$$

Then

$$\langle v^{n+1} - v^n, v^* - v^{n+1}\rangle + \langle \bar{v}^n - v^n, v^{n+1} - \bar{v}^n\rangle + \alpha\langle \nabla_2\Phi(\bar{v}^n, \bar{v}^n) + \nabla\varphi(\bar{v}^n), v^* - \bar{v}^n\rangle$$
$$+(\alpha/2)\langle \bar{\lambda}^n, G(v^*, v^*) - G(\bar{v}^n, \bar{v}^n)\rangle + \alpha^2(L_1 + L_2 C)^2|\bar{v}^n - v^n|^2 \ge 0.$$

We put $w = \bar{v}^n$ in inequality (86)

$$\langle \nabla_2 \Phi(\bar{v}^n, \bar{v}^n) + \nabla \varphi(\bar{v}^n), \bar{v}^n - v^* \rangle + (1/2)\langle \lambda^*, G(\bar{v}^n, \bar{v}^n) - G(v^*, v^*)\rangle \geq 0.$$

Add the last two inequalities to obtain

$$\langle v^{n+1} - v^n, v^* - v^{n+1}\rangle + \langle \bar{v}^n - v^n, v^{n+1} - \bar{v}^n\rangle$$

$$+(\alpha/2)\langle \bar{\lambda}^n - \lambda^*, G(v^*, v^*) - G(\bar{v}^n, \bar{v}^n)\rangle + \alpha^2 (L_1 + L_2 C)^2 |\bar{v}^n - v^n|^2 \geq 0. \qquad (97)$$

Consider first the inequalities from (92) and (93). Put $\lambda = \lambda^*$ in (93)

$$\langle \lambda^{n+1} - \lambda^n, \lambda^* - \lambda^{n+1}\rangle - (\alpha/2)\langle G(\bar{v}^n, \bar{v}^n), \lambda^* - \lambda^{n+1}\rangle \geq 0 \qquad (98)$$

and $\lambda = \lambda^{n+1}$ in (92).

$$\langle \bar{\lambda}^n - \lambda^n, \lambda^{n+1} - \bar{\lambda}^n\rangle + (\alpha/2)\langle G(\bar{v}^n, \bar{v}^n) - G(v^n, v^n), \lambda^{n+1} - \bar{\lambda}^n\rangle$$

$$-(\alpha/2)\langle G(\bar{v}^n, \bar{v}^n), \lambda^{n+1} - \bar{\lambda}^n\rangle \geq 0, \qquad (99)$$

The second term in this inequality can be estimated by means of (90), and then we add both inequalities (98) and (99)

$$\langle \lambda^{n+1} - \lambda^n, \lambda^* - \lambda^{n+1}\rangle + \langle \bar{\lambda}^n - \lambda^n, \lambda^{n+1} - \bar{\lambda}^n\rangle$$

$$+(\alpha/2)^2 L_3^2 |\bar{v}^n - v^n|^2 - (\alpha/2)\langle G(\bar{v}^n, \bar{v}^n), \lambda^* - \bar{\lambda}^n\rangle \geq 0,$$

Using the relations $\langle \bar{\lambda}^n, G(v^*, v^*)\rangle \leq 0$, $\langle \lambda^*, G(v^*, v^*)\rangle = 0$, we rewrite the latter inequality in the form

$$\langle \lambda^{n+1} - \lambda^n, \lambda^* - \lambda^{n+1}\rangle + \langle \bar{\lambda}^n - \lambda^n, \lambda^{n+1} - \bar{\lambda}^n\rangle$$

$$+((\alpha/2))^2 L_3^2 |\bar{v}^n - v^n|^2 + (\alpha/2)\langle G(v^*, v^*) - G(\bar{v}^n, \bar{v}^n), \lambda^* - \bar{\lambda}^n\rangle \geq 0, \qquad (100)$$

We add inequalities (97) and (100)

$$\langle v^{n+1} - v^n, v^* - v^{n+1}\rangle + \langle \bar{v}^n - v^n, v^{n+1} - \bar{v}^n\rangle$$

$$+\langle \lambda^{n+1} - \lambda^n, \lambda^* - \lambda^{n+1}\rangle + \langle \bar{\lambda}^n - \lambda^n, \lambda^{n+1} - \bar{\lambda}^n\rangle$$

$$+\alpha^2 (L_1 + L_2 C)^2 |\bar{v}^n - v^n|^2 + (\alpha/2)^2 L_3^2 |\bar{v}^n - v^n|^2 \geq 0.$$

By means of identity (33) we expand the first four scalar products into the sum of squares

$$\frac{1}{2}|v^{n+1} - v^*|^2 + \frac{1}{2}|\lambda^{n+1} - \lambda^*|^2 + \frac{1}{2}|v^{n+1} - \bar{v}^n|^2 + \frac{1}{2}|\bar{v}^n - v^n|^2 + \frac{1}{2}|\lambda^{n+1} - \bar{\lambda}^n|^2 + \frac{1}{2}|\bar{\lambda}^n - \lambda^n|^2$$

$$\alpha^2((L_1 + L_2 C)^2 + (1/2)^2 L_3^2)|\bar{v}^n - v^n|^2 \leq \frac{1}{2}|v^n - v^*|^2 + \frac{1}{2}|\lambda^n - \lambda^*|^2. \qquad (101)$$

Then

$$|v^{n+1} - v^*|^2 + |\lambda^{n+1} - \lambda^*|^2 + |v^{n+1} - v^n|^2 + d|\bar{v}^n - v^n|^2$$
$$+|\lambda^{n+1} - \lambda^n|^2 + |\bar{\lambda}^n - \lambda^n|^2 \le |v^n - v^*|^2 + |\lambda^n - \lambda^*|^2, \tag{102}$$

where $d = 1 - 2\alpha^2((L_1 + L_2C)^2 + (1/2)^2L_3^2) > 0$, by virtue of (89). Summing (102) from $n = 0$ up to $n = N$ to get

$$|v^{N+1} - v^*|^2 + |\lambda^{N+1} - \lambda^*|^2 + \sum_{k=0}^{k=N} |v^{k+1} - \bar{v}^k|^2$$

$$+d\sum_{k=0}^{k=N} |\bar{v}^k - v^k|^2 + \sum_{k=0}^{k=N} |\lambda^{k+1} - \lambda^k|^2 \sum_{k=0}^{k=N} |\bar{\lambda}^k - \lambda^k|^2 \le |v^0 - v^*|^2 + |\lambda^0 - \lambda^*|^2.$$

From the obtained inequality it follows the boundedness of the trajectory

$$|v^{N+1} - v^*|^2 + \frac{1}{2}|\lambda^{N+1} - \lambda^*|^2 \le |v^0 - v^*|^2 + \frac{1}{2}|\lambda^0 - \lambda^*|^2,$$

and

$$\sum_{k=0}^{\infty} |v^{k+1} - \bar{v}^k|^2 < \infty, \quad \sum_{k=0}^{\infty} |\bar{v}^k - v^k|^2 < \infty, \quad \sum_{k=0}^{\infty} |\lambda^{k+1} - \bar{\lambda}^k|^2 < \infty, \sum_{k=0}^{\infty} |\bar{\lambda}^k - \lambda^k|^2 < \infty,$$

and, consequently, tend to zero of quantities

$$|v^{n+1} - \bar{v}^n|^2 \to 0, \quad |\bar{v}^n - v^n|^2 \to 0, \quad |\lambda^{n+1} - \bar{\lambda}^n|^2 \to 0, |\bar{\lambda}^n - \lambda^n|^2 \to 0, \quad n \to \infty.$$

Since the sequence v^n, λ^n is bounded, there exist a point v', λ' such that $v^{n_i} \to v', \quad \lambda^{n_i} \to \lambda'$? $n_i \to \infty$. We consider inequalities (92),(93) for all $n_i \to \infty$ and, passing to a limit we get

$$\langle \nabla_2\Phi(v', v') + \nabla\varphi(v') + \nabla_2 G^T(v', v')\lambda', w - v' \rangle \ge 0 \quad \forall w \in \Omega,$$

$$\langle -G(v', v'), \lambda - \lambda' \rangle \ge 0 \quad \forall \lambda \ge 0.$$

The inequalities obtained coincide to (76). Then $v' = v^* \in \Omega^*, \quad \lambda' = \lambda^* \ge 0$, i.e., any limit point of v^n, λ^n is an equilibrium solution to the problem. The monotonicity condition of decreasing value $|v^n - v^*| + |\lambda^n - \lambda^*|$ provides of uniqueness of limit point, i.e. the convergence of $v^n \to v^*, \quad \lambda^n \to \lambda^*$ as $n \to \infty$. The theorem is proved.

References

[1] Antipin, A.S. (1995) The Convergence of Proximal Methods to Fixed Points of Extremal Mappings and Estimates of Their Rate of Convergence, *Comp.Maths.Math.Phys*, **35**, No.5, 539-551.

[2] Antipin, A.S. (1997) Equilibrium programming: Gradient-type methods, *Automation and Remote Control.* No.8, 125-137 (Russian).

[3] Antipin, A.S. (1998) Splitting of a gradient approach for solving extreme inclusions, *Comp.Maths.Math.Phys*, **38**, No.7, 1069-1082.

[4] Antipin, A.S. (2000) Gradient-Type Method for Equilibrium Programming Problems with Coupled Constraints, *Yugoslav Journal of Operations research*, **10**. No.2, 1-15.

[5] Antipin, A.S. (2001) Differential equations for equilibrium problems with coupled constraints, *Nonlinear Analysis*, **47**, 1833-1844.

[6] Antipin, A.S. (2001) Gradient approach of computing fixed points of equilibrium problems. *Journal of Global Optimization*,24 (3): 285-309.

[7] Antipin, A. (2002), *Gradient and Extragradient Approaches to Bilinear Equilibrium Programming*, Computing Center RAS, Moscow (in Russian).

[8] Bakushinsky, A. and Goncharsky, A.(1994) *Ill-Posed Problems: Theory and Application.* Kluwer, Dordrecht etc.

[9] Brown, G.W. (1951), Iterative Solutions of Games by Fictitions Play. *Activity Analysis of Production and Allocation, Cowles Commission for Research in Economics* John Wiley and Sons. NY. **13**, 374-376

[10] Aubin, J.-P. and Frankowska, H. (1990) *Set Valued Analysis.* Birkhauser, Boston.

[11] Hansen, T.H. and Scarf, H. (1974) On the approximation of Nash equilibrium points in an N-person noncooperative game *SIAM Journal of Applied Mathematics* **26**, 622-637.

[12] Lemke,C.E. and Howson, J.T. (1964), Equilibrium points of bimatrix games, *SIAM Review* **12**, 45-78.

[13] Mills,H. (1960) Equilibrium points in finite games, *J.Soc.Indust.Appl.Math.* **8**, No.2, 397-402.

[14] Nash, J.F., Jr. (1950) Equilibrium points in n-person games, *Proc. Nat.Acad.Sci.* **36**, 48-49.

[15] Nikaido, H. and Isoda, K. (1955) Note on non-cooperative convex game, *Pacific J. Math.* **5**, Supp. 1, 807-815.

[16] Robinson, J. (1951) An Iterative Method of Solving a Game, *Ann. Math.* **54**, 296-301.

[17] Rosen, J.B. (1965) Existence and uniqueness of equilibrium points for concave N-person games, *Econometrica* **33**, 3, 520-534.

Optimization and Optimal Control, pp. 29 – 35
P.M. Pardalos, I. Tseveendorj, and R. Enkhbat, Editors

A Multilevel Construction Algorithm for the Traveling Salesman Problem

N. Bouhmala (noureddine.bouhmala@hive.no),
T. Natvig, and M. Jakobsen
Vestfold University College
Department of computer science, Norway.

Abstract

In this paper, we introduce a multilevel construction algorithm for the traveling salesman problem. The resulting algorithm starts by coarsening the original problem into a sequence of smaller problems using a coarsening scheme. Thereafter a tour is determined at the smallest problem and is projected back to the original problem by going through a refinement phase at each intermediate level. We investigate the quality of the solution produced based on two different choices made during the coarsening and the refinement phases. A standard set of benchmark instances is used in order to assess the performance of the algorithm. Experimental results indicate that the proposed algorithm produces tours of better quality than those produced by the Clark-Wright algorithm at lower cost.

Keywords: Traveling Salesman Problem, Graph Coarsening, Multilevel Construction Algorithm, Combinatorial Optimization.

1 Introduction

The traveling salesman problem (TSP) is one of the classical problems of combinatorial optimization. It is concerned with a salesman who must visit a number of cities once and return to the city of origin. In this paper we shall concentrate on the Euclidean symmetric TSP in which the distance from city c_i to city c_j is the same as the distance from city c_j to city c_i. The problem is defined with N cities having coordinates in the 2D plane and a symmetric distance matrix $D = [d(c_i, c_j)]$ which gives the Euclidean distance between any two cities c_i and c_j. The goal in the TSP is to find an ordering π of the cities that minimizes the expression 1.

$$L(\pi) = \sum_{i=1}^{N-1} d(c_{\pi(i)}, c_{\pi(i+1)}) + d(c_{\pi(N)}, c_{\pi(1)}). \tag{1}$$

The TSP belongs in the class of combinatorial optimization problems known as NP-hard [3] and has been extensively studied due to its simplicity and applicability. Problems having the TSP structure occur, for example, in circuit board drilling applications [10], instances arising in VLSI fabrication [9], and sometimes as a subproblem for the vehicle routing problem [2]. The simplicity of the problem coupled with its intractability makes it an ideal platform for exploring new algorithmic techniques. Due to the combinatorial explosion nature of the TSP, the researchers have proposed various approaches to solve this problem with the aim usually being just to find near optimal solutions. This has led to the development of construction algorithms (i.e., algorithms that generate starting tours from scratch) and local search algorithms. The reader may refer to [7] for a comprehensive survey of TSP methods. The remainder of this paper is organized as follows. Section 2 describes the multilevel construction algorithm. Section 3 compares the performance of the multilevel construction algorithm with that of the Clark-Wright algorithm. Section 4 contains conclusions and suggestions for future research.

2 Multilevel Contraction Algorithm (MCA)

Multilevel techniques have been applied to the graph partitioning problem [1] [5] [8] [12] and have proved to be effective in producing high quality solutions at a lower cost. Recently, these techniques have been proposed for the traveling salesman problem [13] with good results. The multilevel paradigm is a simple procedure which at its heart involves recursive coarsening to produce smaller and smaller problems that are easier to solve than the original one. In a multilevel framework, the original problem P_0 is coarsened using a coarsening procedure into a sequence of smaller problems P_1,P_2,....P_m such that $|P_0| > |P_1| > |P_2|..... > |P_m|$. Once the coarsening process terminates, a solution is computed for the problem P_m, and this solution is projected back to P_0 by going through intermediate improved solutions for P_{m-1},P_{m-2},...P_1,P_0. A common feature that characterizes these multilevel algorithms, is that any solution in any of the coarsened problems is a legitimate solution to the original problem. Thus at any stage after having generated a solution at the lowest level, the current solution could simply be extended through all problem levels to obtain a solution of the original problem. The basic structure of MCA is as follows: During the coarsening process, a coarsening scheme is used to match pairs of cities to form new ones. These new cities together with their coordinates are used to define a new and smaller problem and recursively iterate the coarsening process until the size of the problem reaches some desired threshold. The coarsening process terminates when the problem is reduced to two cities. A the lowest level a tour is easily computed, however, unlike previous multilevel algorithms, the tour generated at the lowest level is not a feasible solution to the original problem due the manner coarsening is performed. Finally, the tour computed at the lowest level is projected back towards the original problem. Since each city c_k of P_{l+1} contains at most two cities c_i and c_j of P_l, obtaining a tour for P_l from P_{l+1} is done by replacing each city c_k of P_{l+1} by the two merged cities c_i and c_j of P_{l-1}. The two cities c_i and c_j are then inserted on the tour based on two strategies which will be explained when dealing with the uncoarsening phase. A legitimate tour is constructed once the refinement process reaches the original problem. The multilevel construction algorithm consists of the following three phases.

- **Coarsening Phase:**
  The problem P_0 is reduced into a sequence of smaller problems P_1,P_2,....P_m such that $|P_0| > |P_1| > |P_2|..... > |P_m|$. It will require at least $O(logN/N')$ coarsening phases to coarsen the problem down to N' cities.

- **Solution Phase:**
  A tour is generated for the problem $|P_m|$.

- **Uncoarsening Phase:**
  The tour found for the problem P_m is projected back to P_0 by going through intermediate improved tours for P_{m-1},P_{m-2},...P_1,P_0.

2.1 Coarsening Phase

During the coarsening phase, a sequence of smaller problems each with with fewer cities is constructed. In what follows, we describe two coarsening methods that have being implemented.

- **Random Method (RM)** Assume that the original problem P_0 is to be coarsened. The cities are visited in random order. If a city c_i has not being matched yet, then a randomly unmatched city c_j is selected. If such a city c_j exists, then a new city c_k is created. This new city which belongs to the problem P_1 will be positioned at the center of the line joining the two matched cities. The unmatched cities are simply copied over to the next level.

- **Nearest Method (NM)** The cities are again visited in random manner. However, instead of matching a city c_i with a randomly unmatched city, we match the city c_i with an unmatched city c_j such that $d(c_i, c_j)$ is minimum over all valid unmatched cities.

2.2 Uncoarsening Phase

During the uncoarsening phase, the tour found for P_m is projected back to the original problem by going through the problems P_{m-1},P_{m-2}...P_1,P_0. Since each city of P_m is the result of matching at most a pair of cities of P_{m-1}, obtaining a tour for P_{m-1} from P_m is done by simply replacing each city of P_m with the matched cities of P_{m-1}. The two matched cities are then inserted on the tour at the positions that cause the least increase in the tour length. In what follows, two simple insertion methods are described:

- **Simple Insert (SI)** Assume that the city c_k of P_{m+1} is the result of collapsing the two cities c_i and c_j of P_m. The SI method inserts the cities c_i and c_j on the tour between the predecessor and the successor of the city c_k in the order that causes the least increase in tour length.
- **Minimal Insert (MI)** Instead of restricting the insertion of the cities c_i and c_j of P_m on the tour between the predecessor and the successor of c_k, the MI method tries to insert the two merged cities c_i and c_j on the tour at the positions that lead to the least increase in tour length.

3 Experimental Results

In the experiments conducted, we used problems from the publicly available library of TSP problems, TSPLIB [11], in order to assess the performance of the multilevel construction algorithm and that of the Clark-Wright algorithm. All the experiments were performed on a Pentium II/450 with 128 MB of memory. All times reported are in seconds. The set of problems used include eighteen instances ranging from 100 to 15112 cities. Tables 1 and 2 show the results of the experiments. The first column on both tables indicates the problem instance and its size, (e.g., d695 has 695 cities). The solution quality was measured by the percentage excess above the best known solution (or optimal solution if known). The results shown in the two first columns in Table 1 are based on averaged results over five runs. Table 1 shows the quality of the solution produced by MCA based on different choices made during the coarsening and the refinement phases. The best results were achieved using the combination of the NM coarsening method and the MI refinement method. The MI produces better results than the SI regardless of the coarsening strategy used. The MI does up to 68% better than the SI when RM is used as the coarsening method, and up to 10% when NM is used. The results also show that as far as the coarsening is concerned, NM does better than RM. The improvement lies within 71% when the SI is used as the refinement method, and within 10% when the MI is used.

Table 2 compares the performance of the MCA that uses the NM during the coarsening phase and the MI during the refinement phase against the CW algorithm. The results show that MCA produces better results than CW. The improvement lies

Problem Instance	RM-SI	RM-MI	NM-SI	NM-MI
d657	83%	14%	16%	7%
pr1002	88%	17%	15%	8%
pr2392	93%	18%	18%	8%
fl3795	90%	22%	19%	12%
rl5934	92%	27%	21%	14%

Table 1: Comparing the quality of the solution produced by different variants of the multilevel construction algorithm.

	CW		MCA	
Problem Instance	**Q**	**T(s)**	**Q**	**T(s)**
kroA100	8%	0.03	2%	0.01
kroA150	7%	0.05	3%	0.02
kroA200	9%	0.07	6%	0.03
lin318	8%	0.13	6%	0.05
d493	10%	0.29	5%	0.13
d657	9%	0.31	7%	0.15
pr1002	9%	0.37	7%	0.19
kro1084	8%	0.25	7%	0.22
d1291	12%	0.57	11%	0.35
fl1400	9%	0.53	3%	0.41
fl1577	9%	1.12	9%	0.61
vm1748	11%	1.17	7%	0.73
pr2392	13%	2.4	9%	1.40
fl3795	13%	5.6	12%	3.71
rl5934	11%	17.10	12%	9.97
rl11849	10%	116.11	12%	55.12
usa13509	12%	238.94	8%	107.32
d15112	10%	345.33	8%	128.75

Table 2: Various problem instances used in evaluating the multilevel construction algorithm and the Clark-Wright algorithm.

within 6%. For the two instances rl5934 and rl1849 where CW does better, it is marginally better (1%). Furthermore, the time required by MCA is smaller than that required by CW. From Table 2, we see that MCA is usually 2 times faster.

4 Conclusions and Future Research

In this paper, a multilevel construction algorithm is introduced. The resulting algorithm progressively coarsens the problem, initializes a tour and then uses an insertion method to construct a solution on each of the coarsened problems. The experiments have shown that the choices made during the coarsening and the refinement phases have great effect on the overall quality of the solution. The multilevel construction algorithm that uses NM for coarsening and MI for refinement produces better results than CW and requires less time. Further work include a faster implementation of MCA, the use of different coarsening schemes and the possibility of including guided local search during the refinement phase.

References

[1] N. Bouhmala. A Multilevel Algorithm for the Graph Partitioning Problem. *International Conference On Optimization Techniques and Applications,* Vol.1, pp.259-264, 1998.

[2] T. Christofides. Vehicle Routing in the Traveling Salesman Problem. *Lawler, Lenstra, Rinooy Kan and Shmoys, eds., Jhon Wiley,* pp.431-448, 1985.

[3] M.R. Garey and D.S. Johnson. Computers and Intractability: A Guide to the theory of NP-Completeness. *Freeman, San Fransisco, 1979.*

[4] J. Grefenstette, R. Gopal, R. Rosmaita, and D. Gucht. Genetic algorithms for the traveling salesman problem. *In Proceedings of the Second International Conference on Genetic Algorithms. Lawrence Eribaum Associates, Mahwah, NJ, 1985.*

[5] B. Hendrickson and R. Leland. A Multilevel Algorithm for Partitioning Graphs. *Technical Report SAND93-1301,* Sandia National Laboratories, 1993.

[6] J.H. Holland. Adaptation in natural and artificial systems. University of Michigan Press. Ann Arbor, MI, 1975.

[7] D.S. Johnson and L. A. McGeoch. The traveling salesman problem: a case study. In E.Aarts and J.K. Lenstra, editors, *Local Search in Combinatorial Optimization, pages 215-310. Wiley, Chichester, 1997.*

[8] G. Karypis and V. Kumar. A fast and High Quality Multilevel Scheme for Partitioning Irregular Graphs. *Technical report TR 95-035, Computer Science Department,* University of Minnesota, Minneapolis, MN 55455, 1995.

[9] B. Korte. Applications of Combinatorial Optimization. Talk at the 13th International Mathematical Programming Symposium, Tokyo, 1988.

[10] J.D. Litke. An Improved Solution to the Traveling Salesman Problem with Thousands of Nodes.*Communications of the ACM, Vol.27, No.12, pp.1227-1236, 1984.*

[11] G. Reinelt. The Traveling Salesman: Computational Solutions for TSP Applications, Lecture Notes in Computer Science, 840, Springer-Verlag, 1994.

[12] C. Walshaw, C. Cross, and M. Everett. A Localised Algorithm for Optimizing Unstructured Partitions. *Int. Journal. Supercomputer Appl,* Vol. 9, No.4, pp. 280-295, 1995.

[13] C. Walshaw. A Multilevel Approach to the Traveling Salesman Problem. *Accepted for Oper. Res., (Originally published as Univ.Greenwich Tech. Rep. 00/IM/63), 2000.*

Optimization and Optimal Control, pp. 37–58
P.M. Pardalos, I. Tseveendorj, and R. Enkhbat, Editors

Fair Network Resource Allocation and Link Pricing: A Numerical Study

Mustapha Bouhtou (Mustapha.Bouhtou@rd.francetelecom.com)
France Télécom R&D, DAC/OAT,39-40, rue du Général Leclerc, 92131, Issy-Les-Moulineaux, France.

Madiagne Diallo (Madiagne.Diallo@prism.uvsq.fr)
Laboratoire PRISM, Université de Versailles, 45 Av. des Etats-Unis, 78035 Versailles, France.

Laura Wynter (lwynter@us.ibm.com)
IBM Watson Research Center P.O. Box 704, Yorktown Heights, NY 10598, USA.

Abstract

In this paper, we present an analysis of a restricted bi-level optimization model proposed by [19] applied to telecommunication network pricing. This framework provides a generalization of telecommunication resource allocation/shadow price-based schemes such as those of [9] and [23], and permits the optimization of link prices with an aim towards revenue maximization. The analysis of the model is based on numerical experiments that we conducted for a variety of networks with varying parameters. We show that in practice the goal of maximizing operator revenue is not always attainable. We present some such cases, as well as valid alternative objectives for these instances.

Keywords: Internet Pricing, Network Equilibrium, Revenue Maximization, Fair Allocation, Proportional Fairness, Bilevel Program, Traffic Management.

1 Introduction

The evolution of the telecommunications sector has been characterized by the rapid increase in flow levels as well as the increase in applications making use of telephone and Internet media for transmission. Until recently, the provision of telecommunications services was limited to the national telephone companies. Today, following the de-regulation of the telecommunications markets in most countries, the user is faced with a wide range of choices. A simple connection can involve more than two operators. Because of this multiplicity, it is becoming more important to each operator to examine the pricing of its services. Furthermore, the fact that users have many options available has motivated researchers to consider the user's desires simultaneously with those of the telecom operator in order to optimize the operator's objectives.

Traditionally, models of telecom pricing set prices independently of consumer behavior [1], [6], and [5]. A number of recent methods, however, take into account the user by considering his or her willingness to pay, or some QoS requirements, through either a non-linear utility function [7], [9], [10], a system of price-equations [3], [2], [8], [25], or a pre-defined maximum budget [24].

Models that take into account user behavior, in particular, [7], [9], [10] and [23], through the incorporation of user utility functions, often define link prices as Lagrange multipliers. While the existence of the prices is thus guaranteed under the hypotheses of the models, the uniqueness is not guaranteed, and in fact, is usually not satisfied, as we shall see in this paper. Furthermore, defining link prices in terms of Lagrange multipliers has one very important consequence. Namely, the price on a link is zero until that link is congested. This property may seem at odds with a provider's goal of revenue maximization; indeed, it was derived rather from the concept of fairness, especially proportional fairness allocation [7]. We shall show in this paper that the two notions, that of a fair allocation and of revenue maximization, are not necessarily incompatible.

The objective of this work is to apply optimization models and methods to the management of telecommunications networks through pricing. It is of interest for network resource allocation models and methods to take into account the goal of revenue maximization simultaneously with the maximization of the users satisfaction on the resources provided across the network. To the best of our knowledge, these two elements do not appear to have been dealt with jointly in the literature up to now.

In the next section, we recall the frameworks of Proportional Fairness [7] Resource Allocation and Pricing, and Toll Optimization [19]. In Section 3, we provide the numerical study and analysis. Section 4 concludes the paper.

2 Fair Resource Allocation and Revenue Optimization Models

In some cases, we would like for the prices not to be tied to Lagrange multipliers, but rather to be freely optimized so as to obtain the highest possible revenue. This implies the resolution of two models, which are linked by their common variables: on the one hand, the network operator problem includes prices and total resources allocation, and on the other hand, the users' model determines the level of usage taking into account the different links prices with respect to resources availability.

Mathematically, this combination of two linked models is formalized as a bi-level, or hierarchical, program. It allows the operator to set network prices and service levels which take into account their effect on the user response. The resulting problem is a highly complex non-convex optimization problem, for which real-time solutions cannot be expected, despite recent advances on computational methods.

In other cases, however, it is desirable for link prices to be defined by Lagrangian multipliers. Proportional Fairness Resource Allocation and Pricing [7], [9], [10] advocates this point of view. In particular, a proportionally fair resource allocation h^* is one that satisfies for all users, $r \in R$, with a willingness to pay w_r, the inequality

$$\sum_{r \in R} w_r \frac{(h_r{}^* - h_r)}{h_r{}^*} \geq 0 \tag{1}$$

for all feasible allocations h_r. This inequality amounts to the optimality condition of the following optimization program:

$$\max \sum_{r \in R} w_r \log h_r \tag{2}$$

such that h_r is feasible with respect to user demand, the capacity constraints on the network links are satisfied, and flows are non-negative. Kelly and colleagues then interpreted the Lagrangian multipliers on the link capacity constraints as the added costs that would drive the allocation to proportional fairness, without imposing them explicitly.

Our approach generalizes the method of pricing suggested by the Proportional Fairness model. In particular, we consider a special case of the bilevel or hierarchical program that has a single, globally optimal, set of network prices. Furthermore, the problem is solvable in polynomial time (for real valued flows), as opposed to its more complex, bi-level, counterpart. To do so, we have adopted the framework of Larsson and Patriksson [19] defined for applications in the transport field. Next, we recall this toll optimization framework.

The model combines two problems. When applied to our setting, one must first solve optimally a non-linear resource allocation problem (such as the Proportional

Fairness allocation model). Then, one uses the optimal solution and the Lagrange multipliers of this first problem to maximize the operator's revenue function over a restricted set of prices.

Let $G(N, A)$ denote a strongly connected network, where N is the set of nodes and A the set of links. Each link a has associated with it a capacity $u_a \in R_+ \cup \{\infty\}$, $a \in A$. For certain ordered pairs of nodes, $s = (o, q) \in S \subset N \times N$, there is a demand for communication from o to q, known as d_s. The flow associated with the set of all such demands over the network can be expressed as a flow on each link a, or on each origin-to-destination route. We shall assume here that several routes serve the same origin-to-destination pair. The set of all routes on the network will be referred to as R. This may represent the case of several physical routes that a request may take, or the presence of more than one network operator serving that pair.

Indicator matrices can be used to convert from the flow on links to flow on routes. Let Δ be the link-route indicator matrix, where $\Delta_{ar} = 1$ if link a is present along on route $r \in R$, and 0 otherwise. Let Γ be the od-route indicator matrix, where $\Gamma_{rs} = 1$ if route r serves the od-pair $s \in S$, and 0 otherwise.

Here, we shall suppose in the remainder of this paper that the users on each od pair are homogeneous in their quality of service (QoS) requirements. Adding user differentiation and/or multiple QoS classes within each od-pair is straightforward, but complicates the notation.

The method of [19] works as follows. Consider a canonical network optimization problem with capacity constraints:

$$\min \quad f(x) \tag{3}$$

subject to

$$\begin{aligned} \Gamma y &= d && (4)\\ \Delta y &\leq u && (5)\\ d, y &\geq 0, && (6) \end{aligned}$$

where the flow on a link, a, is $x_a = \sum_{r \in R} y_r \Delta_{ar}$. Then, the system based on the Karush-Kuhn-Tucker (KKT) optimality conditions for the problem (3)–(6) as expressed in [19] is given below:

$$T(\lambda, \pi) = \begin{cases} \Delta^T[\nabla f(x^*) + \lambda] - \Gamma \pi \geq 0, \\ [\nabla f(x^*) + \lambda]^T x^* - d^T \pi = 0, \\ \lambda^T (x^* - u) = 0, \\ \lambda \geq 0 \end{cases} \tag{7}$$

where π is the Lagrange multiplier vector for constraints (4). Line 2 of Equation set (7) can be obtained by writing the following linear programming formulation of (3)–(6), solving for x^*:

$$\min_x \quad \nabla f(x^*)^T x \tag{8}$$

subject to constraints (4)–(6). The function $\nabla f(x^*)$ is the value of the gradient of the function f *evaluated at the optimal solution value*, x^*. While this model is clearly not solvable in practice, since the function value $f(x^*)$ at x^* is not known, it permits writing a linear programming dual to the original nonlinear problem (3)–(6). Taking the Lagrangian relaxation of the capacity constraints (5) and adding that into the linear programming objective function (8), we obtain

$$\begin{aligned} &\min_x \quad \nabla f(x^*)^T x + \lambda^T (x - u) = \\ &\min_x \quad \nabla f(x^*)^T x + \lambda^T x. \end{aligned} \tag{9}$$

Then, the linear programming dual of (9) with constraints (4) and (6) is:

$$\max_\pi \quad d^T \pi \tag{10}$$

subject to

$$\Gamma^T \pi \;\leq\; \Delta^T [\nabla f(x^*) + \lambda]. \tag{11}$$

Strong (linear programming) duality gives the term analogous to that of Line 2 of Equation set (7), that is:

$$[\nabla f(x^*) + \lambda]^T x^* - d^T \pi = 0. \tag{12}$$

For a fixed, optimal, value of x^*, the System (7) describes a polyhedron in λ and π. Furthermore, it can be observed that the multiplier vector λ will not necessarily be unique in network flow problems.

Indeed, the requirement of linear independence of binding constraint gradients is not always satisfied in networks, as the constraint gradients take the form of indicator matrices for the graph, which are not constructed so as to have linear independent rows. When these multipliers are not unique, then, the authors in [19] proposed optimizing a secondary objective over the set of those multiplier values. That is, given the System (7), they posed the following secondary optimization problem:

$$\max_\lambda \quad \phi(\lambda) \tag{13}$$

$$\text{subject to} \quad \lambda \in T \cap P, \tag{14}$$

where in our setting T is the set of link prices, λ, given by the System (7), and $\phi : \Re^{|A|} \mapsto \Re$ is coercive, and preferably convex. Since the feasible set is polyhedral, when P is bounded, this secondary, or network manager's optimization problem is easy to solve using standard linear or nonlinear programming algorithms (depending on whether ϕ is a linear or nonlinear function of λ).

The network manager's objective can take various forms. Larsson and Patriksson [19], in the context of transport planning, discussed at length the network objective of trying to achieve a particular flow of (transport users) by adding prices judiciously; this includes trying to adjust link flows from their user-equilibrium values to a system optimum (using marginal cost pricing) or to some other flow (which then changes the prices that are produced by the model), or even allowing prices to be positive or negative, in which case, travel subsidies are considered (negative prices).

The interest in the model of [19] stems from the possibility to optimize network revenue through prices, and to do so using a restricted, and therefore simpler, model than the general bi-level program. Furthermore, the model has the advantage of possessing a unique solution, under very mild and natural conditions.

In [19], it was claimed that the maximization of operator revenue in the model they propose, will very often be unbounded, leading to infinitely large prices; in other words, they claim that the revenue maximization problem is usually ill-defined for this model. This outcome is not, however, valid when the network is constructed so as to always contain a non-tolled path, for each user, that is, linking every origin-destination pair. In the case of Internet networks, this assumption willl always be verified, as explained further below. When the network is constructed as such, link prices cannot be made arbitrarily large as long as the free path has sufficient capacity. A non-trivial solution (that is, non-zero revenue) arises when the free paths offer lower quality of service than the paths tolled by the network operator.

However, in order to be of use in practice, it is necessary to explore the extent to which one can optimize revenue over the polyhedral set of prices, in System (7).

3 Numerical Experience

The purpose of the tests presented in this section was to evaluate the extent to which link prices that are tied to Lagrange multipliers can be optimized toward revenue maximization. We were therefore interested in evaluating whether the non-uniqueness of the prices that is theoretically possible in the context of the network flow problem arises in practice, and whether it can be optimized sufficiently to allow for revenue maximization.

To that end, by solving the fair network resource allocation problem via an augmented Lagrangian algorithm, relaxing the link capacity constraints, an optimal "fair" link flow, x^* is obtained, along with an initial multiplier, or price, vector λ. With this optimal flow x^* to the (for example, proportional) fairness problem in hand along with λ, it is possible to explore the polyhedron of feasible $\lambda(x^*)$, and, in principle, optimize over that set. We are thus interested in to what extent, *in practice* this secondary, revenue maximizing optimization is feasible.

The results that follow were all obtained using the Network Equilibrium Toolbox, *CiudadSim* found at

`http://www-rocq.inria.fr/scilab/CiudadSim`

in Scilab [22] to solve the fair resource allocation problems and the linear programming solver of Scilab to solve the linear revenue maximization problems in λ.

In the remainder of this section, we provide numerical experience with the models of the previous section. The networks were constructed to represent several possible choices for the user. For each user, characterized by an origin-destination pair, there is always a "best effort" choice, that is, a path with generally higher delay (low QoS) but very high capacity relative to demand levels and consequently not subject to pricing by the network operator. The term "best effort" comes from the current allocation strategy of the Internet: usage is free but no quality guarantees are provided; this service is known as "best effort" service since capacity is allocated to a user "as best as possible".

In addition, to the best effort path available to each user, there is at least one path, or provider, offering service with higher QoS, meaning here lower delay; consequently capacity on those links is lower. The delay functions we chose were quadratic. In our setting we are interested in a particular form of (13). In particular the revenue maximization problem

$$\max_{\lambda \in T \cap P} \phi(\lambda) = (x^*)^T \lambda. \tag{15}$$

Note that, through the set P, this may represent a joint maximization of all providers revenues, or the maximization over a subset of link prices, corresponding to one of several possible providers on the network.

In the first set of examples, the polyhedral set of prices T will be shown to be a singleton; in other words, the vector of shadow price, λ, is unique, and therefore no further optimization is possible. In the second set of examples, the set of prices T is not a singleton, and, furthermore, is bounded, thereby permitting further optimization by way of a secondary optimization problem. We thus examine the extent to which operator revenue can be maximized while maintaining the fair allocation. In the last set of examples, we show the possibility of T to be unbounded in the direction of revenue maximization. In other words the objective of global revenue maximization over all non "best-effort" links is not well defined. In this case, we show how one can maximize other objectives such as the revenue over restricted sets of links or routes.

3.1 Network Instances with Unique Prices

The first example is performed on a telecommunications network with 5 nodes and 7 arcs, provided in Figure 1. The dashed lines indicate the origin and the destination of a user's request along with the respective demand values, while, the numbers above the solid links indicate the respective optimal link flows. The network data is provided in Table 1. There were two users in this example: one requesting 40 units of service from node 1 to node 5, the other requesting 50 units from node 2 to node 5.

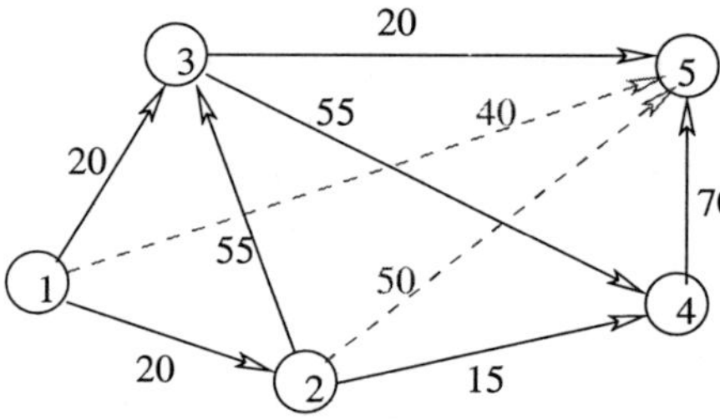

Figure 1: Network with Fair Flows

Link #	1	2	3	4	5	6	7
Link Definition	(1,3)	(1,2)	(2,3)	(2,4)	(3,4)	(3,5)	(4,5)
Link Capacity	20	100	100	15	100	20	100
Link Delay	$3{x_1}^2$	$8x_2 + 2{x_2}^2$	$7x_3 + 2{x_3}^2$	$8{x_4}^2$	$9x_5 + 2{x_5}^2$	$9{x_6}^2$	$8x_7 + 2{x_7}^2$
Fair Link Flow x^*	20	20	55	15	55	20	70
Initial Link Price λ	105	0	0	116	0	87	0

Table 1: Network data for Figure 1

Using the augmented Lagrangian method of [16], we obtained the fair resource allocation illustrated in the following network and in Table 1. This solution provides a feasible vector of prices denoted Initial Link Prices λ (cf. Table 1). The initial revenue obtained using the initial link prices λ is thus

$$\textbf{Initial Revenue} = (\mathbf{x}^*)^{\mathbf{T}}\lambda = \mathbf{5580}.$$

Given the optimal link flow vector x^* (cf. Table 1), we can explicitly express the polyhedron of shadow prices, $T(\lambda, \pi)$, guaranteeing the fair allocation, x^*, as follows:

$$T(\lambda,\pi)=\begin{cases}\lambda_1 \quad +\lambda_6 \quad -\pi_{15} \geq -240\\ \lambda_1 \quad +\lambda_5 \quad +\lambda_7 \quad -\pi_{15} \geq -327\\ +\lambda_2 \quad +\lambda_3 \quad +\lambda_6 \quad -\pi_{15} \geq -345\\ +\lambda_2 \quad +\lambda_3 \quad +\lambda_5 \quad +\lambda_7 \quad -\pi_{15} \geq -432\\ +\lambda_2 \quad +\lambda_4 \quad +\lambda_7 \quad -\pi_{15} \geq -316\\ +\lambda_3 \quad +\lambda_6 \quad -\pi_{25} \geq -297\\ +\lambda_3 \quad +\lambda_5 \quad +\lambda_7 \quad -\pi_{25} \geq -384\\ +\lambda_4 \quad +\lambda_7 \quad -\pi_{25} \geq -268\\ 20\lambda_1 +20\lambda_2 +55\lambda_3 +15\lambda_4 +55\lambda_5 +20\lambda_6 +70\lambda_7 -40\pi_{15} -50\pi_{25} = -30900\\ -80\lambda_2 \; -45\lambda_3 \; -45\lambda_5 \; -30\lambda_7 = 0\\ \lambda_1, \lambda_2, \lambda_3, \lambda_4, \lambda_5, \lambda_6, \lambda_7 \geq 0\end{cases}$$

We are interested in checking if the Lagrangian multipliers (shadow prices) of the resource allocation problem are unique or not, that is, whether or not this set T of prices is a singleton. If the Lagrangian multipliers are not unique, the secondary optimization (15) over those prices may produce prices giving higher revenue to the network operator(s).

To do so, we examine the system obtained from the KKT necessary conditions: the Lagrange multipliers are unique if the active inequality and the equality constraints gradients are linearly independent.

From constraints (4) and (5), observe that the equality and inequality constraints gradients are respectively Γ and Δ. Thus we can easily identify in Δ the active constraints, as they correspond, in the arc-route-matrix Δ, to the rows of the saturated arcs. Let us denote the matrix composed of the active and the equality constraints by I. From the network data in Table 1, we verify that arcs 1, 4, and 6 are saturated, thus the inequality constraints corresponding to those arcs are active. The matrices Δ, Γ and I are provided below.

$$\Delta=\begin{vmatrix}1&1&0&0&0&0&0&0\\0&0&1&1&1&0&0&0\\0&0&1&1&0&1&1&0\\0&0&0&0&1&0&0&1\\0&1&1&0&0&0&1&0\\1&0&0&1&0&1&0&1\\0&1&1&0&1&0&1&1\end{vmatrix}\quad \Gamma=\begin{vmatrix}1&1&1&1&1&0&0&0\\0&0&0&0&0&1&1&1\end{vmatrix}\quad I=\begin{vmatrix}1&1&0&0&0&0&0&0\\0&0&0&0&1&0&0&1\\1&0&0&1&0&1&0&1\\1&1&1&1&1&0&0&0\\0&0&0&0&0&1&1&1\end{vmatrix}$$

In this example, the rows of matrix I above are indeed linearly independent, since $\text{rank}(I) = 5$. In other words, the set of prices T is a singleton with unique element the vector of initial prices in Table 1. Thus, for this network with the given communication demands, no further revenue maximization is possible over this set. Therefore, the maximal revenue is thus equivalent to the initial revenue, which was 5580.

Figure 2 is the simplest network tested, having only 3 nodes, 3 links, and a single user, from nodes 1 to 3. The network data is summarized in Table 2.

Given the data, and solving for the optimal, fair, resource allocation, in x^*, only link 2 is saturated. Thus, the matrix of active constraint gradients is

$$I = \begin{vmatrix} 0 & 1 \\ 1 & 1 \end{vmatrix}$$

which is clearly linearly independent. Therefore, the multiplier values are all unique: those on links 1 and 3 are zero, and the only possible value of λ_2 is 672.

Link #	1	2	3
Link Definition	(1,2)	(1,3)	(2,3)
Link Capacity	100	20	100
Link Delay Data	$4x_1 + 9{x_1}^2$	$7x_2$	$6x_3 + 3{x_3}^2$
Fair Link Flow x^*	80	20	80
Initial Link Prices λ	0	672	0

Table 2: Network data for Figure 2

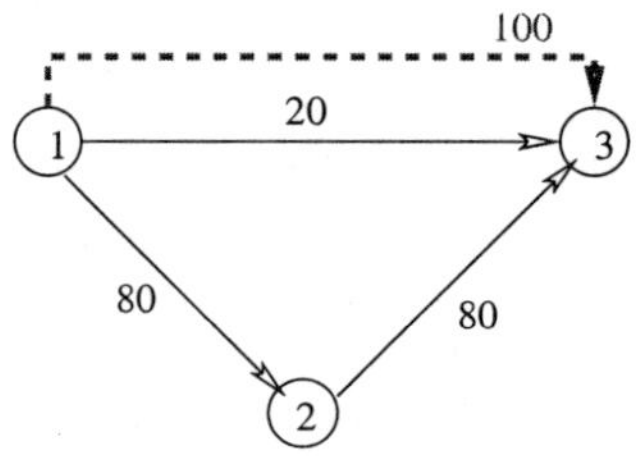

Figure 2: Network with Fair Flows

While a number of networks exhibit unique values of the price vector, it should be recalled that this phenomenon is closely related to the network topology. In networks with very few routes, it is likely that the sub-matrix obtained by extracting the rows of the arc-route indicator matrix Δ which correspond to active capacity constraints, will be linearly independent. This turned out to be the case in some of the examples we constructed. When, however, the number of routes is large, compared to the number of links, and demand is sufficiently high to saturate a large number of links capacities, this linear independence is unlikely to hold. The next examples illustrate the case where link prices are not unique and, furthermore, the set of prices T is bounded.

3.2 Network Instances with Non Unique Finite Prices

The present example makes use of the network of Figure 3, and delay data from Table 3. Again, there were two users represented on this network. The first requires

2.5 units from node 1 to 5; the second requires 3 units from node 2 to 6. As before at least one path with high QoS exists for each user as well as a "best effort" choice for each.

Link #	1	2	3	4	5	6	7	8	9
Link Definition	(1,2)	(1,4)	(2,4)	(2,3)	(2,5)	(3,4)	(3,6)	(4,6)	(6,5)
Link Capacity	100	1.5	100	1.8	1	0.3	1.5	100	100
Link Delay	$5x_1 + 4x_1{}^2$	$9x_2{}^2$	$7x_3 + 9x_3{}^2$	$9x_4{}^2$	$4x_5{}^2$	$1x_6{}^2$	$6x_7{}^2$	$4x_8 + 5x_8{}^2$	$6x_9 + 2x_9{}^2$
Fair Link Flow x^*	1	1.5	1.2	1.8	1	0.3	1.5	3	1.5
Initial Link Price λ	0	0	0	1.3	28.5	0	10.3	0	0

Table 3: Network Data for Figure 3

Again, using the augmented Lagrangian method, we obtained the fair resource allocation illustrated on the network of Figure 3 and in Table 3. As in Section 3.1, this method provides along with the solution a feasible vector of prices, denoted Initial Prices, λ. The initial revenue with respect to the fair link flow and the Initial prices, λ is

$$\textbf{Initial Revenue} = (\mathbf{x}^*)^{\mathbf{T}}\lambda = \mathbf{46.3}.$$

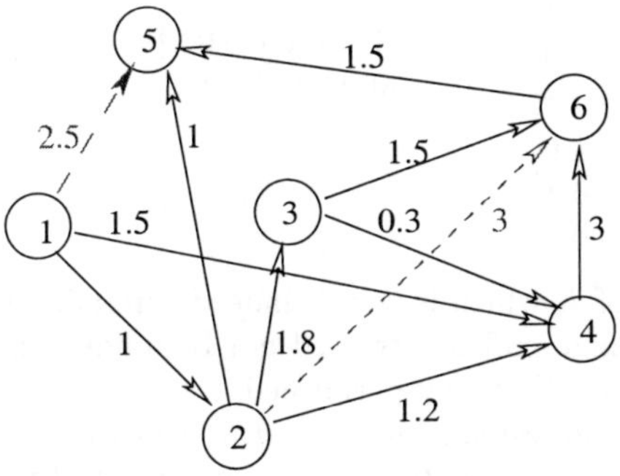

Figure 3: Network with Fair Flows

With the fair link flow x^* of Table 3 at hand, we write explicitly the set of fair link prices, $T(\lambda, \pi)$, for the present example as follows:

$$T(\lambda, \pi) =$$

$$\left\{\begin{array}{ccccccccccccc}
\lambda_1 & & & & +\lambda_5 & & & & & -\pi_{15} & & \geq & -13 \\
\lambda_1 & & & +\lambda_4 & & & +\lambda_7 & & +\lambda_9 & -\pi_{15} & & \geq & -43.2 \\
\lambda_1 & & +\lambda_3 & & & & & +\lambda_8 & +\lambda_9 & -\pi_{15} & & \geq & -54.8 \\
\lambda_1 & & & +\lambda_4 & & +\lambda_6 & & +\lambda_8 & +\lambda_9 & -\pi_{15} & & \geq & -53.5 \\
 & +\lambda_2 & & & & & & +\lambda_8 & +\lambda_9 & -\pi_{15} & & \geq & -41.5 \\
 & & +\lambda_3 & & & & & +\lambda_8 & & & -\pi_{26} & \geq & -36.8 \\
 & & & +\lambda_4 & & & & & & & -\pi_{26} & \geq & -25.2 \\
 & & & +\lambda_4 & & +\lambda_6 & +\lambda_7 & +\lambda_8 & & & -\pi_{26} & \geq & -35.5 \\
\lambda_1 & +1.5\lambda_2 & +1.2\lambda_3 & +1.8\lambda_4 & +1\lambda_5 & +0.3\lambda_6 & +1.5\lambda_7 & +3\lambda_8 & +1.5\lambda_9 & -2.5\pi_{15} & -3\pi_{26} & = & -167.86 \\
-99\lambda_1 & & -98.5\lambda_3 & & & & & -97\lambda_8 & -98.5\lambda_9 & & & = & 0 \\
\lambda_1, & \lambda_2, & \lambda_3, & \lambda_4, & \lambda_5, & \lambda_6, & \lambda_7, & \lambda_8, & \lambda_9 & & & \geq & 0
\end{array}\right.$$

When the multipliers are not unique, the polyhedron of prices T need not be a singleton, and we may be able to maximize the network manager's revenue further than that which was obtained from the initial Lagrange multipliers λ.

Again, we examine the KKT necessary conditions, denoting by I the set of the active and the equality constraints gradients. It is easy to verify that the capacity constraints relative to arcs 2, 4, 5, 6, and 7 are active since those arcs are saturated. Using the arc-route and od-route matrices, respectively Δ and Γ, we construct the set I.

For the present example, the matrices Δ, Γ and I are:

$$\Delta = \begin{vmatrix} 1&1&1&1&0&0&0&0 \\ 0&0&0&0&1&0&0&0 \\ 0&0&1&0&0&1&0&0 \\ 0&1&0&1&0&0&1&1 \\ 1&0&0&0&0&0&0&0 \\ 0&0&0&1&0&0&0&1 \\ 0&1&0&0&0&0&1&0 \\ 0&0&1&1&1&1&0&1 \\ 0&1&1&1&1&0&0&0 \end{vmatrix} \qquad \Gamma = \begin{vmatrix} 1&1&1&1&1&0&0&0 \\ 0&0&0&0&0&1&1&1 \end{vmatrix} \qquad I = \begin{vmatrix} 0&0&0&0&1&0&0&0 \\ 0&1&0&1&0&0&1&1 \\ 1&0&0&0&0&0&0&0 \\ 0&0&0&1&0&0&0&1 \\ 0&1&0&0&0&0&1&0 \\ 1&1&1&1&1&0&0&0 \\ 0&0&0&0&0&1&1&1 \end{vmatrix}$$

We compute the rank of I, which is $6 < 7$; indeed, summing rows 4 and 5 in I gives row 2. Thus, the Lagrange multipliers need not be unique and an optimization over the set of prices could be effectively performed.

In this case, we are interested in *the extent to which revenue can be maximized by performing the secondary optimization (15) over the set of prices, T*. In order to do so, we evaluate the size of the polyhedron of feasible prices, T, solving the following pair of optimization problems

$$\min_{\lambda \in T} < e^i, \lambda >, \tag{16}$$

and

$$\max_{\lambda \in T} < e^i, \lambda > . \tag{17}$$

We denote e^i the ith coordinate vector, i.e, $i = 1, \ldots, 11$, where the 11 variables are the 9 price multipliers, 1 for each arc, and the two demand constraints multipliers, π_{15} and π_{26}. All of the solutions to the above problems were finite; the set T is therefore bounded in all coordinate directions.

The next step is then to solve the revenue maximization problem over the polyhedral (and, in this case, bounded) set of multipliers, T. That is, given the fair allocation x^*, the revenue maximization problem is:

$$\max_{\lambda \in T} \ \phi(\lambda) = (x^*)^T \lambda$$

which gives rise to the optimal prices summarized in Table 4.

Thus, we obtain the maximal revenue

$$\textbf{Max Revenue} = (\mathbf{x}^*)^{\mathbf{T}} \lambda^* = \mathbf{79.54}.$$

This value is substantially higher than the initial revenue of 46.3. Recall that the initial link prices were obtained by solving the fair resource allocation problem and taking prices to be the Lagrange multiplier values output by the augmented Lagrangian algorithm. The optimal link prices were obtained by maximizing revenue over the set of Lagrange multipliers. The advantage of solving the secondary linear program is significant in this example.

Link #	1	2	3	4	5	6	7	8	9
Optimal Link Prices λ^*	0	13.3	0	1.3	41.8	0	10.3	0	0

Table 4: Optimal Link Prices for Network of Figure 3

We next provide two other network instances where the polyhedral set of prices is also bounded and not a singleton, thus permitting a secondary optimization to maximize revenue. These networks are represented in Figures 4 and 5. The example of Figure 4 is the largest network studied with 10 nodes and 16 links, where two demands are to be satisfied: one from nodes 5 to 1 requiring 8 units of flow and the other from nodes 3 to 6 with a demand of 13 units. The network data is summarized in the Table 5.

As in all examples, the link delay functions are quadratic of the form $ax_i + bx_i{}^2$, where x_i is the flow on the link i, a and b are positive. In Table 5. containing the network data, we provide the values of the constants a and b, as (a, b).

In this example, we have from the fair resource allocation and corresponding multiplier values that

$$\textbf{Initial Revenue} = 52.91.$$

Link #	1	2	3	4	5	6	7	8	9	(10)	11	12	13	14	15	16
Link Definition	(2,5)	(2,7)	(2,8)	(3,2)	(3,10)	(4,2)	(5,4)	(5,6)	(5,9)	(7,1)	(7,5)	(8,6)	(9,1)	(9,4)	(10,4)	(10,8)
Link Capacity	100	100	10	7	100	100	100	100	3.7	100	2	10	3.7	2	100	7
Link Delay Data	(2,8)	(3,2)	(3,0)	(6,0)	(3,8)	(4,4)	(10,4)	(6,4)	(13,0)	(2,2)	(12,0)	(2,0)	(12,0)	(2,0)	(4,5)	(6,0)
Fair Link Flow x^*	3	4.3	4	7	6	4.3	4.3	3	3.7	4.3	0	10	3.7	0	0	6
Initial Link Prices λ	0	0	3.7	4.3	0	0	0	0	0.61	0	0	0.34	0.58	0	0	0

Table 5: Network Data for Figure 4

The optimal link prices obtained by solving the revenue maximization problem over λ is found in Table 6, and the maximal revenue is

$$\textbf{Max Revenue} = 100.44.$$

Again, a significant increase in network revenue can be obtained by solving the secondary optimization problem over the set of prices, given by the Lagrange multipliers.

Optimal Link Prices λ^*	0	0	0	8	0	0	0	0	0	0	0	4	1.2	0	0	0

Table 6: Optimal Link Prices for the Network of Figure 4

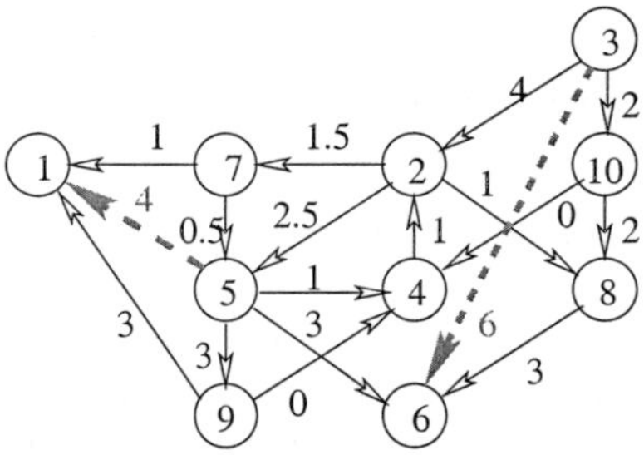

Figure 4: Network with Fair Flows

We turn now to a slight modification on the capacities and delay function constants for the Network of Figure 2. The modified data is given in Table 7. Namely, solving the fair resource allocation problem with the modified data, we have now that arcs 1 and 3 are saturated, rendering the matrix of active constraint gradients linearly *dependent.* As we have seen already, this allows, in principle, for a marge of manoeuvre in setting the multipliers. The initial revenue, read from Table 7, is equal to $20*101.5+20*101.5 = 4060$.

However, optimizing the multiplier values with the aim of revenue maximization, we obtain the optimal link prices below.

Link #	1	2	3
Link Definition	(1,2)	(1,3)	(2,3)
Link Capacity	20	100	20
Link Delay Data	(7,0)	(6,3)	(7,0)
Fair Link Flow x^*	20	80	20
Initial Link Prices λ	101.5	0	101.5

Table 7: Network data for Figure 5

Link#	1	2	3
Optimal Link Prices λ^*	203	0	0

In other words, while the multiplier values are not unique, and the number of links with non-zero prices can vary, the route prices, in this example, were unique, as is, therefore, the provider's revenue which is equal to 4060.

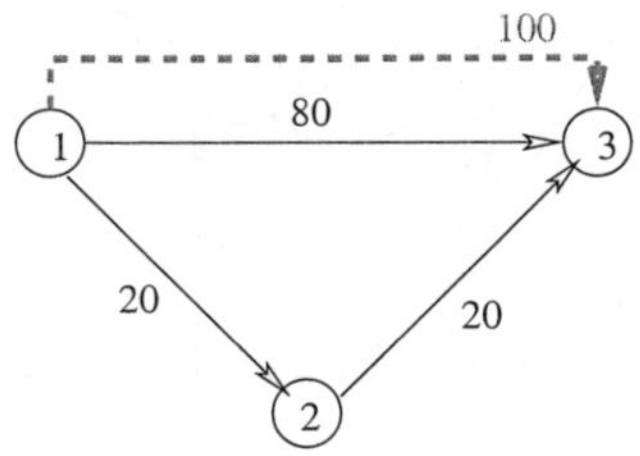

Figure 5: Network with Fair (Optimal) Flows

3.3 Unbounded Price Set and Alternative Optimizations

In this last set of examples, we illustrate the possibility of T to be unbounded in the direction of the profit maximization objective. Note that the present example was obtained by making small changes in the demand values, capacities and delay functions on the data of the Network of Figure 4. Recall that in the example with the original network data of that figure, the set of prices was a singleton.

In Table 8, we provide the values of the constants a and b of the delay functions. The network is provided in Figure 6. Two users are represented; the first requests 4 units from node 5 to 1 while the second user requests 6 units from node 3 to 6. At least one high QoS choice is available to each user, along with "a best effort" option.

By solving the user problem with the augmented Lagrangian scheme we find the fair link flow x^* in Table 8. The Lagrange multipliers obtained jointly with the fair

Link #	1	2	3	4	5	6	7	8	9	(10)	11	12	13	14	15	16
Link Definition	(2,5)	(2,7)	(2,8)	(3,2)	(3,10)	(4,2)	(5,4)	(5,6)	(5,9)	(7,1)	(7,5)	(8,6)	(9,1)	(9,4)	(10,4)	(10,8)
Link Capacity	100	100	10	4	100	100	100	100	10	100	0.5	3	3	5	100	2
Link Delay Data	(12,6)	(10,7)	(16,10)	(0,12)	(8,10)	(6,8)	(4,10)	(4,10)	(0,9)	(10,8)	(0,10)	(0,8)	(0,10)	(0,9)	(11,12)	(0,9)
Fair Flow x^*	2.5	1.5	1	4	2	1	1	3	3	1	0.5	3	3	0	0	2
Ini. Prices λ	0	0	0	0	0	0	0	0	0	0	1.5	11	9.5	0	0	28

Table 8: Network Data for Figure 6

resource allocation x^* are the shadow prices λ in the same table. For these prices the initial revenue is:

$$\textbf{Initial Revenue} = \mathbf{118.24}.$$

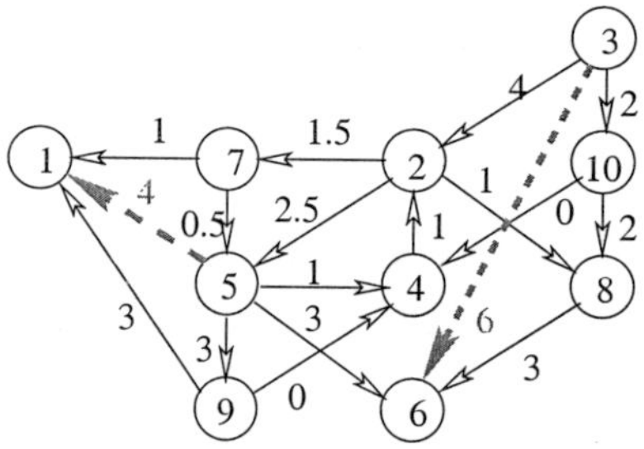

Figure 6: Network with Fair (Optimal) Flows

Using the fair link flow x^*, we express the polyhedron of shadow prices T as follows:

$$T(\lambda,\pi) =$$

$$\begin{pmatrix} & +\lambda_2 & & & & +\lambda_6 & +\lambda_7 & & & +\lambda_{10} & & & & & & \\ & & & & & & & & +\lambda_9 & & & & +\lambda_{13} & & & \\ +\lambda_1 & & & +\lambda_4 & & & & +\lambda_8 & & & & & & & & \\ & +\lambda_2 & & +\lambda_4 & & & & +\lambda_8 & & & +\lambda_{11} & & & & & \\ & & +\lambda_3 & +\lambda_4 & & & & & & & & +\lambda_{12} & & & & \\ & & & & +\lambda_5 & & & & & & & +\lambda_{12} & & & & +\lambda_{16} \end{pmatrix} \Delta + \begin{pmatrix} 0 \\ 0 \\ -\pi_{36} \\ -\pi_{36} \\ -\pi_{36} \\ -\pi_{36} \end{pmatrix} + \begin{pmatrix} -\pi_{51} \\ -\pi_{51} \\ 0 \\ 0 \\ 0 \\ 0 \end{pmatrix} \geq \begin{pmatrix} -66.5 \\ -57 \\ -109 \\ -107.5 \\ -99 \\ -70 \end{pmatrix}$$

$$\begin{pmatrix} 2.5 & +1.5 & +1 & +4 & +2 & +1 & +1 & +3 & +3 & +1 & +0.5 & +3 & +3 & & & +2 \\ -97.5 & -98.5 & -9 & & -98 & -99 & -99 & -97 & -7 & -99 & & & & -5 & -100 & \end{pmatrix} \Delta + \begin{pmatrix} -4\pi_{51} \\ 0 \end{pmatrix} + \begin{pmatrix} -6\pi_{36} \\ 0 \end{pmatrix} = \begin{pmatrix} -801.75 \\ 0 \end{pmatrix}$$

$$\lambda = (\lambda_1,\lambda_2,\lambda_3,\lambda_4,\lambda_5,\lambda_6,\lambda_7,\lambda_8,\lambda_9,\lambda_{10},\lambda_{11},\lambda_{12},\lambda_{13},\lambda_{14},\lambda_{15},\lambda_{16}) \geq 0$$

where, Δ indicates the arc-route matrix of the network of Figure 6. For the present example, the arc-route Δ, od-route Γ, and, active and equality gradients constraints I matrices are:

$$\Delta=\begin{vmatrix}0&0&1&0&0&0\\1&0&0&1&0&0\\0&0&0&0&1&0\\0&0&1&1&1&0\\0&0&0&0&0&1\\1&0&0&0&0&0\\1&0&0&0&0&0\\0&0&1&1&0&0\\0&1&0&0&0&0\\1&0&0&0&0&0\\0&0&0&1&0&0\\0&0&0&0&1&1\\0&1&0&0&0&0\\0&0&0&0&0&0\\0&0&0&0&0&0\\0&0&0&0&0&1\end{vmatrix}\qquad \Gamma=\begin{vmatrix}1&1&0&0&0&0\\0&0&1&1&1&1\end{vmatrix}\qquad I=\begin{vmatrix}1&1&0&0&0&0\\0&0&1&1&1&1\\0&0&1&1&1&0\\0&0&0&1&0&0\\0&0&0&0&1&1\\0&1&0&0&0&0\\0&0&0&0&0&1\end{vmatrix}$$

As in the previous example, we applied the same test to determine whether the multipliers are unique or not and to provide an idea of the size of the polyhedron T (summing up rows 3 and 7 gives rise row 2, thus I is LD). In the present example, it happens that at least one of the problems

$$\min_{\lambda} < e^i, \lambda >,$$

and

$$\max_{\lambda} < e^i, \lambda >,$$

is unbounded; namely for $i = 4, 9, 16$, where $\lambda \in T$ and e^i is the ith coordinate vector, i.e, $i = 1,\ldots., 18$. The 18 variables are the 16 price multipliers, 1 for each arc, and the two demand constraints multipliers, π_{51} and π_{36}. Thus we see that it is not possible to consider global profit maximization as secondary optimization problem. However, consider selecting a restricted set of arcs such as:

$$S_1 = \{11, 12, 13, 14\}.$$

Then the secondary optimization (15) will attempt to maximize revenue over this smaller set of links. This is of interest if, for example, those links are owned by one operator and a remaining links owned by other providers. Alternatively, they may represent a subset of links of one operator which are critical to a particular subnetwork under study. In this case, we solve:

$$\max_{\lambda\in T} \phi(\lambda) = \sum_{i\in S_1} x_i^* \lambda_i$$

We observe that the optimal solution of $\phi(\lambda)$ is finite and equal to the initial values of the λ_i, $i \in S_1$.

Similarly, it is often of interest to consider a particular set of routes. Consider, for instance the set R of routes r_j where

$$R = \{r_1, r_2, r_3\}.$$

and $r_1 = \{3, 12\}, r_2 = \{11, 7, 6, 3\}, r_3 = \{14, 6\}$. Letting S_2 denote the links included in route set R, we solve, in this case,

$$\max_{\lambda \in T} \phi(\lambda) = \sum_{i \in S_2} x_i^* \lambda_i.$$

Again, we obtain a finite value for the maximal revenue, but again, this value is unique for the selected routes, and is equal to 33.75.

We next provide another example with an unbounded link price polyhedra. It consists of a network (Figure 7) with 7 nodes and 13 links with three user demands. Demands are from nodes 2 to 6, 4 to 2 and 3 to 4 with respective quantities 40, 5, and 5. The network data, the resulting fair link flows and the initial link prices obtained with the augmented Lagrangian routine are summarized in Table 9. In this example, although the polyhedral set is unbounded, it is possible to maximize revenue over subsets of arcs or routes, contrary to the above instance in which prices on subsets of links were unique.

Link #	1	2	3	4	5	6	7	8	9	11	12	13
Link Definition	(1,6)	(1,5)	(2,1)	(3,2)	(3,5)	(3,7)	(4,5)	(5,6)	(5,7)	(5,2)	(6,3)	(7,4)
Link Capacity	100	1	100	8	2	100	100	1	2	100	4	100
Link Delay Data	(4,4)	(6,0)	(7,5)	(3,0)	(6,0)	(9,5)	(8,6)	(6,0)	(6,0)	(8,5)	(7,0)	(6,7)
Fair Link Flow x^*	39	1	40	0	2	3	5	1	2	5	0	5
Initial Link Price λ	0	74	0	0	4	0	0	74	4	0	0	0

Table 9: Network Data for Figure 7

In this instance, the initial revenue, as obtained through the initial Lagrange multipliers, is

$$\textbf{Initial Revenue} = 164.$$

Applying the KKT conditions shows that the multipliers are not unique, and, furthermore, solving the pair of optimization problems (16)–(17) shows that T is unbounded in the direction of the objective $(x^*)^T \lambda$.

Contrary to the previous example of Figure 6, however, here, we may choose a restricted set of arcs, such as

$$S_1 = \{2, 4, 7, 9\},$$

and optimize the revenue over this set. From Table 9, we compute the initial revenue:

$$\textbf{Initial Revenue over } \mathbf{S_1} = \mathbf{82}.$$

Solving the following revenue maximization problem over S_1, we then obtain:

$$\max_{\lambda \in T} \phi(\lambda) = \sum_{i \in S_1} x_i^* \lambda_i$$

which gives

$$\textbf{Max Revenue} = \textbf{904},$$

where $\lambda^*_{i \in S_1} = (148, 8, 148, 8)$.

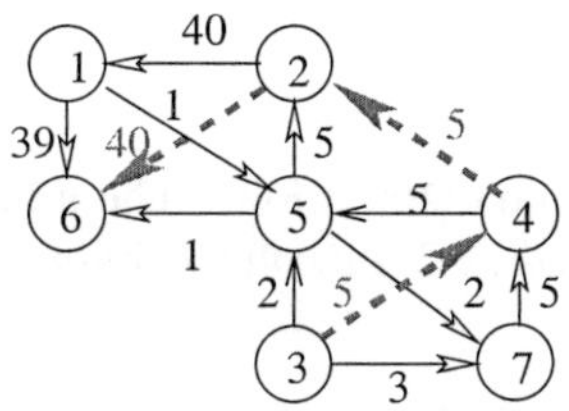

Figure 7: Network with Fair Flows

4 Perspectives and Related Results from the Literature

This paper presented the application of a special case of the bilevel programming problem for the solution of a set of fair prices on a telecommunication network, where a network manager optimizes a revenue maximization objective over a particular set. The set is determined by the optimality conditions for a multi-commodity flow problem over the network, subject to capacity constraints. A restriction on the prices states formally that the link prices must be Lagrange multipliers for the capacity constraints, and will therefore be strictly positive only when capacity is reached. This approach generalizes some popular approaches to pricing through fairness, and other Lagrangian schemes for solving telecommunications network flow problems. It draws a parallel between a large body of telecommunications network modeling and the more general bi-level mathematical programming model. A natural next stage of this work would involve developing a characterization of the telecommunication networks that exhibit sufficiently large Lagrange multiplier sets so as to permit considerable revenue maximization. This work also leads naturally into a study of the more general bi-level programming model, by freeing the prices of the complementarity constraints that define them to be Lagrange multipliers.

The numerical exercises have shown that at least, on the quadratic cost fair resource allocation problem, the method of optimizing a secondary objective over the set of Lagrange multipliers, or shadow prices, is worthwhile. Indeed, we have seen that in most instances, revenue maximization is possible for some subnetwork, if not for the entire network, and for some network data. Furthermore, the complexity of the secondary optimization is lower than that of the fair resource allocation problem as it is a linear program, rather than a nonlinear program. An interesting topic for future research would be to test whether the solution to this two-step procedure may come quite close to the true bi-level optimization solution. In other words, the solution to the system described here could be used as a good approximation to the true optimal pricing problem in which prices are not restricted to be Lagrange multipliers.

References

[1] L. Anania and R. J. Solomon (1997) Flat: The minimalist price, *In L. W. McKnight and J. P. Bailey, editors, Internet Economics*, 91–118, Cambridge, Massachusetts, MIT Press.

[2] R. Cocchi et al. (1991) A study of priority pricing in multiple service class networks, *Proc. Sigcomm*, 123–130, Oxford University Press.
Available from ftp://ftp.parc.xerox.com/pub/net-research/pricing2.ps.Z.

[3] R. Cocchi et al. (1993) Pricing in computer networks: motivation, formulation, and example, *ACM/IEEE Trans. Net.*, 1:614–627.
Available from ftp://ftp.parc.xerox.com/pub/net-research/pricing1.ps.Z.

[4] Luiz A. DaSilva (2000) Pricing for 'qos'-enabled networks: a survey, *IEEE Communications Surveys and Tutorials*, 3(2):2–8, Second Quarter.
Available from URL: http://www.comsoc.org/pubs/surveys/.

[5] M. Falkner, M. Devetsikiotis, and I. Lambadaris (2000) An overwiew of pricing concepts for broadband ip networks, *IEEE Communications Surveys and Tutorials*, 3(2):8–13, Second Quarter.
Available from URL: http://www.comsoc.org/pubs/surveys/.

[6] P. C. Fishburn and A. M. Odlyzko (2000) Dynamic behavior of differential pricing and quality of service options for the internet, *In Proc. First Intern. Conf. on Information and Computation Economies (ICE-98)*, 128–139, ACM Press.
Available from URL http://www.research.att.com/~amo.

[7] R. J. Gibbens and F. P. Kelly (1999) Resource pricing and the evolution of congestion control, *Automatica*, 35:1969–1985.
Available from URL http://www.statslab.cam.ac.uk/~frank/evol.html.

[8] A. Gupta et al. (1997) Priority pricing of integrated services networks, *In L. W. McKnight and J. P. Bailey, editors, Internet Economics*, 323–352, Cambridge, Massachusetts, MIT Press.

[9] F. P. Kelly (1997) Charging and rate control for elastic traffic, *European Trans. Telecommunications*, 8:33–37.
Available from URL http://www.statslab.cam.ac.uk/~frank/elastic.html.

[10] F. P. Kelly, A. K. Maulloo, and D. K. H. Tan (1998) Rate control for communication networks: shadow prices, proportional fairness, and stability, *Journal of the Operational Research Society*, 49:237–252.
Available from URL http://www.statslab.cam.ac.uk/~frank/rate.html.

[11] Y. A. Korilis, A. A. Lazar, and A. Orda (1997) Achieving network optima using stackelberg routing strategies, *IEEE/ACM Transactions on Networking*, 5(1):161–173.
Available fron URL http://citeseer.nj.nec.com/220769.html.

[12] Y. A. Korilis, T. A. Varvarigou, and S. R. Ahuja (1998) Incentive-compatible pricing strategies in noncooperative networks, *Conference on Computer Communications*, 439–446. IEEE/INFOCOM.

[13] T. Larsson and M. Patriksson (1992) Simplicial decomposition with disaggregated representation for the traffic assignment problem, *Transportation Science*, 26:4–17.
Available from the URL http://www.math.chalmers.se/~mipat/traffic.html.

[14] T. Larsson and M. Patriksson (1994) An augmented lagrangean scheme for capacitated traffic assignment problems, *In N. Bhouri F. Boillot and F. Leurent, editors, Proceedings of the 2nd Meeting of the EURO Working Group on Urban Traffic and Transportation, Actes INRETS no. 38*, 163–199, Paris, France.
Available from the URL http://www.math.chalmers.se/~mipat/traffic.html.

[15] T. Larsson and M. Patriksson (1994) On side constrained models of traffic equilibria, *In F. Giannessi and A. Maugeri, editors, Proceedings of the International School of Mathematics "G. Stampacchia" 19th Course on Variational Inequalities and Network Equilibrium Problems*, 169–178, Plenum Press, New York, USA.
Available from the URL http://www.math.chalmers.se/~mipat/traffic.htmL.

[16] T. Larsson and M. Patriksson (1995) An augmented lagrangean dual algorithm for link capacity side constrained traffic assignment problems, *Transportation Research*, 29B:433–455.
Available from the URL http://www.math.chalmers.se/~mipat/traffic.html.

[17] T. Larsson and M. Patriksson (1995) Equilibrium characterizations of solutions to side constrained asymmetric traffic assignment models, *In J. Barcelo, editor, 3rd Meeting of the EURO Working Group on Transportation*, 143–164, Barcelona, Spain.
Available from the URL http://www.math.chalmers.se/~mipat/traffic.html.

[18] T. Larsson and M. Patriksson (1997) Traffic management through link tolls—an approach utilizing side constrained traffic equilibrium models, *Rendiconti del Circolo Matematico di Palermo Serie II*, 48:147–170.
Available from the URL http://www.math.chalmers.se/~mipat/traffic.html.

[19] T. Larsson and M. Patriksson (1998) Side constrained traffic equilibrium models—traffic management through link tolls, *In P. Marcotte and S. Nguyen, editors, Chapter 7 in Equilibrium and Advanced Transportation Modelling*, 125–151, Kluwer Academic Publishers, Boston, MA.
Available from the URL http://www.math.chalmers.se/~mipat/traffic.html.

[20] T. Larsson and M. Patriksson (1999) Side constrained traffic equilibrium models—analysis, computation and applications, *Transportation Research*, 33B:233–264.
Available from the URL http://www.math.chalmers.se/~mipat/traffic.html.

[21] T. Larsson, M. Patriksson, and W. L. Zhuang (1997) A dual scheme for traffic assignment problems, *Optimization*, 42:323–358.
Available from the URL http://www.math.chalmers.se/~mipat/traffic.html.

[22] Scilab Scientific Computing Library (2000), www-rocq.inria.fr/scilab/.

[23] S. H. Low and D. E. Lapsley (1999) Optimization flow control, 1: Basic algorithm and convergence. *IEEE/ACM Transactions on Networking*, 7(6):861–875.

[24] J. K. Mackie-Mason and H. R. Varian (1995) Pricing congestible network resources, *IEEE JSAC*, 13(7):1141–1149.

[25] H. Mendelson and S. Whang (1990) Optimal incentive–compatible priority pricing for the $m/m/1$ queue, *Operations Research*, 38:870–883.

[26] Jan A. Van Mieghem (2000) Price and service discrimination in queuing systems – incentive compatibility of gcu scheduling, *Management Science*, 46(9):1249–1267.

Optimization and Optimal Control, pp. 59 – 68
P.M. Pardalos, I. Tseveendorj, and R. Enkhbat, Editors

Nonlinear Phenomena in Economics

S. Budnyam (budnyam@icb.edu.mn)
Institute of Commerce and Business,
Ulaanbaatar, Mongolia

Abstract

This paper is devoted to nonlinear phenomena in economics which arise in a monopoly.

Keywords: microeconomics, linear system, monopoly, adaptive search, dynamic system, chaos

1 Introduction

Traditional microeconomic theory deals with two basic market types : Perfect competition and monopoly. In the case of perfect competition individual firms are assumed to be so small in comparison with the entire market that they cannot noticeably influence market price on their own ;

The single monopolist, facing a lot of these households, on the other hand, is assumed to deliberately choose to limit the quantity supplied so as to keep price sufficiently high to yield monopoly profit.

As to the general form of a demand function, it is assumed to be monotonically downward sloping, demand decreasing as price increases. As there are always substitutes for most goods, like driving as an alternative to using the railway, there would be a maximum price above which nobody would demand anything. Likewise, there would be a maximum saturation demand when price does down to zero. In most cases the demand curve is just a downward sloping straight line, but this is too simplistic. The elasticity of demand would in general, vary over different sections of the demand curve, giving it a convex/ concave outline. The full analysis of such cases in graphs was given in the pioneering work by Joan Robinson in 1933, who recognized the existence of multiple local equilibria (fig 1).

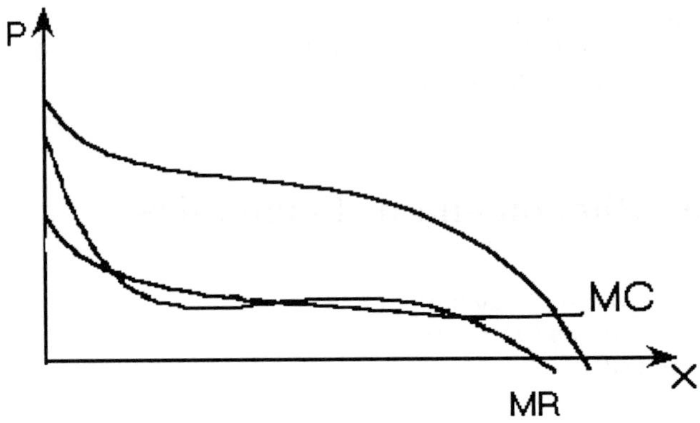

Fig.1 Demand curve, marginal revenue and cost curves

2 The Model

Such a case, illustrated in fig.1, which could in fact have been obtained from Joan Robinson's book, can be represented by the truncated Taylor series:

$$p = A - Bx + Cx^2 - Dx^3 \tag{1}$$

where p denotes commodity price, x denotes the quantity demanded, and A, B, C, D are some positive constants.

In order that function (1) be invertible, i.e. there be just one quantity demand corresponding to each price, the curve must be downward-sloping. The least steepest slope is at the point of inflection, defined by:

$$d^2p/dx^2 = 2C - 6Dx = 0 \tag{2}$$

Here the slope is:

$$dp/dx = -B + 2Cx - 3Dx^2 = 0 \tag{3}$$

or, substituting from (2) for $x = C/(3D)$ we obtain

$$\frac{dp}{dx} = -B + \frac{C^2}{3D} < 0 \tag{4}$$

The demand curve is downward-sloping, as indicated by the imposed sign requirement, provided that the condition:

$$C^2 < 3BD \tag{5}$$

holds.

Economists usually analyse the monopolist's behavior, using the marginal revenue, MR, i.e. the derivative of total revenue, $d(px)/dx = p + xdp/dx$. For the demand curve (1) the marginal revenue curve would be:

$$MR = A - 2Bx + 3Cx^2 - 4Dx^3 \tag{6}$$

which, in fact, lacks a unique inverse, though the demand curve has one. The marginal revenue curve is displayed along with the demand curve in Fig.1. The lack of an inverse is an interesting case, because it provides an opportunity for multiple equilibria.

Next we want the marginal revenue curve to have a positive derivative at the inflection point.

This point is defined by:

$$\frac{d^2MR}{dx^2} = 6C - 24DX = 0 \tag{7}$$

The slope is:

$$\frac{dMR}{dx} = -2B + 6Cx - 12Dx^2 \tag{8}$$

or, substituting from (7) for $x = C/4D$, we get

$$\frac{dMR}{dx} = -2B + \frac{3C^2}{4D} > 0. \tag{9}$$

The condition for an upward slope at the point of inflection, i.e. for the marginal revenue curve (6) not to have a unique inverse is:

$$C^2 > \frac{8}{3}BD \tag{10}$$

so, from (5) and (10) there is in fact a latitude of choice for

$$C^2 \in \left(\frac{8}{3}BD, \frac{9}{3}BD\right), \tag{11}$$

such that the demand curve (1) has a unique inverse but the marginal revenue curve (6) does not. Along with the marginal revenue curve, economists use the marginal cost curve, denoted MC.

It is the derivative of the total cost as a function of the quantity supplied. Typically the marginal cost curve is assumed to be concave from the above, with marginal cost first decreasing with increasing supply and eventually increasing. The truncated Taylor series for the marginal cost would be:

$$MC = E - 2Fx + 3Gx^2, \tag{12}$$

where E, F and G are positive. The integers are introduced for later convenience in integration.

The standard analysis would involve the condition for maximum profit, $MR = MC$, i.e. from (6) and (12)

$$(A - E) - 2(B - F)x + 3(C - G)x^2 - 4Dx^3 = 0 \quad (13)$$

Whenever the curves look like those in Fig.1 this equation has three different real roots. In theory, the monopolist would calculate these by solving the cubic, and then evaluate second order derivatives:

$$-2(B - F) + 6(C - G)x - 12Dx^2 \quad (14)$$

to check that the intermediate root is a profit minimum, whereas the higher and lower ones represent genuine maximum. Finally, the monopolist would choose between the two local maxima to identity the global profit maximum.

3 Adaptive Search

In reality, however the monopolist does not know more than a few points on the demand function. The marginal revenue curve, representing the derivative of an incompletely known function, would not be known except in terms of interpolation through recently visited points.

Given this, the task would be to design an algorithm for the maximum of unknown profit function:

$$\Pi(x) = (A - E)x - (B - F)x^2 + (C - G)x^3 - Dx^4 \quad (15)$$

This would be quite easy, provided the monopolist knew that the function was quartic.

The simplest way is to estimate the difference of marginal costs and revenues from the two last visited points in the profit function, and proceed to move in the direction of increasing profits.

Denoting the next last visited points by x and y respectively, and step length by δ, we get the next point as:

$$y + \delta \frac{\Pi(y) - \Pi(x)}{y - x} \quad (16)$$

In order to avoid unnecessary numerical computations the denominator in the quotient may be factored out as :

$$\begin{aligned} P(x, y) = \ & \tfrac{\Pi(y)-\Pi(x)}{y-x} = (A - E) - (B - F)(x + y) + \\ & +(C - G)(x^2 + xy + y^2) - D(x^3 + x^2y + xy^2 + y^3) \end{aligned} \quad (17)$$

4 Numerical Results

Iteration of this search process may lead to any of the two local profit maxima, to oscillating processes, or to chaos, depending on the values of the coefficients A through G, and the step size δ. The chaotic attractors may be unique, or else coexist with different attraction basins. In the following computer simulations, as well as in drawing Fig.1. the following values of the coefficients were used: A= 5.6, B=2.7, C=0.62, D=0.05, E=2, F=0.3 and G=0.02. With these coefficients the profit function (15) would become:

$$\Pi(x) = 3.6x - 2.4x^2 + 0.6x^3 - 0.05x^4 \tag{18}$$

Equating its derivative to zero yields the equation:

$$\Pi'(x) = 3.6 - 4.8x + 1.8x^2 - 0.2x^3 \tag{19}$$

which could also have been obtained by substituting the numerical coefficients in condition (13) for MR=MC. Equation (19) has three real roots:

$$x = 3 \quad and \quad x = 3 \pm \sqrt{3} \tag{20}$$

The search process with the coefficients introduced is defined by the map:

$$x_{t+1} = f(x_t, y_t) \tag{21}$$

$$y_{t+1} = g(x_t, y_t) \tag{22}$$

where

$$f(x, y) = y \tag{23}$$

and

$$g(x, y) = y + \delta P(x, y) \tag{24}$$

where $P(x, y)$ denotes the third order polynomial (17) which approximates the marginal profit, with the numerical values A through G substituted.

5 Fixed Points and Cycles

Before proceeding with the simulations, we note that the fixed points of the iterative mapping (21) - (22), not unexpectedly, are given by the equations

$$x = y \quad and \quad 3.6 - 4.8x + 1.8x^2 - 0.2x^3 = 0 \tag{25}$$

and that they are the maxima and the minimum of the profit function. To find out the stability of these points we calculate the Jacobian

$$\frac{\partial(f,g)}{\partial(x,y)} = \delta(2.4 + 1.8x - 0.3x^2) \tag{26}$$

We have taken $x = y$ in the fixed points. Of those we know that the profit minimum $x = 3$ is unstable, but the profit maximum at

$$x = 3 \pm \sqrt{3} \tag{27}$$

need further discussion. Substituting any of those roots in the Jacobian (26) we get

$$\frac{\partial(f,g)}{\partial(x,y)} = \frac{3}{5}\delta \tag{28}$$

Loss of stability occurs for both roots when the Jacobian unitary, i.e. at the common value:

$$\delta \quad = \quad 5/3 \tag{29}$$

This can be at the first branching point in the bifurcation diagram in Fig.2, where the phase variable, x (or y) is plotted against the parameter δ.

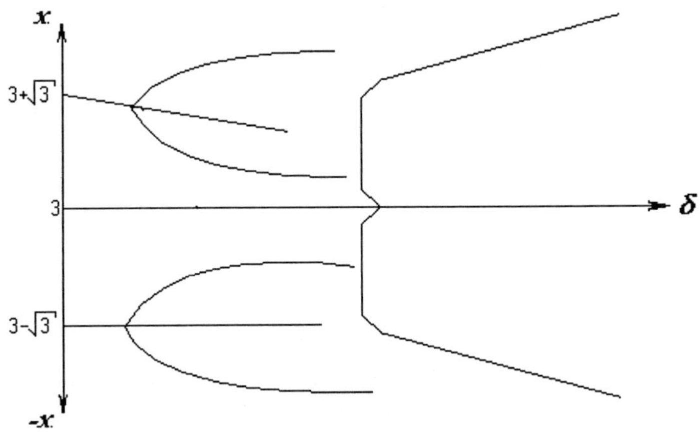

Fig.2. Bifurcation diagrams.

We see the two alternative fixed points coexist at parameter δ. Than 5/3. After that, the fixed points are replaced by cycles, over a quite extensive interval, and chaotic bands, with each it's own basin of attraction. After a certain point the attractions merge in a single one. Fig.3. displays the largest Lyapunov exponent, plotted against the step size parameter. The most striking feature is that, once chaos sets in definitely, the Lyapunov exponent remains positive, with no windows of order at all.

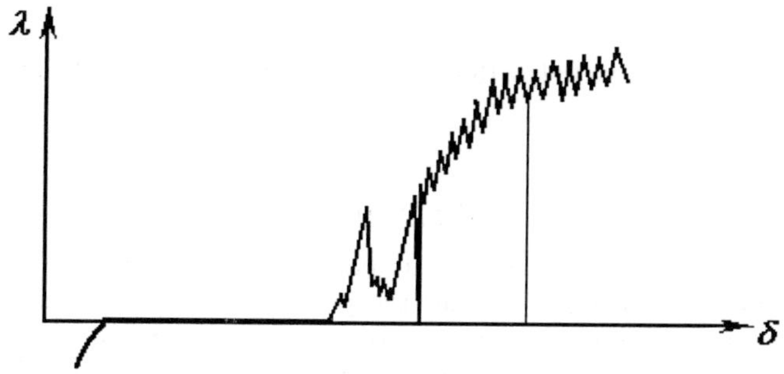

Fig.3. The largest Lyapunov exponent

The cycles shown in Fig.2. seemingly are of period three, but this is deceptive. For each of the two coexistent cycles, three different values are taken on, so let us denote those of one set a, b, and c. They are, however only passed in the sequence $a, b, c, d,\ a, b, c, ...$, so that one of the values is taken on twice as often as the other two. Recall that our system is two- dimensional, and we are accordingly dealing with a process where the points $(a, b), (b, c), (c, b)$, and (b, a) are visited over and over so the process is of period four, not three.

To see some details of the periodic solution, let us locate period four point. After four iterations we are back again, so the following four equations must be fulfilled:

$$z = y + \delta\frac{\Pi(y) - \Pi(x)}{y - x} \tag{30}$$

$$w = z + \delta\frac{\Pi(z) - \Pi(y)}{z - y} \tag{31}$$

$$x = w + \delta\frac{\Pi(w) - \Pi(z)}{w - z} \tag{32}$$

$$y = x + \delta\frac{\Pi(x) - \Pi(w)}{x - w} \tag{33}$$

Multiplying through by the denominators, and adding (30)-(33) pairwise, we arrive at the following two equations:

$$(z - y)(x - y + z - w) = \delta(\Pi(x) - \Pi(z)) \tag{34}$$

$$(w - z)(x - y + z - w) = \delta(\Pi(w) - \Pi(y)) \tag{35}$$

Whereas adding all four of them, (30)- (33) makes all the profit functions in the right-hand side cancel, and we end up with a perfect square that has to equal zero:

$$(x - y + z - w)^2 = 0 \tag{36}$$

Using this last condition in (34) - (35) we get a pair of useful conditions:

$$\Pi(x) = \Pi(z) \tag{37}$$

$$\Pi(y) = \Pi(w) \tag{38}$$

So, profits in the sequence x, y, z, w are pair wise equal with a lag of two periods. This may happen in two ways. Either the points visited are themselves equal, as conjectured for the value visited twice as often as the other two. Or, they may be located sufficiently widely apart as to fall on different rising and falling branches of the quartic profit function. We now turn to the numerical example to get more substance, and substitute expressions (18) for the pair wise equal profits in (37)-(38). The equations then factor into:

$$(x - z)(x + z - 6)((x - 3)^2 + (z - 3)^2 - 6) = 0 \tag{39}$$

$$(y - w)(y + w - 6)((y - 3)^2 + (w - 3)^2 - 6) = 0 \tag{40}$$

These give us three possibilities each. Taking z and w for an example, they may be equal, their sum can be equal to 6, or they may be located on a circle centered around the point $(3, 3)$. Among these the middle option is not relevant for a periodic point that moves entirely above or below the line at 3 in fig.2.

Any sum of two amplitudes is strictly smaller or larger than 6, depending on which of the alternative cycles we consider. Actually, the middle option refers to the unstable fixed point at 3 of fig.2.

Obviously, both $x = z$ and $y = w$ cannot hold simultaneously, as we would then no longer deal with a four- period, but with a two –period point, which, by the way, like the three- period point, can be shown to be impossible. So, what we actually locate are the now like wise unstable fixed points $3 \pm \sqrt{3}$, in addition to the aforementioned one at 3.

In the same way we cannot have both pairs located on the circles $(x-3)^2+(z-3)^2 = 6$ and $(y - 3)^2 + (w - 3)^2 = 6$, because, after a somewhat more the derivation, we end up at the same conclusion, i.e. that we are dealing with one of the unstable fixed points. So, two points in the four cycle must be equal, say x and z, and two, say y and w, must be on the circle.

These considerations, along with (36), provide us with three equations:

$$x = z \tag{41}$$

$$(y - 3)^2 + (w - 3)^2 = 6 \tag{42}$$

$$x + z = y + w \tag{43}$$

In addition, we have equation (30), which, using the fact that $x = z$ according to (41), becomes:

$$x = y + \delta \frac{\Pi(y) - \Pi(x)}{y - x} \tag{44}$$

Equations (41) - (44) are sufficient for calculating the four values of the cycle. For instance with $\delta = 2$, according to Fig.2. definitely has a cycle, we get:

$$x = z = 3 \pm \frac{\sqrt{65}}{5} \tag{45}$$

$$y = 3 \pm \left(\frac{\sqrt{65}}{5} - \frac{\sqrt{10}}{5} \right) \quad w = 3 \pm \left(\frac{\sqrt{65}}{5} + \frac{\sqrt{10}}{5} \right) \tag{46}$$

The minus sign applies to the lower half of Fig.2. the plus sign to the upper half. These values agree with the results from simulation.

6 Chaos

We see from Figs.(2) and (3) that chaos appears once the four- period point loses stability. It is therefore of particular interest to locate the parameter value at which this happens. This can be done by analytical means, though the procedure is a little bit tedious.

The Jacobion determinant at the periodic point, using the values specified in (45)-(46), becomes:

$$j = \left(\frac{\delta}{20} \right)^4 \left(\frac{4\sqrt{26}}{5} - 10 \right) \left(\frac{4\sqrt{26}}{5} + 10 \right) \left(\frac{8\sqrt{26}}{25} - \frac{54}{5} \right) \left(\frac{8\sqrt{26}}{25} + \frac{54}{5} \right) \tag{47}$$

and loss of stability occurs when the Jacobion becomes unitary, i.e, when

$$j = \frac{1345}{51549} \delta^4 = 1 \tag{48}$$

Calculating the critical step length parameter in (48) we get:

$$\delta = 2.48813 \tag{49}$$

The method of critical lines

Let us now find out how the attractor, could be studied using the concepts of critical lines and absorbing areas. Note that (26) and (47) as derived above will not do because they were obtained for the fixed point and the four period cycle respectively.

Rather we need the general expression derived from (23)-(24):

$$\begin{vmatrix} f_x & f_y \\ g_x & g_y \end{vmatrix} = \begin{vmatrix} 0 & 1 \\ \delta p_x & 1 + \delta p_y \end{vmatrix} = -\delta p_x = 0 \tag{50}$$

Note also that we put the determinant equal to zero, not to unity as before, because we are now interested in the boundary line along which the image is folded, which is an issue different from that of determining the stability of a periodic point. We may also note the interesting fact that fold (50) is independent of the step length parameter δ, as we must have $p_x = 0$.

According to equation (17) which defines the polynomial $P(x, y)$, and using the numerical values stated above, we get by simple differentiation and a slight rearrangement:

$$3(x-3)^2 + 2(x-3)(y-3) + (y-3)^2 = 6 \tag{51}$$

This is the equation of an ellipse tilted counterclockwise by 45^0 and centered at the point $x = y = 3$. Accordingly, inside the ellipse there are three preimages to the map (23)-(24), outside there is just one.

References

[1] Abraham, R.H and Shaw, C.D, (1992) *Dynamics, The Geometry of Behavior.* Second edition, Addison-Wesley, Reading MA.

[2] Abraham, R.H, Gardini, L. and Mira C., (1997) *Chaos in Discrete Dynamical Systems*, TELOS, Springer-Verlag, NY.

[3] Chiarella, C. (1990) *The Elements of Nonlinear Theory of Economics Dynamics, Lecture Notes in Economics and Mathematical systems 343.*, Springer-Verlag.

[4] Tonu Puu (2000) *Attractors, Bifurcations, and Chaos - Nonlinear Phenomena in Economics*, Springer-Verlag.

Optimization and Optimal Control, pp. 69 – 77
P.M. Pardalos, I. Tseveendorj, and R. Enkhbat, Editors

Necessary Optimality Conditions and Methods of Global Optimization

V.P. Bulatov (bulatov@isem.sei.irk.ru)
Institute of Energy Systems, Irkutsk, Russia

D. Ganhyuag (gans@icb.edu.mn)
Institute of Commerce and Business, UlaanBaatar, Mongolia

Abstract.

We consider the problem of finding all real roots of systems of nonlinear equations. We present a global optimization technique to solve this problem. Different approaches are described and the connection of the problem under consideration with the classical necessary optimality conditions is discussed.

Keywords: Systems of nonlinear equations, global optimization, necessary optimality conditions

1 Statement of the problem and method of solution

Consider the problem of finding all points $\overline{x} \in R^0$ such that

$$q_i(\overline{x}) = 0, q \in E^n, \tag{1}$$

$q_i(x)$ are convex continuously differentiable functions, R^0 is a convex closed bounded set, and E^n is the nth dimensional Euclidean space.

Roots $\overline{x} \in R^0$ will be called different if matrix $\nabla q(\overline{x})$ has rank n and of multiplicity $k : 1 \leq k < n$ if rank $\nabla q(\overline{x})$ is equal to $n - k$. It is also assumed that the number of roots of system (1) over R^0 is finite, i.e. the roots of system (1) are isolated.

The following auxiliary problem is used to find the current root x^k

$$\min\{\|x\| : x \in R\}, \tag{2}$$

$$R = \{x \in E^n : q_i(x) = 0, x \in L, i = \overline{1,m}, m \leq n\}, \tag{3}$$
$$L = \{x : f_j(x) \leq 0, j = \overline{1,l}\}.$$

We associate to problem (2)-(3) the following one

$$\min\{\psi(x) = \|x\| + N\sum_{i=1}^{m} q_i(x) : x \in R^0\}, \tag{4}$$

$$R^0 = \{x : q_i(x) \geq 0, f_j(x) \leq 0, i = \overline{1,m}, j = \overline{1,l}\}, N > 0. \tag{5}$$

Denote by

$$\Phi(x) = \left\{ \begin{array}{c} q(x) \\ f(x) \end{array} \right\},$$

where $q \in E^m, f \in E^l, \Phi \in E^{m+l}$. Let R^0 satisfies regularity conditions, X be the set of minimum points of $\varphi(x) = \|x\|$ over R^0 and R_λ be the set of optimal solutions of the dual problem. Then the following theorem is true.

Theorem 1. *Assume that*
1) there exists $\varepsilon > 0$ such that $\forall x\ X, \lambda \in R_\lambda$

$$\|\nabla\Phi(x)^T\lambda\| \geq \varepsilon\|\lambda\|;$$

2) $\|\nabla\varphi(x)\| \leq c < \infty$. Then there exists N^* such that $\forall N > N^*$ any minimum point of $\psi(x)$ over R^0 satisfies $q(x) = 0$.

Assume that a sequence $\{x^j\}$ such that $q_i(x^j) = 0, \forall j = \overline{1,k}$ is constructed. Determine the cone L^k at point x^k

$$L^k = \{x : \nabla q(x^k)^T x \leq \nabla q(x^k)^T x^k\}.$$

Equations of the edges of the cone L^k have the following form

$$x^j = x^k - \lambda^j s^{kj}, \lambda^j > 0, \tag{6}$$

where s^{kj} are columns of the matrix inverse to matrix $\nabla q(x^k)$ (since according to the assumption $|\nabla q(x^k)| \neq 0$). From equations $\sum_{i=1}^{n} q_i(x^k - \lambda^j s^{kj}) = 0$ find λ^{kj} and the corresponding points $x^{kj} = x^k - \lambda^{kj}s^{kj}$ of intersection of rays (6) with surface $\sum_{i=1}^{n} q_i(x) = 0$. Determine cutting half space $\{x : \beta^{kT}x \leq t_k\}$, which cuts off the point x^k. The next solution x^{k+1} of the system of nonlinear equations is found from solving the problem

$$\min\{\psi(x) : x \in R^{k+1}\},$$
$$R^{k+1} = \{x : x \in R^k, \beta^{kT}x \leq t_k\}.$$

Let S^k be convex closed set such that $S^k \supset R = \{x : q_i(x) = 0, i = \overline{1,n}\}$ and all solutions of system (1) are boundary points of S^k. Assume that there exist $\lambda^{kj}(0 < \lambda^{kj} < \infty)$ and the corresponding points $x^{kj} = x^k - \lambda^{kj}s^{kj}$ of intersection of rays (6) with the boundary of set S^k. Draw the plane $\{x : \beta^{kT}x = t_k\}$ through the points x^{kj} and determine the cutting half space $\{x : \beta^{kT}x \leq t_k\}$, which cuts off the point x^k. The next solution x^{k+1} of system of nonlinear algebraic equations is found from solving the following problem

$$\min\{\psi(x) : x \in R^{k+1}\},$$

$$R^{k+1} = \{x : x \in R^k, \beta^{kT}x \leq t_k\}.$$

From geometrical consideration we have

Lemma. *Intersection of cone L^k and half space $\{x : \beta^{kT}x > t_k\}$ does not contain solutions of system (1) different from x^k.*

Let us give some examples of set S^k.

$$S = S^k(\lambda) = \{x : \sum_{i=1}^{n} \lambda_i q_i(x) \leq 0, \forall \lambda_i > 0\}$$

or

$$S = S^k(1) = \{x : \sum_{i=1}^{n} q_i(x) \leq 0\}.$$

The described cutting plane constructions are similar to those developed in integer programming and can be used in iteration processes of global optimization in combination with the necessary optimality conditions. Consider, for example, the following boolean programming problem

$$\min\{\varphi(x) : x \in R, x_i \in \{0,1\}, i = \overline{1,n}\}, R \subset E^n.$$

Determine

$$e = (1, ..., 1) \in E^n, R^0 = \{x \in E^n : 0 \leq x \leq 1\}, q_i(x) = x_i - x_i^2.$$

It is obvious that the constraints $q_i(x) = x_i - x_i^2 = 0$ are equivalent to $x_i \in \{0,1\}, i = \overline{1,n}, x \in R^0$. Associate to the initial problem the following one

$$\min\{\varphi(x) + N\sum_{i=1}^{n} q_i(x) : x \in R, q_i(x) \geq 0, i = \overline{1,n}\}, N > 0.$$

If $\varphi(x)$ is a concave function and R is determined by the system of linear inequalities $R = \{x : Ax \leq b\}$, then the latter problem can be rewritten as follows

$$\min\{\varphi(x) + Nx^T(e - x) : x \in R \cap R^0\}.$$

This problem is the problem of minimization of a concave function over a polytope. It is well known that there exist N^* such that for all $N > N^*$ the set of all global minima of the latter problem coincide with the set of all optimal integer solutions of the initial problem. Moreover, it is possible to get a constructive estimate on N based on the input data.

2 Connection to the necessary optimality conditions

Let us apply now the described technique for solving a more general mathematical programming problem

$$\max\{\varphi(x) : q_j(x) \leq 0, j = \overline{1,m}\}. \tag{7}$$

Here $\varphi(x)$ is a convex smooth function, $q_i(x)$ are smooth functions, max is understood in the global sense. Rewrite the constraints in (7) as equalities $q_i(x) + y_i^2 = 0$, associate to problem (7) the Lagrange function and write down the necessary optimality conditions for the Lagrange problem. Then the equivalent to (7) problem will have the following form

$$\max\{\varphi(x) : \nabla\varphi(x) + \lambda^T \nabla q(x) = 0, q_i(x) + y_i^2 = 0, \lambda_j y_j = 0, j = \overline{1,m}\}. \tag{8}$$

In new notations

$$\frac{\partial \varphi}{\partial x_i} + \sum_{j=1}^{m} \lambda_j \frac{\partial q_j}{\partial x_i} = \psi_i, i = \overline{1,n}, q_j(x) + y_j^2 = \psi_j, j = \overline{n+1, n+m},$$

$$2\lambda_j y_j = \psi_j, j = \overline{n+m+1, n+2m}$$

we have

$$\max\{\varphi(x) : \psi_j(x,y,\lambda) = 0, j = \overline{1, n+2m}\}. \tag{9}$$

The feasible set of the latter problem is given by a closed system of nonlinear algebraic equations. Assume that we are looking for the solutions belonging to convex compact set R^0. Assume also, that the set of solutions of the system $\psi(x,y,\lambda) = 0$ is finite over R^0, i.e. we consider isolated roots. Then the initial problem (7) is equivalent to problem of maximization of a convex function $\varphi(x)$ over the finite set $R = \{x : x \in R^0, \psi_j(x,y,\lambda) = 0, j = \overline{1, n+2m}\}$.
In a similar way as above we associate to problem (9) the following auxiliary problem of finding a local maximum of a convex function

$$\max\{\varphi(x) + N \sum_{j=1}^{n+2m} \psi_j(x,y,\lambda) : x \in R\}, \tag{10}$$

$$R = \{x, y, \lambda : \psi_j(x,y,\lambda) \leq 0, x \in R^0\}, N > 0. \tag{11}$$

Due to the Theorem 1 for sufficiently large N we have $\psi_j(x^*, y^*, \lambda^*) = 0, j = \overline{1, n+2m}$ at any locally optimal solution x^*, y^*, λ^* of problem (10)-(11).
Assume that problem (10)-(11) is nondegenerate, i.e. $|\nabla\psi(x^*, y^*, \lambda^*)| \neq 0$ for any solution and vector-function $\psi(x,y,\lambda)$ is jointly convex. It is not difficult to see that a local solution of problem (10)-(11) can be found by solving a sequence of problems

of maximization of linear functions over R.

Let us apply now the described method for solving problem (9). Assume that point $z^k = (x^k, y^k, \lambda^k)$ is found from the solution of the problem similar to (10)-(11):

$$\max\{\varphi(x) + N \sum_{j=1}^{n+2m} \psi_j(x, y, \lambda) : x, y, \lambda \in R^k\},$$

$$R^k = \{(x, y, \lambda : \psi_j(x, y, \lambda) \leq 0, x \in R^0, s^{iT} z \leq t_i, j = \overline{1, n+2m}, i = \overline{1, k}\}.$$

Let $\beta_k = \max\{\varphi(x^k), \beta_{k-1}\}$. Determine the cone L^k at point z^k

$$L^k = \{z : \nabla\psi(x^k, y^k, \lambda^k)^T z \leq \nabla\psi(x^k, y^k, \lambda^k) z^k\}. \tag{12}$$

Due to the assumption that the matrix $\nabla\psi(x^k, y^k, \lambda^k)$ is nondegenerate, equations of the edges of cone (12) have the form

$$z^j = z^k - \lambda^j s^{kj}, j = \overline{1, n+2m}, z^k = (x^k, y^k, \lambda^k), \lambda^j > 0, \tag{13}$$

where s^{kj} are columns of the matrix inverse to $\nabla\psi(z^k)$.

From equations $\psi(z^k - \lambda^j s^{kj}) = \beta_k$, where $\psi(z) = \varphi(x) + N \sum_{j=1}^{n+2m} \varphi_j(z)$, determine parameters λ^{kj} and the corresponding points z^{kj} of intersection of rays (13) with the boundary of set $\{z : \psi(z) \leq \beta_k\}$. Due to the nondegeneracity $\lambda^{kj} > 0, \forall j$. Draw a plane through points z^{kj} and construct half space $\{z : (s^{k+1})^T z \leq t_{k+1}$ which does not contain z^k. Obtain the next solution z^{k+1} from solving the problem

$$\max\{\psi(z) : z \in R^k, (s^{k+1})^T z \leq t_{k+1}\},$$

and $\psi_j(z^{k+1}) = 0, j = \overline{1, n+2m}$. In a finite number of steps we obtain the solution of the initial problem (9).

Remark. *It is assumed that $\psi_j(x, y, \lambda)$ are jointly convex.*

The assumption is not too strong as it can be seen from the first glance. If $\psi_j(x)$ and $q_i(x)$ in the initial problem are given by power series then by introduction of new variables and dimension enlargement one can construct a problem of type (9) with convex functions in the left hand sides of the system of constraints and the constructed problem will be equivalent to the initial one.

3 Systems of algebraic polynomials

Let us describe how systems with algebraic polynomials in the left hand sides can be reduced to the equivalent systems with convex functions in the left hand sides.

Consider the problem of finding all real roots of the equation

$$a_m x^m + a_{m-1} x^{m-1} + \ldots + a_3 x^3 + a_2 x^2 + a_1 x + a_0 = 0. \tag{14}$$

Introduce new notations

$$x^{m-1} = y_{m-1}, x^{m-2} = y_{m-2}, ..., x^3 = y_3, x^2 = y_2. \tag{15}$$

Substituting (15) in (14) we obtain the following bilinear system

$$a_m x y_{m-1} + a_{m-1} x y_{m-2} + ... + a_3 x y_2 + a_2 x^2 + a_1 x + a_0 = 0, \tag{16}$$

$$x y_{m-2} - y_{m-1} = 0, x y_{m-3} - y_{m-2} = 0, ..., x y_2 - y_3 = 0. \tag{17}$$

Using the equality $\frac{1}{4}ab = (a+b)^2 - (a-b)^2$ we have from (16)-(17)

$$a_m[(x+y_{m-1})^2 - (x-y_{m-1})^2] + a_{m-1}[(x+y_{m-2})^2 - (x-y_{m-2})^2] + ...$$

$$a_3[(x+y_2)^2 - (x-y_2)^2] + 4a_2x^2 + 4a_1x + 4a_0 = 0,$$

$$(x+y_{m-2})^2 - (x-y_{m-2})^2 - 4y_{m-1} = 0,$$

$$(x+y_{m-3})^2 - (x-y_{m-3})^2 - 4y_{m-2} = 0,$$

$$\dots\dots\dots\dots\dots\dots$$

$$(x+y_2)^2 - (x-y_2)^2 - 4y_3 = 0. \tag{18}$$

By substitution $(x-y_{m-1})^2 = z_{m-1}, (x-y_{m-2})^2 = z_{m-2}, ..., (x-y_2)^2 = z_2$ we finally have the following system of $2(m-2)+1$ quadratic algebraic equations with convex functions in the left hand sides

$$a_m(x+y_{m-1})^2 + a_{m-1}(x+y_{m-2})^2 + ... + a_3(x+y_2)^2 + 4a_2x^2 + 4a_1x-$$

$$a_m z_{m-1} - a_{m-1} z_{m-2} - ... - a_3 z_2 + 4a_0 = 0,$$

$$(x-y_{m-1})^2 - z_{m-1} = 0, (x-y_{m-2})^2 - z_{m-2} = 0, ..., (x-y_2)^2 - z_2 = 0,$$

$$(x+y_{m-2})^2 + z_{m-2} - 4y_{m-1} = 0, (x+y_{m-3})^2 - z_{m-3} - 4y_{m-2} = 0, ...,$$

$$(x+y_2) - z_2 - 4y_3 = 0.$$

Consider now a system of nonlinear algebraic equations determined by multivariable polynomials in the left hand sides

$$\sum_{i_p=1}^{n} \sum_{i_{p-1}=1}^{n} ... \sum_{i_1=1}^{n} a^j_{i_1 i_2 ... i_p} x_{i_1} x_{i_2} ... x_{i_p} + ... \sum_{i_2=1}^{n} \sum_{i_1=1}^{n} a^j_{i_1 i_2} x_{i_1} x_{i_2} +$$

$$\sum_{i_1=1}^{n} a^j_{i_1} x_{i_1} + q^0_j = 0, j = \overline{1,n}. \tag{19}$$

Similar to (15) introduce the notations

$$\sum_{i_3=1}^{n} \sum_{i_2=1}^{n} a^j_{i_1 i_2 i_3} x_{i_1} x_{i_2} = y^j_{i_1},$$

....

$$\sum_{i_p=1}^{n} \sum_{i_{p-1}=1}^{n} \dots \sum_{i_2=1}^{n} a^j_{i_1 i_2 \dots i_p} x_{i_2} x_{i_3} \dots x_{i_p} = y^j_{i_{p-1}}. \tag{20}$$

Equation (19) can be rewritten in the bilinear form

$$q_j^0 + \sum_{i_1=1}^{n} a^j_{i_1} x_{i_1} + \sum_{i_2=1}^{n} \sum_{i_1=1}^{n} a^j_{i_1} x_{i_1} x_{i_2} +$$

$$\sum_{i_1=1}^{n} x_{i_1} y^j_{i_1} + \dots + \sum_{i_1=1}^{n} x_{i_1} y^j_{i_{p-1}} = 0.$$

In the same way as in (17) equations (20) can be rewritten in the bilinear form too. Using the equality $4x_{i_k} y_{i_p} = (x_{i_k} + y_{i_p})^2 - (x_{i_k} - y_{i_p})^2$ and denoting by $(x_{i_k} - y_{i_p})^2 = z_{i_k i_p}$ we again obtain a system of nonlinear quadratic equations with convex functions in the left hand sides.

It is not difficult to see that the systems are equivalent, i.e. if $\overline{x}_1, \dots, \overline{x}_n$ is a solution of system (19) then there exist $\overline{z}_{i_k i_p}$ and $\overline{y}_{i_k}$ such that $(\overline{x}_i, \overline{z}_{i_k i_p}, \overline{y}_{i_k})$ is a solution of the quadratic system and vise versa.

Roots of system (19) can be distributed in a quite exotic way and in general a solution found at the current iteration can not be separated by a plane from other solutions. This can lead to obtaining the same solution at the next iteration.

If the problem consists in finding all isolated solutions of system with convex functions in the left hand sides over a compact set R^0 then the iteration process can be constructed in the following way. Let x^0 be a solution of system $q_j(x) = 0$ over R^0. Then x^0 is an extreme point of the convex set $\{x : x \in R, q_j(x) \leq 0\}$ and it can be separated by plane $\{x : \alpha^{0T} x = \beta_0\}$ from the other solutions. Determine the set $R^1 = \{x : x \in R^0, \alpha^{0T} x \leq \beta_0\}$ which does not contain x^0. Find x^1 as a solution of the system $q_j(x) = 0$ over R^1 and separate x^1 as an extreme point of the set $\{x : x \in R^1, q_j(x) \leq 0, j = \overline{1,n}\}$ from the other solutions. We stop when the current set R^k does not contain solutions of the system $q_j(x) = 0$.

Below we present some results of finding all roots of system of quadratic equations. The corresponding programs were developed in GAMS. The suggested approach was compared with the Newton method combined with multistart. For systems containing 2-3 equations both programs found all solutions, but the combined Newton method takes more time. For problems with higher dimensions the combined Newton method failed to find all solutions (in all cases no more than half of all solutions). The maximum number of roots found by the suggested above approach were 1000. All test problems were randomly generated.

Example. The systems with the following left-hand sides was considered

$$g_1(x) = 1.031242x_1^2 + 0.214315x_2^2 + 0.482032x_3^2 + 1.739011x_4^2 - 10,$$

$$g_2(x) = 0.381442x_1^2 + 0.327203x_2^2 + 0.970965x_3^2 + 1.038026x_4^2 - 10,$$

$$g_3(x) = 2.057346x_1^2 + 0.407672x_2^2 + 0.184593x_3^2 + 0.985299x_4^2 - 10,$$
$$g_4(x) = 0.765527x_1^2 + 0.255545x_2^2 + 0.902573x_3^2 + 1.086937x_4^2 - 10.$$

Below in the table the following notations are used:
itr - number of iterations;
def - total deviation;
T - time in sec.;
N_1 - number of starts for finding a feasible point;

	x_1	x_2	x_3	x_4	N_1	itr	def	T
1	-1.343	2.687	2.101	-1.602	1	4	0	0.63
2	-1.343	-2.687	-2.101	1.602	1	4	0	0.62
3	-1.343	2.687	2.101	1.602	1	4	0	0.63
4	1.343	2.687	-2.101	1.602	1	4	0	0.63
5	-1.343	-2.687	2.101	-1.602	1	4	0	0.62
6	-1.343	-2.687	-2.101	-1.602	1	4	0	0.64
7	1.343	-2.687	-2.101	1.602	2	4	0	0.72
8	-1.343	2.687	-2.101	-1.602	1	4	0	0.67
9	1.343	2.687	2.101	-1.602	1	4	0	0.65
10	-1.343	-2.687	2.101	1.602	2	4	0	0.71
11	1.343	-2.687	2.101	1.602	1	4	0	0.66
12	1.343	-2.687	-2.101	-1.602	1	4	0	0.63
13	-1.343	2.687	-2.101	1.602	5	4	0	1.00
14	1.343	2.687	2.101	1.602	14	4	0	1.89
15	1.343	2.687	-2.101	-1.602	14	4	0	1.86
16	1.343	-2.687	2.101	-1.602	1	4	0	0.63

References

[1] V.P. Bulatov (1977), *Imbedding Methods in Problems of Optimization*, Nauka, Novosibirsk, 1977.

[2] V.P. Bulatov and T.I. Belykh (2001), "Concave Programming and Numerical Methods of finding Real Roots of Nonlinear Systems", *DAN*, vol. 376, N2, pp.171-174

[3] V.P. Bulatov (2000), "Global Optimization and Numerical Methods for Finding All Real Roots of Systems of Nonlinear Algebraic Equations with Convex Functions in the Left-Hand Sides", *Computational Mathematics and Mathematical Physics*,vol. 60, N3, pp.348-355

[4] V.P. Bulatov and D. Ganhuyag (2000), "Global Opimization and Methods for Finding All Real Roots of Systems of Nonlinear Algebraic Equations", Abstracts

of the International Conference ".Discrete Analysis and Operations Research", Novosibirsk, pp.18-21

[5] A. Brooke, D. Kendrick, A. Meeraus (1996), *GAMS. A User's Guide*, GAMS Development Corporation

Optimization and Optimal Control, pp. 79 – 102
P.M. Pardalos, I. Tseveendorj, and R. Enkhbat, Editors

On Some Theory, Methods and Algorithms for Concave Programming

R. Enkhbat (renkhbat46@yahoo.com)
Department of Mathematics of the School of Economic Studies
National University of Mongolia
P.O.Box 46/635,
Ulaanbaatar,
MONGOLIA

Abstract

We are concerned with concave programming or the convex maximization problem, which has many applications and is known as NP-hard problem [26]. In this paper, we survey several results in concave programming that have appeared in the literature since 1987. The proposed methods and algorithms for solving the problem are based on the global optimality conditions first obtained by Strekalovsky in 1987 [28]. We show that some of the results can be applied to a nonconvex optimal control problem with a convex terminal objective function.

Keywords: concave programming, algorithm, simple set, approximation

1 Introduction

We consider the problem of maximizing a convex function over an arbitrary set $D \subset R^n$:

$$f(x) \longrightarrow \max, \quad x \in D, \tag{1.1}$$

Problem (1.1) has many applications in economics and engineering. Convex maximization (equivalently, concave programming) problem as a special case of problem (1.1), where D is a polytope, is originated from Tuy [30]. The cutting plane method for this case was proposed in [30]. There are many iterative algorithms

[2, 16–20, 24, 25, 30, 31] for solving the convex maximization problem by resorting to the cutting plane and/ or branch and bound techniques. For most of them, it is common that auxiliary problems are linear programming, and these methods don't use global optimality conditions.

Unlike convex minimization problem, problem (1.1) is multiextremal and from the viewpoint of computational complexity, it is NP-hard. Classical local optimality conditions can't guarantee always finding a global solution.

We consider, in this paper, mainly methods and algorithms for solving problem (1.1) developed using the global optimality condition in [28]. Our approach uses concepts of approximation of level sets for the objective function, first and second order permissive sets, and simple set. We also consider the quasiconvex maximization problem as a generalized form of problem (1), and present its optimality conditions [6].

The paper is organized as follows. In section 2, we give existing optimality condions for problem (1.1). Section 3 describes methods and algorithms based on resolving sets. In section 4 we give a overview of methods for the quadtratic case. In section 5, methods and algorithms fot the strongly convex case with the use of simple sets are described. In section 6, we report computational experiments on the proposed algorithms.

2 Convex Maximization Problem

2.1 Global Optimality Conditions

Consider the convex maximization problem,

$$f(x) \longrightarrow \max, \quad x \in D, \tag{2.1}$$

where $f : R^n \to R$ is a convex function and $D \subset R^n$ is a nonempty arbitrary subset of R^n. First global optimality conditions for this problem was obtained in 1987 by Strekalovsky [28].

Theorem 2.1 ([28]). Given an arbitrary set D, suppose that a point $z \in D$ satisfies

$$-\infty < \inf_{R^n} f < f(z) < +\infty$$

and $L_{f(z)}(f) = \{x \in R^n \mid f(x) \leq f(z)\} \subset intdom\ f$ is compact. Then z is a global maximizer of problem (1.1) if and only if

$$\begin{aligned} &\langle c, x - z \rangle \leq 0 \quad \text{for all } x \in \overline{conv}\ D \text{ and } c \in \partial f(z), \\ &\langle v, x - z \rangle \leq 1 \quad \text{for all } x \in \overline{conv}\ D \text{ and } v \in S(f, z), \end{aligned} \tag{2.2}$$

where

$$\begin{aligned} S(f,z) &= \left\{ v \in R^n \;\middle|\; \begin{array}{l} \exists y \in R^n : \; y \neq z, \; y \in E_{f(z)}(f), \\ \exists \alpha > 0 : \; \alpha v \in \partial f(y), \; \langle v, x - z \rangle = 1. \end{array} \right\}, \\ \partial f(z) &= \{c \in R^n \mid f(x) - f(z) \geq \langle c, x - z \rangle, \;\; x \in R^n\}, \\ intdom\ f &= \{x \in R^n \mid f(x) < +\infty\}, \\ E_{f(z)}(f) &= \{y \in R^n \mid f(y) = f(z)\}, \end{aligned}$$

and $\overline{conv}\ D$ is the closed convex hull of D. ■

Another equivalent formulation of condition (2.2) under some relaxed assumptions was given in [29].

Theorem 2.2 ([29]). Let a point $z \in D$ satisfy $-\infty \leq \inf_D f < f(z)$. Then z is a global maximizer of (2.1) if and only if

$$\partial f(x) \subset N(x|D) \text{ for all } x \in E_{f(z)}(f), \tag{2.3}$$

where $N(x|D)$ is the normal cone to D at x defined as:

$$N(x|D) = \{c \in R^n \mid \langle c, y - x \rangle \leq 0, \;\; y \in D\}.$$

■

It is clear that condition (2.3) can be written as follows:

$$\langle c, x - y \rangle \leq 0 \text{ for all } x \in D, c \in \partial f(y) \text{ and } y \in E_{f(z)}(f).$$

When D is closed and convex then there is another related result given by Hiriart-Urruty and Ledaev.

Theorem 2.3 ([15]). Let a point $z \in D$ satisfy $-\infty \leq \inf_D f < f(z)$. Then z is a global maximizer of (2.1) if and only if

$$\partial f(x) \subset N(x|D) \text{ for all } x \in D \text{ and } x \in E_{f(z)}(f) \tag{2.4}$$

■

A new global optimality condition which implies (2.3) was given in [32].

Theorem 2.4([32]) Let a point $z \in D$ be such that $-\infty \leq \inf_D f < f(z)$. Then z is a global maximizer of (2.1) if and only if

$$\partial f(x) \cap N(x|D) \neq \varnothing \text{ for all } x \in D \text{ and } x \in E_{f(z)}(f), \tag{2.5}$$

■

Also, there is a different global optimality condition from (2.3)- (2.5) obtained by Hiriarty-Urrity in 1989 which is formulated as:

Theorem 2.5 ([14]). If D is closed set in problem (2.1), then $z \in D$ is a global maximizer of (2.1) if and only if

$$\partial_\varepsilon f(z) \subset N_\varepsilon(z \mid D) \text{ for all } \varepsilon > 0, \tag{2.6}$$

where

$$\partial_\varepsilon f(z) = \{c \in R^n \mid f(x) \geq f(z) + \langle c, x - z\rangle - \varepsilon \text{ for all } x \in R^n\},$$
$$N_\varepsilon(z \mid D) = \{c \in R^n \mid \langle c, y - z\rangle \leq \varepsilon \text{ for all } y \in D\}.$$

■

Now consider the problem of maximizing a quasiconvex function over a set $D \subset R^n$ as a generalized form of problem (2.1):

$$f(x) \longrightarrow \max, \quad x \in D, \tag{2.7}$$

where $f: R^n \to R$ is a quasiconvex and differentiable function and $D \subset R^n$ is a nonempty arbitrary subset of R^n. Then an optimality condition for this problem is given by the following theorem:

Theorem 2.6 ([6]). Let z be a solution of problem (2.7) then

$$\langle f'(y), x - y\rangle \leq 0 \text{ for all } y \in E_{f(z)}(f) \text{ and } x \in D. \tag{2.8}$$

If, in addition, $f'(y) \neq 0$ holds for all $y \in E_{f(z)}(f)$, then condition (2.8) is sufficient for $z \in D$ being a solution to problem (2.7). ■

There are some algorithms and methods for solving problem (2.1) based on global optimality condition (2.6) using support functions and copositivity techniques [3] .

In this paper we will restrict ourselves to consideration of algorithms and methods based on global optimality condition (2.3).

3 Methods and Algorithms

3.1 Resolving set

First attempt to construct an algorithm for solving problem (2.1) based on condition (2.3) using a notion of resolving set was done in [7].

Definition 3.1 A set

$$R(z) = \{y^1, y^2, \ldots, y^m \mid f(y^i) = f(z), \quad i = 1, 2, \ldots, m\}$$

is called a resolving set of problem (2.1) for a point $z \in D$, if the inequality

$$\max_{1 \leq i \leq m} \sup_{x \in D} \langle f'(y^i), x - y^i\rangle \leq 0$$

implies that z is a solution to the problem. ■

In problem (2.1), resolving sets exist for some special case of D and f, but it is hard to construct a resolving set for a general case.

Example 3.1 ([7]). Let in problem (2.1) $f(x) = \| x \|^2$ and $D = \{x \in R^n \mid a_i \leq x_i \leq b_i,\ b_i > 0 > a_i, i = 1, 2, ..., n\}$. Then a resolving set for a point $z \in D$ is

$$R(z) = \{y^1, y^2, \ldots, y^n \mid y^i = z - 2z_i e^i, \quad i = 1, 2, \ldots, n\},$$

where $e^1 = (1, 0, ..., 0)$, $e^2 = (0, 1, 0, ..., 0), \ldots,$ $e^n = (0, \ldots, 0, 1)$ constitute the natural basis for R^n.

Lemma 3.1([7]). Let a point $z \in D$ be a local maximizer of the above problem such that

$$\exists \tilde{x} \in D : \ f(\tilde{x}) > f(z)$$

then there exists a point $u \in D$ and a number $j \in \{1, 2, \ldots, n\}$ such that

$$\langle f'(y^j), u - y^j \rangle > 0.$$

■

Example 3.2 ([7]). Now consider the problem

$$f(x) = \| x \|^2 \rightarrow max, \ x \in D$$

$$D = \{x \in R^n \mid \langle Ax, x \rangle \leq c\},$$

where A is an $n \times n$ positive definite matrix, and $c \in R,\ c > 0$.

Lemma 3.2([7]). Let a point $z \in D$ ($z \neq 0$) satisfy the condition

$$\langle z, x - z \rangle \leq 0 \text{ for all } x \in D,$$

$$\exists \tilde{x} \in D : \ f(\tilde{x}) > f(z).$$

Then there exist $u \in D$ and $y \in R^n : f(y) = f(z)$ such that $\langle f'(y), u - y \rangle > 0$, where $y = \frac{z}{\|z\|}h,\ u = y + \varepsilon\frac{h}{\|h\|},\ \varepsilon > 0$ and h is an eigenvector corresponding to the largest eigenvalue of matrix A. ■

Example 3.3

$$f(x) = \| x \|^2 \rightarrow max, \ x \in D$$

$$D = \{x \in R^n \mid \sum_{i=1}^{n} \frac{x_i}{a_i} \leq 1, \ x_i \geq 0, \ i = 1, 2, \ldots, n\}$$

$$a_i > 0, \ i = 1, 2, \ldots, n$$

$$R(z) = \{y^i \in R^n \mid y^i = \| z \| e^i, \ i = 1, 2, \ldots, n\}$$

Lemma 3.3([7]). Let a point $z \in D$ be a local maximizer of the above problem such that

$$\exists \tilde{x} \in D : \ f(\tilde{x}) > f(z)$$

then there exists a point $u \in D$ and a number $j \in \{1, 2, \ldots, n\}$ such that $\langle f'(y^j), u - y^j \rangle > 0$. ■

3.2 Algorithm and its Convergence

Assume that there exists a resolving set for problem (2.1). In addition, we suppose that the problem of maximizing a linear function on D is solvable. Then we have the following algorithm.

Algorithm 1.

⇝ **Step 1** Choose a point $x^0 \in D$. Set $k := 0$.

Step 2 Find a stationary $z^k \in D$ using one of the gradient methods starting with an initial approximation point x^k.

Step 3 Construct the resolving set for z^k, *i.e.*,

$$R(z^k) = \{y^1, y^2, ..., y^m \mid f(y^i) = f(z^k), \quad i = 1, 2, ..., m\}$$

Step 4 Solve

$$\langle f'(y^i), x\rangle \; \rightarrow \; \max, \quad x \in D$$

for each $i = 1, 2, ..., m$. Let u^i be a solution of this problem.

Step 5 Find an integer j such that

$$\theta_k \triangleq \langle f'(y^j), u^j - y^j\rangle = \max_{1\le i\le m} \langle f'(y^i), u^i - y^i\rangle$$

Step 6 If $\theta_k \le 0$ then stop, and z^k is a solution. Otherwise, go to next step.

Step 7 $x^{k+1} := u^j, k := k + 1$ and go to Step 2 .

Convergence of the algorithm is given by

Theorem 3.1([8]). Assume that problem (2.1) has a finite number of stationary points. Then the sequence of points z^k produced by Algorithm 1 converges to a solution of problem (2.1) in finite number of steps. ■

4 Quadratic Convex Maximization Problem

4.1 Problem Statement and Optimality Condition

Consider the quadratic maximization problem.

$$f(x) = \langle Cx, x\rangle + \langle d, x\rangle + q \; \longrightarrow \; \max, \quad x \in D, \tag{4.1}$$

where C is a positive semidefinite $(n\times n)$ matrix, and $D \subset R^n$ is a nonempty arbitrary subset of R^n. A vector $d \in R^n$ and a number $q \in R$ are given. Then optimality condition (2.3) has the following form due to next theorem.

Theorem 4.1 ([6]). Let $z \in D$ be such that $f'(z) \neq 0$. Then z is a solution of problem (4.1) if and only if

$$\langle f'(y), x - y\rangle \leq 0 \text{ for all } y \in E_{f(z)}(f) \text{ and } x \in D, \tag{4.2}$$

where $E_c(f) = \{y \in R^n \mid f(y) = c\}$.

Proof. Necessity. Assume that z is a global maximizer of problem (4.1) and let $y \in E_{f(z)}(f)$ and $x \in D$. It is clear $f'(y) \neq 0$. Then the convexity of f implies that

$$0 \geq f(x) - f(z) = f(x) - f(y) \geq \langle f'(y), x - y\rangle$$

Sufficiency. Suppose, on the contrary, that z is not a solution to problem (4.1); i.e., there exists an $u \in D$ such that $f(u) > f(z)$. It is clear that the closed set $L_{f(z)}(f) = \{x \in R^n \mid f(x) \leq f(z)\}$ is convex. Since $int\ L_{f(z)}(f) \neq \emptyset$ then there is the projection y of u on $L_{f(z)}(f)$ such that

$$\|y - u\| = \min_{x \in L_{f(z)}(f)} \|x - u\|.$$

Clearly,

$$\|y - u\| > 0 \tag{4.3}$$

holds because $u \notin L_{f(z)}(f)$. Moreover, this y can be considered as a solution of the following convex minimization problem:

$$g(x) = \frac{1}{2}\|x - u\|^2 \longrightarrow \min, \quad x \in L_{f(z)}(f).$$

Applying the Lagrange method to this latter problem, we obtain the following optimality condition at the point y:

$$\begin{cases} \lambda_o \geq 0, \ \lambda \geq 0, \ \lambda_o + \lambda > 0 \\ \lambda_o g'(y) + \lambda f'(y) = 0 \\ \lambda(f(y) - f(z)) = 0 \end{cases} \tag{4.4}$$

If $\lambda_o = 0$, then (4.4) implies that $\lambda > 0$, $f(y) = f(z)$ and $f'(y) = 0$. Hence we conclude that z must be a global minimizer of f over R^n, which contradicts the assumption in the theorem. If $\lambda = 0$, then we have $\lambda_o > 0$ and $g'(y) = y - u = 0$, which also contradicts (4.3). So, without loss of generality, we can set $\lambda_o = 1$ and $\lambda > 0$ in (4.4) to obtain

$$\begin{aligned} &y - u + \lambda f'(y) = 0, \quad \lambda > 0, \\ &f(y) = f(z). \end{aligned}$$

From this we conclude that $\langle f'(y), u - y\rangle > 0$, which contradicts (4.2). This last contradiction implies that the assumption that z is not a solution of problem (4.1) must be false. This completes the proof. ∎

4.2 Approximation Techniques of the Level Set

Furthermore, to construct a numerical method for solving problem (4.1) based on optimality conditions (4.2), we assume that D is a bounded polyhedral set and C is a symmetric positive defined $n \times n$ matrix. Then problem (4.1) can be written as follows.

$$f(x) = \langle Cx, x\rangle + \langle d, x\rangle + q \longrightarrow \max, \quad x \in D, \tag{4.5}$$

where $D = \{x \in R^n \mid Ax \leq b\}$, A is an $r \times n$ matrix, and $b, d \in R^n$, $q \in R$.

Now introduce the definitions [9].

Definition 4.1 The set $E_{f(z)}(f)$ defined by

$$E_{f(z)}(f) = \{y \in R^n \mid f(y) = f(z)\}$$

is called the level set of f at z.

Definition 4.2 The set A_z^m defined by

$$A_z^m = \{y^1, y^2, \dots, y^m \mid y^i \in E_{f(z)}(f), \quad i = 1, 2, \dots, m\} \tag{4.6}$$

is called the approximation set to the level set $E_{f(z)}(f)$ at the point z.
When we construct A_z^m, we specify the elements $y^i \in R^n$, $z \in D$, and the integer m. Assume that A_z^m is given. Then for each $y^i \in A_z^m$, $i = 1, 2, \dots, m$ solve the auxiliary problem

$$\langle f'(y^i), x\rangle \longrightarrow max, \quad x \in D. \tag{4.7}$$

Let u^i, $j = 1, 2, \dots, m$ be solutions of those problems, which always exist due to compactness of D:

$$\langle f'(y^i), u^i\rangle = \max_{x \in D}\langle f'(y^i), x\rangle. \tag{4.8}$$

Define θ_m as follows:

$$\theta_m = \max_{i=1,2,\dots,m}\langle f'(y^i), u^i - y^i\rangle. \tag{4.9}$$

There are some properties of A_z^m and θ_m.

Lemma 4.1([9]). If there is a point $y^i \in A_z^m$ for $z \in D$ such that $\theta_m = \langle f'(y^i), u^i - y^i\rangle > 0$, where $u^i \in D$ satisfies $\langle f'(y^i), u^i\rangle = \max_{x \in D}\langle f'(y^i), x\rangle$,
then

$$f(u^i) > f(z)$$

holds.
The idea of approximation of the level set with eigenvectors was described in [9]. Let l^j and λ^j, $j = 1, 2, \ldots, n$, be orthogonal eigenvectors of the matrix C; i.e., $Cl^j = \lambda^j l^j$, $j = 1, 2, \ldots, n$, such that

$$\langle d, l^i \rangle \geq 0, \quad i = 1, 2, \ldots, n. \tag{4.10}$$

Define the elements $y^j \in R^n$, $j = 1, 2, \ldots, 2n$ as follows.

$$y^j = x^* + \alpha_j l^j, \tag{4.11}$$
$$y^{n+j} = 2x^* - y^j, \; j = 1, 2, ..., n, \tag{4.12}$$

where $f(x^*) = \min\limits_{x \in R^n} f(x)$ and $\alpha_j = \frac{1}{\|l^j\|}(\frac{f(z)-f(x^*)}{\lambda^j})^{\frac{1}{2}}$, $j = 1, 2, \ldots, n$. We can check that $f(y^j) = f(z)$, $j = 1, 2, \ldots, 2n$ and therefore $y^j \in A_z^{2n}$.

Since C is a symmetric and positive definite matrix, then $\lambda^j > 0, j = 1, 2, \ldots, n$.

Algorithm 2

Input: A convex quadratic function f and a polyhedral set D.

Output: An approximate solution x to problem (4.5); i.e., an approximate global maximizer of f over D.

⇝ **Step 1** Choose a point $x^0 \in D$. Set $k := 0$.

Step 2 Find a local maximizer $z^k \in D$ using one of the gradient methods starting with an initial approximation point x^k.

Step 3 Construct an approximation set $A_{z^k}^{2n}$ at the point z^k by (4.11) and (4.12).

Step 4 For each $y^i \in A_{z^k}^{2n}$, $i = 1, 2, \ldots, 2n$ solve the problem

$$\langle f'(y^i), x \rangle \longrightarrow max, \quad x \in D.$$

Let u^i, $i = 1, 2, \ldots, 2n$ be solutions, i.e.,

$$\langle f'(y^i), u^i \rangle = \max_{x \in D} \langle f'(y^i), x \rangle.$$

Step 5 Find a number $j \in \{1, 2, \ldots, 2n\}$ such that

$$\theta_{2n}^k = \langle f'(y^j), u^j - y^j \rangle = \max_{i=1,2,\ldots,2n} \langle f'(y^i), u^i - y^i \rangle.$$

Step 6 If $\theta_{2n}^k \leq 0$ halt. In this case z^k is an $\epsilon = \epsilon^k$ global maximizer, where $\epsilon^k = f(\bar{x}) - f(z^k)$ and

$$\bar{x} = \sum_{i=1}^{n} \alpha_i^k l^i, \ \alpha_i^k = \frac{1}{||l^i||}(\frac{f(z^k) - f(x^*)}{\lambda^i})^{\frac{1}{2}}, \quad i = 1, 2, \ldots, n.$$

Step 7 Set $x^{k+1} := u^j, \ k := k + 1$ and go to step 2.

As we have seen from Algorithm 2, the approximation sets $A_{z^k}^{2n}$ can not always provide a global maximizer in problem (4.5). For example, if in step 6 of the algorithm, $\theta_{2n}^k \leq 0$ holds then we restrict ourselves to finding an approximate solution which might be very far from the global maximizer of problem (4.5). However, under some assumptions a convergence of the algorithm is guaranteed by the following proposition.

Theorem 4.2([9]). Assume that in Algorithm 2, $\theta_{2n}^k > 0$ for all $k = 0, 1, \ldots$. Then the sequence $\{z^k\}, (k = 0, 1, \ldots)$ generated by the algorithm converges to a global solution of problem (4.5) in a finite number of steps.

4.3 The Second Order Permissive Set

In order to improve Algorithm 2, it is helpful to define another approximation set based on the A_z^m of (4.6). That is,

$$B_z^m = \{\bar{y}^1, \bar{y}^2, \ldots, \bar{y}^m \mid \bar{y}^j = x^* + \bar{\alpha}_j(u^j - x^*) \in E_{f(z)}(f), \ \ j = 1, 2, \ldots, m\}, \tag{4.13}$$

where $f(x^*) = \min\limits_{x \in R^n} f(x), \ \bar{\alpha}_j = (\frac{f(z) - f(x^*)}{\langle C(u^j - x^*), u^j - x^* \rangle})^{\frac{1}{2}}, \ j = 1, 2, \ldots, m, \ \langle f'(y^i), u^i \rangle = \max\limits_{x \in D} \langle f'(y^i), x \rangle$ for each $y^j \in E_{f(z)}(f), \ \ j = 1, 2, \ldots, m$.

$$\theta_m = \langle f'(y^j), u^j - y^j \rangle = \max_{i=1,2,\ldots,m} \langle f'(y^i), u^i - y^i \rangle > 0.$$

Since $C > 0$, the function f of (4.5) is strongly convex having the unique global minimizer x^* of f over R^n which satisfies $2Cx' + d = 0$. Hence, we can compute x' as $x' = -2C^{-1}d$. The use of x' is justified by the relationship between A_z^m and B_z^m in the following lemma.

Lemma 4.2([9]). Let $f(z) \neq f(x^*)$. If $\theta_m > 0$ then there exist a $j \in \{1, 2, \ldots, m\}$ and $v \in D$ such that $\bar{y}^j \in B_z^m$ satisfies $\langle f'(\bar{y}^j), v - \bar{y}^j \rangle > 0$.

In analogy with θ_m for A_z^m, introduce $\bar{\theta}_m = \{\bar{y}^1, \bar{y}^2, \ldots, \bar{y}^m\}$ for the set B_z^m as follows:

$$\bar{\theta}_m = \max_{i=1,2,\ldots,m} \langle f'(\bar{y}^i), v^i - \bar{y}^i \rangle$$

where v^i is defined by$\langle f'(\bar{y}^i), v^i\rangle = \max_{x\in D}\langle f'(\bar{y}^i), x\rangle$.

Definition 4.3 The approximation set B_z^m defined at $z \in D$ by (4.13) is said to be the second order permissive if the inequalities

$$\bar{\theta}_m \leq 0$$

implies that z is a global maximizer of problem (4.5).

Assume that there exist a second order permissive set. Then we can construct the following algorithm for (4.5).

Algorithm 3

Input: A convex quadratic function f and a polyhedral set D.

Output: A global maximizer of f over D.

$\rightsquigarrow$ **Step 1** Choose a point $x^0 \in D$. Set $k := 0$.

Step 2 Find a local maximizer $z^k \in D$ using one of the gradient methods starting with an initial approximation point x^k.

Step 3 Construct an permissive approximation set $A_{z^k}^m$ at the point z^k.

Step 4 For each $y^i \in A_{z^k}^m,\ i = 1, 2, \ldots, m$ solve the problem

$$\langle f'(y^i), x\rangle \longrightarrow max, \quad x \in D.$$

Let $u^i,\ i = 1, 2, \ldots, m$ be solutions, i.e.,

$$\langle f'(y^i), u^i\rangle = \max_{x\in D}\langle f'(y^i), x\rangle.$$

Step 5 Find a number $j \in \{1, 2, \ldots, m\}$ such that

$$\theta_m^k = \langle f'(y^j), u^j - y^j\rangle = \max_{i=1,2,\ldots,m}\langle f'(y^i), u^i - y^i\rangle.$$

Step 6 If $\theta_m^k > 0$ then $x^{k+1} := u^j,\ k := k + 1$ and go to step 2.

Step 7 Construct an second order permissive set $B_{z^k}^m$ at the point z^k by (4.13).

Step 8 For each $\bar{y}^i \in B^m_{z^k}$, $i = 1, 2, \ldots, m$ solve the problem

$$\langle f'(\bar{y}^i), x\rangle \longrightarrow max, \quad x \in D.$$

Let v^i, $i = 1, 2, \ldots, m$ be solutions, i.e.,

$$\langle f'(\bar{y}^i), v^i\rangle = \max_{x \in D}\langle f'(\bar{y}^i), x\rangle.$$

Step 9 Find a number $s \in \{1, 2, \ldots, m\}$ such that

$$\bar{\theta}^k_m = \langle f'(\bar{y}^s), v^s - \bar{y}^s\rangle = \max_{i=1,2,\ldots,m} \langle f'(\bar{y}^i), v^i - \bar{y}^i\rangle.$$

Step 10 If $\bar{\theta}^k_m > 0$ then $x^{k+1} := v^s$, $k := k + 1$ and go to step 2. Otherwise z^k is a global maximizer and terminate. ■

The convergence of this Algorithm is given by the following theorem.

Theorem 4.3 ([9]). The sequence $\{z^k\}, (k = 0, 1, \ldots)$ generated by the algorithm 3 converges to a global solution of problem (4.5) in a finite number of steps.

4.4 A Maximization Algorithm over a Box

Consider the following quadratic maximization problem with box constraints.

$$f(x) = \langle Cx, x\rangle + \langle d, x\rangle + q \longrightarrow \max, \quad x \in D \subset R^n, \tag{4.14}$$

$$D = \{x \in R^n \mid a_i \leq x_i \leq b_i, \quad i = 1, 2, \ldots, n\},$$

where C is a symmetric positive semidefinite $n \times n$ matrix and the vectors $a, b, d \in R^n$ and a number $q \in R$ are given.

Let $z = (z_1, z_2, \ldots, z_n)$ be a local maximizer of problem (4.14). Then $z_i = a_i \vee b_i$, $i = 1, 2, \ldots, n$. In order to construct an approximation set A^m_z take the following steps.
Define points $v^1, v^2, \ldots, v^{n+1}$ by formulas

$$v^k_i = \begin{cases} z_i & \text{if } i \neq k, \\ a_k & \text{if } z_k = b_k, \\ b_k & \text{if } z_k = a_k, \quad i, k = 1, 2, \ldots, n, \end{cases} \tag{4.15}$$

and

$$v^{n+1}_i = \begin{cases} a_i & \text{if } z_i = b_i, \\ b_i & \text{if } z_i = a_i, \quad i = 1, 2, \ldots, n. \end{cases} \tag{4.16}$$

Clearly,

$$\|v^{n+1} - z\| > \|v^k - z\|, \quad k = 1, 2, \dots, n,$$
$$\sum_{i=1}^{n}(a_i - b_i)^2 = \|v^{n+1} - z\|^2.$$

Denote by h^i vectors $h^i = v^i - z,\ i = 1, 2, \dots, n+1$. Note that $\langle h^k, h^j \rangle = 0,\ k \neq j,\ k, j = 1, 2, \dots, n$. Define the approximation set A_z^{n+1} by

$$A_z^{n+1} = \{y^1, y^2, \dots, y^{n+1} \mid y^i \in E_{f(z)}(f),\ y^i = z - \alpha_i h^i,\ i = 1, \dots, n+1\}, \quad (4.17)$$

where $\alpha_i = \frac{\langle 2Cz+d, h^i \rangle}{\langle Ch^i, h^i \rangle},\ i = 1, 2, \dots, n+1$.

Then an algorithm modified from Algorithms 2 and 3 for solving (4.14) is described in the following.

Algorithm 4

Input: A convex quadratic function f and a box set D.

Output: An approximate solution x to problem (4.14); i.e., an approximate global maximizer of f over D.

⇝ **Step 1** Choose a point $x^0 \in D$. Set $k := 0$.

Step 2 Find a local maximizer $z^k \in D$ the projected gradient method starting with an initial approximation point x^k.

Step 3 Construct an approximation set $A_{z^k}^{n+1}$ at the point z^k by formulas (4.15), (4.16) and (4.17).

Step 4 For each $y^i \in A_{z^k}^{n+1},\ i = 1, 2, \dots, n+1$ solve the problems

$$\langle f'(y^i), x \rangle \longrightarrow max, \quad x \in D,$$

which have analytical solutions $u^i,\ i = 1, 2, \dots, n+1$ found as

$$u_s^i = \begin{cases} a_s & \text{if } (2Cy^i + d)_s \leq 0, \\ b_s & \text{if } (2Cy^i + d)_s > 0, \end{cases}$$

where $i = 1, 2, \dots, n+1$ and $s = 1, 2, \dots, n$.

Step 5 Find a number $j \in \{1, 2, \ldots, n+1\}$ such that

$$\theta^k_{n+1} = \langle f'(y^j), u^j - y^j \rangle = \max_{i=1,2,\ldots,n+1} \langle f'(y^i), u^i - y^i \rangle.$$

Step 6 If $f(u^j) \geq f(z^k)$ then $x^{k+1} := u^j,\ k := k+1$ and go to step 1.

Step 7 Find a $y \in E_{f(z^k)}(f)$ such that

$$y = z^k - \frac{\langle 2Cz^k + d, u^j - z^k \rangle}{\langle C(u^j - z^k), u^j - z^k \rangle}(u^j - z^k).$$

Step 8 Solve the problem

$$\langle f'(y), x \rangle \longrightarrow max, \quad x \in D.$$

Let v be solution, i.e.,

$$\langle f'(y), v \rangle = \max_{x \in D} \langle f'(y), x \rangle.$$

Step 9 If $\langle f'(y), v - y \rangle > 0$ holds, then $x^{k+1} := v,\ k := k+1$ and return to step 2. Otherwise,halt; z^k is an approximate maximizer. ■

5 Maximization of Strongly Convex Functions

5.1 Algorithm over a Simple Set

We introduce the following definition of simple set [5].
Definition 5.1 A set D in R^n is called a simple set if the following conditions hold:

a) D is compact.

b) The problem of maximizing a linear function over D is solvable by a "simple" method.

Consider the convex maximization problem as a particular case of problem (2.1),

$$f(x) \longrightarrow \max, \quad x \in D, \tag{5.1}$$

where $f: R^n \to R$ is strongly convex and $D \subset R^n$ is a simple set. Then the global optimality condition of Theorem 2.2 can be stated:

Theorem 5.1 A point $z \in D$ with $f'(z) \neq 0$ is a solution of problem (5.1) if and only if the following condition hold:

$$\langle f'(y), x - y \rangle \leq 0 \text{ for all } y \in E_{f(z)}(f) \text{ and } x \in D,$$

where and $E_c(f) = \{y \in R^n \mid f(y) = c\}$.
The proof is completely analogous to the proof of Theorem 4.1. ■

Define the auxiliary function $\pi(y)$ as follows:

$$\pi(y) = \max_{x \in D} \langle f'(y), x - y \rangle, \quad y \in R^n. \tag{5.2}$$

There are some properties of $\pi(y)$.

Lemma 5.1([5]). The function $\pi(y)$ is continuous on R^n.

Theorem 5.2([5]). The directional derivative of $\pi(y)$ at any point $y \in R^n$ in any direction $h \in R^n$ of norm 1 exists, and is given by

$$\frac{\partial \pi(y)}{\partial h} = \max_{x \in D(y)} \langle f''(y)h, x \rangle - \langle f''(y)y + f'(y), h \rangle.$$

Before presenting an algorithm for solving problem (5.1), it is useful to restate Theorem 5.1 in a convenient way via the following function $\theta(z)$ defined for $z \in D$.

$$\theta(z) = \max_{y \in E_{f(z)}(f)} \pi(y),$$

where $\pi(y) = \max_{x \in D} \langle f'(y), x - y \rangle$. Since f is strongly convex, the set $E_{f(z)}(f)$ is compact [21]. Thus, clearly, $\theta(z) < +\infty$. Also, we can note that

$$\pi(y) \leq \theta(z) \text{ for all } y \in E_{f(z)}(f).$$

Theorem 5.3 Let $z \in D$ and $f'(z) \neq 0$. If $\theta(z) = 0$ then the point z is a solution to problem (5.1).

Algorithm 5

Input: A strongly convex function f and a simple set D.

Output: A solution x to problem (5.1); i.e., a global maximizer of f over D.

⇝ **Step 1** Choose a point $x^0 \in D$ such that $f'(x^0) \neq 0$. Set $k := 0$.

Step 2 Solve the following problem

$$\pi(y) \longrightarrow \max, \quad y \in E_{f(x^k)}(f).$$

Let y^k be a solution of this problem, i.e.,

$$\pi(y^k) = \max_{y \in E_{f(x^k)}(f)} \max_{x \in D} \langle f'(y), x - y \rangle.$$

Let $\theta(x^k) := \pi(y^k)$, and let x^{k+1} be a solution satisfying

$$\pi(y^k) = \langle f'(y^k), x^{k+1} - y^k \rangle.$$

Step 3 If $\theta(x^k) = 0$ then output $x = x^k$ and terminate. Otherwise set $k := k+1$ and return to Step 2. ∎

The convergence of this Algorithm is given by the following theorem.

Theorem 5.4([5]). Assume that $f : R^n \to R$ be strongly convex and continuously differentiable. Then the sequence $\{x^k\}, (k = 0, 1, ...)$ generated by Algorithm 5 is a maximizing sequence for problem (5.1), that is,

$$\lim_{k\to\infty} f(x^k) = \max_{x\in D} f(x),$$

and every accumulation point of the sequence $\{x^k\}$ is a global maximizer of (5.1).

Algorithm 5 seems to us more theoretical rather than numerical. Note that its efficiency depends on how we can solve successfully the following problem

$$\pi(y) \longrightarrow \max, \quad y \in E_{f(x^k)}(f). \tag{5.3}$$

In general, this problem belongs to a class of nondifferentiable optimization problems. We can use such well known methods as a penalty function method [21, 33, 34], a method of generalized gradients [22], or a method of nondifferentiable optimization [4]. Moreover, if D is strictly convex, then $\pi(y)$ is differentiable [5]. Thus, in principle, we can apply gradient type of methods as well as the Lagrangean method [1]. Hence, we can see that the global optimality condition for problem (5.3) not only is connected to the classical conditions, but also its global algorithm is connected to the classical methods.

Also, there is a numerical method [23] for such problem as (5.3) with the equality constraint, based on the set covering and finite set methods, which find a solution with a given (ε) accuracy. In general, in Algorithm 5, we can use a numerical method providing ε approximate solution to problem (5.3), in order to find an approximate solution to problem (5.1) with a given accuracy. To make Algorithm 5 numerically implementable for this purpose, we need adapt it as follows.

Algorithm ϵ-MAX

Input: A strongly convex function f and a simple set D, and a sequence $\{\varepsilon_k\}$ such that $\varepsilon_k > 0$ for all k and $\lim_{k\to\infty} \sum_{k=0}^{\infty} \varepsilon_k < +\infty$, and $\varepsilon > 0$.

Output: A global maximizer x of f over D.

⇝ **Step 1** Choose a feasible solution $x^o \in D$ and a sequence $\{\varepsilon_k\}$ such that $f'(x^o) \neq 0$. Set $k := 0$.

Step 2 Find an ε_k solution $\bar{y}^k$ of the problem

$$\pi(y) \longrightarrow \max, \quad y \in E_{f(x^k)}(f),$$

that is, $\pi(\bar{y}^k) = \max_{x \in D}\langle f'(\bar{y}^k), x - \bar{y}^k\rangle \geq \max_{y \in E_{f(x^k)}(f)} \pi(y) - \varepsilon_k$ and $f(\bar{y}^k) = f(x^k)$.
Let $\bar{x}^{k+1} \in D$ be a solution satisfying $\pi(\bar{y}^k) = \langle f'(\bar{y}^k), \bar{x}^{k+1} - \bar{y}^k\rangle$.

Step 3 If $\pi(\bar{y}^k) \leq \varepsilon - \varepsilon_k$ then output $x = x^k$ and terminate. Otherwise set $k := k+1$ and $x^k := \bar{x}^k$ return to Step 2. ■

Theorem 5.5 ([6]). Assume that $f : R^n \to R$ be strongly convex and continuously differentiable. Let a sequence $\{\varepsilon_k\} \subset R$ be such that $\varepsilon_k > 0$ for $k = 0, 1, ...$ and

$$\lim_{k\to\infty} \sum_{k=0}^{\infty} \varepsilon_k + f_* < f(x^0),$$

where $f_* = \min_{x \in R^n} f(x)$. Then the sequence $\{x^k\}$ generated by Algorithm ϵ-MAX is a maximizing sequence for problem (5.1), that is,

$$\lim_{k\to\infty} f(x^k) = \max_{x \in D} f(x),$$

and every accumulation point of the sequence $\{x^k\}$ is a global maximizer of (5.1).

5.2 Nonconvex Optimal Control Problem

Consider the following optimal control problem:

$$J(u) = \phi\Big(x(t_1)\Big) \to \max \tag{5.4}$$

$$\begin{cases} \dot{x}(t) = A(t)x + B(t)u(t) + C(t) \\ x(t_0) = x^0 \end{cases} \tag{5.5}$$

$$u \in V = \Big\{u \in L_2^{(r)}(T)/u(t) \in U,\, t \in T\Big\}, \tag{5.6}$$

where $t \in T = [t_0, t_1]$, $-\infty < t_0 < t_1 < +\infty$, $u(t) = \Big(u_1(t), u_2(t), \ldots, u_r(t)\Big)^T$, $x(t) = \Big(x_1(t), x_2(t), \ldots, x_n(t)\Big)^T$. Here $x^0 \in R^n$ is an initial state, t_0, t_1 and x^0 are given. Matrix functions $A(t)$, $B(t)$ and $C(t)$ are piecewise continuous on $[t_0, t_1]$ with dimensions $(n\times n)$, $(n\times r)$ and $(n\times 1)$ respectively. $U \subset R^r$ is a simple set, $\phi : R^n \to R$ is a strongly convex and differentiable function.

The problem (5.4)-(5.6) belongs to the class of nonconvex optimal control problem and application of Pontryagin's maximum principle [13, 27] can give only stationary processes $\Big(x(\cdot), u(\cdot)\Big)$.

We will show how to apply algorithm 5 to the solution of problem (5.4)-(5.6). It is well known that every admissible control $u \in V$ corresponds to a unique solution $x(t) = x(t, u)$ of the Cauchy problem (5.5) by the formula [33, 34]:

$$x(t,u) = F(t)x^0 + \int_{t_0}^{t} F(t)F^{-1}(\xi)\Big(B(\xi)u(\xi) + C(\xi)\Big)d\xi, \tag{5.7}$$

where $F(t)$ is a fundamental matrix with dimension $(n \times n)$ that satisfies the matrix equation

$$\dot{F} = A(t)F, \quad F(t_0) = E$$

on T with the unit matrix E. Note that a state $x(t, u)$ is an absolutely continuous vector-function of time t, which satisfies the system (5.5) almost everywhere in T [33]. Let us denote by

$$\tilde{D} = \tilde{D}(t_1) = \Big\{y \in R^n / y = x(t_1, u), u \in V\Big\}$$

the reachable set of the control system (5.5)-(5.6). Under the above assumptions the set $\tilde{D} \subset R^n$ is convex and compact [27, 34]. Then we can present problem (5.4)-(5.6) in the following form:

$$\phi(x) \to \max, \quad x \in \tilde{D} \tag{5.8}$$

Let a point x^* be a solution of problem (5.8), then the admissible control $u^* = u^*(t)$, $t \in T$ corresponding to x^* is the global optimal control solution to problem (5.4)-(5.6). Therefore, algorithm 5 can also be used for solving the problem (5.4)-(5.6). As we have done before, write down the auxiliary function $\pi(y)$ for problem (5.8) as follows:

$$\tilde{\pi}(y) = \max_{x \in \tilde{D}} < \phi'(y), x - y > \tag{5.9}$$

We first consider the following linearized optimal control problem

$$< \phi'(y), x > \to \max, \quad x \in \tilde{D}. \tag{5.10}$$

In order to solve problem (5.10), we introduce a conjugate state

$$\psi(t,y) = \Big(\psi_1(t), \dots, \psi_n(t)\Big)^T$$

as a solution of the following differential system:

$$\dot{\psi} = -A^T\psi, \quad \psi(t_1) = -\phi'(y) \text{ for every } y \in R^n \tag{5.11}$$

This system has the unique piecewise differentiable solution $\psi(t) = \psi(t, y)$ defined on T. Then a solution of problem (5.10) will be given by the following

Theorem 5.6([27, 34]). Let $\psi(t) = \psi(t, y)$, $t \in T$ be a solution of conjugate system (5.11) for $y \in R^n$. Then for an admissible control $z(t) = z(t, y)$ to be optimal in problem (5.11) it is necessary and sufficient that the condition $< \psi(t, y), B(t)z(t, y) >= \min_{u \in U} < \psi(t, y), B(t)u >$ be fulfilled for almost every $t \in T$.

Using this theorem we can easily show that the reachable set $\tilde{D} = \tilde{D}(t_1)$ is a simple set. Therefore, the value $\tilde{\pi}(\tilde{y})$ is calculated by the following scheme:

1. Solve problem (5.11) for a given $\tilde{y} \in R^n$. Let $\psi(t) = \psi(t, \tilde{y})$ be a solution of (5.11).
2. Find the optimal control $z = z(t, \tilde{y})$ as a solution of the problem

$$< \psi(t), B(t)u >\to \min, \qquad u \in U$$

at each moment of $t \in T$.

3. Find a solution $\tilde{x} = x(t, z)$ of problem (5.5) for $u = z(t, \tilde{y})$.
4. Find $\tilde{x}(t_1) = \tilde{x}(t_1, z)$ by the formula (5.7) for $t = t_1$.
5. Calculate $\tilde{\pi}(\tilde{y})$ by the formula

$$\tilde{\pi}(\tilde{y}) = \max_{x \in \tilde{D}} < \phi'(\tilde{y}), x - \tilde{y} >=< \phi'(\tilde{y}), \tilde{x}(t_1) - \tilde{y} > .$$

Based on Algorithm 5 now we are ready to present an algorithm for solving problem (5.4)-(5.6).

Algorithm 6

$\leadsto$ **Step 1** Let $k = 0$ and an arbitrary control $u^k \in V$ be given.
Find $x^k = x(t_1, u^k)$ by solving the system (5.5) for $u = u^k$.

Step 2 Solve the problem $\tilde{\pi}(y) \to \max$, $\phi(y) = \phi(x^k)$. Let y_k be the solution of this problem and $z^k = z^k(t, y_k)$ be the a solution of the corresponding problem

$$< \phi'(y_k), x >\to \max, \;\; x \in \tilde{D}$$

or

$$< \psi^k(t), B(t)z^k >= \min_{u \in U} < \psi^k(t), B(t)u >, \;\; t \in T,$$

where

$$\dot{\psi}^k(t) = -A^T(t)\psi^k(t), \quad \psi^k(t_1) = -\phi'(y_k)$$

Step 3 If $\pi(y^k) \leq 0$ then go to step 5. Otherwise, go to step 4.

Step 4 Set $u^{k+1} = z^k(t, y_k)$ and $k := k + 1$. Then go to step 1.

Step 5 Iteration is terminated. $u^* = u^k$ is a global optimal solution of problem (5.4)-(5.6)

Convergence of the algorithm is based on the following statement.

Theorem 5.7([12]). The sequence $\{u^k\} \subset V$ constructed by Algorithm 6 is a maximizing sequence for problem (5.4)-(5.6), i.e.,

$$\lim_{k\to\infty} J(u^k) = \max_{u\in V} J(u)$$

6 Numerical Experiments

Computational experiments on implementation of all proposed algorithms 2-6 were given in [5–12]. The proposed algorithms have been implemented on IBM PC/586 microcomputer in Pascal 7. For example, as a test problem the quadratic assignment problem of the following type

$$\sum_{i=1}^{n}\sum_{j=1}^{n}\sum_{k=1}^{n}\sum_{e=1}^{n} a_{ij}b_{ke}x_{ik}x_{je} \longrightarrow min,$$

subject to

$$\sum_{i=1}^{n} x_{ij} = 1, \quad j = 1,2,\ldots,n,$$
$$\sum_{j=1}^{n} x_{ij} = 1, \quad i = 1,2,\ldots,n,$$
$$x_{ij} \in \{0,1\},\ i,j = 1,2,\ldots,n,$$

has been solved by Algorithm 2 and 3 [11] in its equivalent convex maximization problem form [24]:

$$\langle Qx, x\rangle \longrightarrow max,$$

where

$$D = \{x \in R^{n^2} \mid \sum_{j=1}^{n} x_{ij} = 1, \quad \sum_{i=1}^{n} x_{ij} = 1, \quad x_{ij} \geq 0, \quad i,j = 1,2,\ldots,n\}$$

where $Q = \alpha I - S$ is positive definite for α such that $\alpha > \|S\|_\infty$, and $\|S\|_\infty$ is defined by $\|S\|_\infty = max\{\sum_{j=1}^{n^2} |S_{1j}|, \ldots, \sum_{j=1}^{n^2} |S_{n^2j}|\}$. The $(n^2 \times n^2)$ symmetric matrix S has nonnegative entries which are the products of $a_{ij}b_{ke}$.

Also, the problem of finding the largest eigenvalue of positive definite matrix A was formulated as the following convex maximizing problem

$$\langle Ax, x\rangle \to max, \quad \| x \| \leq 1,$$

and then has been numerically solved by Algorithm 1 [8].
Also, the list of some optimal control test problems solved by Algorithm 6 is the following [12]:

$$(p1) \quad \begin{cases} x_1^2(1) + x_2^2(1) \to \max \\ \dot{x}_1 = u_1 \\ \dot{x}_2 = u_2 \\ x_1(0) = x_2(0) = 0, \quad u(t) \in U, \quad t \in [0,1] \\ U = \{u \in R^2 / -3 \leq u_1 \leq 4, \ -2 \leq u_2 \leq 5\} \end{cases}$$

$$(p2) \quad \begin{cases} x_1^2(1) + x_2^2(1) \to \max \\ \dot{x}_1 = x_2 + u_1 \\ \dot{x}_2 = x_1 + u_2 \\ x_1(0) = x_2(0) = 0, \quad u(t) \in U, \quad t \in [0,1] \\ U = \{u \in R^2 / -3 \leq u_i \leq 4, \ i = 1,2\} \end{cases}$$

$$(p3) \quad \begin{cases} x_1^2(t_1) + x_2^2(t_1) + x_3^2(t_1) \to \max \\ x_1 = -x_1 + x_2 \\ \dot{x}_2 = -x_1 - x_2 - 10u \\ \dot{x}_3 = x_2 \\ x_1(0) = x_2(0) = 0, \quad x_3(0) = -30, \quad T = [0, 2.8] \\ u(t) \in U = \{u \in R / -15 \leq u \leq 15\} \end{cases}$$

The results of the numerical experiments for these problems are shown in the following table 1 [12].

Problems	Local value of ϕ	Global value of ϕ	Computing time (min:sec)
p1	13	41	0:51
p2	53.1448	94.4798	0:56
p3	149.055	263.519	0:6.78

7 Conclusions

Global Optimality conditions developed since 1987 for concave programming have been presented. We have described methods and algorithms for solving the convex maximization problem based on global optimality conditions by Strekalovsky. Also, we have examined a possibility to solve some nonconvex optimal control problem as well as the quadratic assignment problem by the proposed algorithms.

References

[1] Bertsekas, D.P. (1982), Constrained Optimization and Lagrangean Multiplier Methods, *Academic Press*, New York.

[2] Bulatov, V.P. (1977), The Embedding Methods in Extremum Problems, *Nauka*, Novosivirsk.

[3] Bomze, I. and Danninger, G. (1994), A Finite Algorithm for Solving General Quadratic Problem, *Journal of Global Optimization*, 4, pp.1-16.

[4] Demyanov, V.P. and Vasiliev, L.V. (1981), Nondifferentiable Optimization, *Nauka*, Moscow.

[5] Enkhbat, R. (1996), An Algorithm for Maximizing a Convex Function over a Simple Set, *Journal of Global Optimization*, 8, pp.379-391.

[6] Enkhbat, R. and T.Ibaraki. (2001), On the Maximization and Minimization of a Quasiconvex Function, *Technical Report 2000-010*, Kyoto University pp.1-37.

[7] Enkhbat, R. (1990), Algorithms for Global Maximization of Convex Functions over Sets with a Special Structure, *PhD thesis*, Irkutsk University, Irkuts, Russia.

[8] Enkhbat, R, Kamada, M. and Bazarsad, Ya. (1999), A Finite Algorithm for Maximizing a Convex Function over Compact Set, *Journal of the Faculty of Engineering of Ibaraki University*, 46(3), p.1-21.

[9] Enkhbat, R. and T.Ibaraki (2001), Global Optimization Algorithms for General Quadratic Programming, *Journal of the Mongolian Mathematical Society*, No.5, pp.1-35.

[10] Strekalovsky, A.S. and Enkhbat, R. (1990), Global Maximum of Convex Functions on a Arbitrary Sets, *Dep.in VINITI*, No.1063, Irkutsk University.

[11] Enkhbat, R. and Ts.Javzandulam (2000), Global Search Algorithm for the Quadratic Assignment Problem, *Journal of the Mongolian Mathematical Society*, No.4, pp.16-28.

[12] Enkhbat, R. (2000), A Global Method for Some Class of Optimization and Control Problems, *International Journal of Mathematics and Mathematical Sciences*, 23(9), pp.605-616.

[13] Ioffe A.D. and Tihomirov V.M. (1979), Theory of Extremal Problems, *North-Holland*, New York.

[14] Hiriart-Urruty, J.B. (1989), From Convex Optimization to Nonconvex Optimization, *Nonsmooth Optimization and Related Topics*, Plenum, pp.219-239.

[15] Hiriart-Urruty, J.B. and Ledyaev, J.S. (1996), A Note on the Characterization of the Global Maxima of a (tangentially) Convex Function over a Convex Set, *Journal of Convex Analysis*, 3(1), pp.55-61.

[16] Horst, R. (1986), A General Class of Branch and Bound Methods in Global Optimization with some New Approaches for Concave Minimization, *Journal of Optimization Theory and Applications*, 51, pp.271-291.

[17] Horst, R. (1987), Outer Cut Methods in Global Optimization, *Springer- Verlag* Lecture Notes in Economics and Mathematical Systems, 304, pp.28-40.

[18] Horst, R., Thoai, N.V. and Tuy, H. (1987), Outer Approximation by Polyhedral Convex Sets, *Operations Research Spectrum*, 9(3), pp.153-159.

[19] Horst, R. (1987), A New Branch and Bound Approach for Concave Minimization Problems, *Springer- Verlag* Lecture Notes in Computer, Science, 41, pp.330-337.

[20] Horst, R. and Tuy, H. (1990), Global Optimization(Deterministic Approaches), *Springer*, Berlin.

[21] Karmanov, V.G. (1989), Mathematical Programming, *Mir Publisher*, Moscow.

[22] Michaelyevich, V.S., Gupal, V.S. and Norkin, V.I.(1987), Methods of Nonconvex Optimization, *Nauka*, Moscow.

[23] Nefedov, V.N. (1987), Finding the Global Maximum of a Function of Several Variables on a Set Given by Inequality Constraints, *Journal of Numerical Mathematics and Mathematical Physics*, 27(1), pp.35-51.

[24] Pardalos, P.M. and Rosen, J.B. (1987), Constrained Global Optimization: Algorithms and Applications, *Springer- Verlag* Lecture Notes in Computer, Science.

[25] Pardalos, P.M. and Rosen, J.B. (1986), Methods for Global Concave Minimization: A Bibliographic Survey, *SIAM Review*, 28, pp.367-379.

[26] Reiner Horst and Panos M.Pardalos, Edidors: (1995), Handbook of Global Optimization *Kluwer Academic Publishers*, London.

[27] Pontryagin L.S., Boltyanski R.G., Gamkrelidze R.V. and Mischenko E.E. (1962), Mathematical Theory of Optimal Processes, *Interscience Publ. Inc.*, New York.

[28] Strekalovsky, A.S. (1987), On the Global Extremum Problem, *Soviet Math.Doklady*, 292(5), pp.1062-1066.

[29] Strekalovsky, A.S. (1998), Global Optimality Conditions for Nonconvex Optimization, *Journal of Global Optimization* , 12, pp.415-434.

[30] Tuy, H. (1964), Concave Programming under Linear Constraints, *Soviet Math.Doklady*, 159(1), pp.32-35.

[31] Tuy, H. (1991), Normal Conical Algorithm for Concave Minimization over Polytopes, *Mathematical Programming*, 51, pp.229-245.

[32] Teseveendorj, I. (1998), On the Conditions for Global Optimality, *Journal of the Mongolian Mathematical Society*, 2, pp.58-62.

[33] Vasiliev, F.P. (1987), Numerical Methods for Solving Extremum Problems, *Nauka*, Moscow.

[34] Vasiliev, O.V. (1996), Optimization Methods, *World Federation Publishers*, Atlanta.

Optimization and Optimal Control, pp. 103 – 119
P.M. Pardalos, I. Tseveendorj, and R. Enkhbat, Editors

Maximum Clique Regularizations

Dominique Fortin (Dominique.Fortin@lidil.univ-mrs.fr)
Université de la Méditerranée, LIF - Case 901,
163, Avenue de Luminy, 13288, Marseille cedex 9, FRANCE,

Ider Tseveendorj (Ider.Tseveendorj@rd.francetelecom.com)
France Télécom R&D, DAC/Laboratoire Optimisation, Architecture et Trafic,
38-40, rue du Général Leclerc, 92794, Issy les Moulineaux, FRANCE

Abstract

Given a simple graph $G = (\mathcal{V}, \mathcal{E})$, we refine the well known regularization for the maximum clique problem: diagonal offsets are dynamically updated along searching for maximum clique. A DC-algorithm retrieves local optimas by walking around clique and stable sets at the same time and computational experiments report early steps results depending on Perron's eigenvector starting point and policies of weighing diagonal offsets.

Keywords: the maximum clique, global optimization, regularizations.

1 Introduction

We consider a simple graph $G = (\mathcal{V}, \mathcal{E})$ which consists of a finite nonempty set $\mathcal{V}$ of vertices and a finite and nonempty set $\mathcal{E}$ of edges. We assume that the graph has no selfloops and no isolated vertex.

A clique (resp. stable set) in a graph $G = (\mathcal{V}, \mathcal{E})$ is a set of vertices any two of which are adjacent (resp. non adjacent).

A clique is said to be *maximal* if there is no larger clique containing it. A (maximal) clique is said to be a *maximum* clique if it has biggest order among all cliques.

The maximum clique is hard to approximate within $|\mathcal{V}|^{1-\epsilon}$ [12] and the best known approximation guarantee is $O(|\mathcal{V}|/\log^2|\mathcal{V}|)$ [5].

The maximum size of a clique in G is the *clique number* and denoted $\omega(G)$ or simply ω when no confusion arises. All previous notations extend to the weighted case, namely when a *weight* w_u for all $u \in \mathcal{V}$ is provided; then the *weighted clique number* $\omega(G, w)$ is the largest of $\sum_{u\in\sigma} w_u$ among all cliques in G. Clearly $\omega(G) = \omega(G, e)$ where e stands for the all 1s weight vector.

A *complete l-partite* graph is one whose vertex set could be partitioned as l stable sets and every pair of vertices in different parts are adjacent.

We preferably use uppercase letters for matrices so that E denotes the matrix of all 1s, while I is the identity matrix; we freely jump from clique problem to stable set problem and therefore use $\bar{A} = E - I - A$ correspondence to relate underlying graph to its complement.

First, Motzkin and Strauss[17] showed that $\omega(G) = (1 - \langle Az, z\rangle)^{-1}$, where z is a global solution to the following nonconvex quadratic optimization problem:

$$\begin{cases} \text{maximize } \langle Ax, x\rangle, \\ \text{subject to } \langle e, x\rangle = 1, x \geq 0. \end{cases} \qquad (MS)$$

whose solution is known to be blurred by spurious local optima. Then, Bomze[6], introducing a regularization $\frac{1}{2}\langle Ix, x\rangle$ to the objective function:

$$\begin{cases} \text{maximize } \langle (A + \frac{1}{2}I)x, x\rangle, \\ \text{subject to } \langle e, x\rangle = 1, x \geq 0. \end{cases} \qquad (B)$$

sharpened Motzkin-Straus's result to $\omega(G) = (2(1 - \langle (A + \frac{1}{2}I)z, z\rangle))^{-1}$; moreover, z solves (B) globally **iff** z is the characteristic vector of a maximum clique.

In this article, we review diagonal regularization of max clique and sibling max stable at the same time.

The paper is organized as follows. In the remaining of the introduction section, we recall some results from Gibbons et al.[11] and standard 1st and 2nd order necessary optimality conditions. Then, in section 2 we introduce a diagonal regularization $\hat{A} = A + D$ instead of constant diagonal offset $\hat{A} = A + \frac{1}{s}I$ (where s is a lower bound of the clique size) proposed by the authors[10] earlier. In section 3 we use Legendre-Fenchel transform to relate diagonal parameters between clique and sibling stable problem. Then, we provide an algorithm to alternately switch between clique and stable set and report qualitative results to practically check relationships found.

For the problem

$$\begin{cases} \text{maximize } \frac{1}{2}\langle Ax, x\rangle, \\ \text{subject to } \langle e, x\rangle = 1, x \geq 0. \end{cases}$$

we recall both first and second order necessary conditions for optimality:

$$\begin{cases} A\bar{x} + \lambda e + \mu = 0 \\ \langle e, \bar{x}\rangle = 1 \\ \bar{x} \geq 0 \\ \mu \geq 0 \\ \mu_u \bar{x}_u = 0 \end{cases} \qquad 1stOC$$

$$\begin{cases} \langle Ay, y\rangle \leq 0, \text{ for all } y \text{ satisfying} \\ \qquad\qquad\qquad\qquad \langle e, y\rangle = 0, \\ \qquad\qquad y_u = 0 \ \ \forall u : \bar{x}_u = 0, \end{cases} \qquad 2ndOC$$

where $\bar{x}$ fulfills 1stOC.

2 Regularizations

Let A, I be adjacency and identity matrix respectively, define $\hat{A} = A + D$ where $D = \text{diag}(\frac{1}{s_i})$ regularizes adjacency matrix A, for some $s_i \geq 2$ for all i

$$\begin{cases} \text{maximize } \langle \hat{A}x, x\rangle, \\ \text{subject to } \langle x, e\rangle = 1, x \geq 0 \end{cases} \qquad (MRC)$$

Proposition 1 *$\bar{x} = \arg\max(MRC)$ is the barycenter of the characteristic face of some maximal clique σ of size k, namely $\bar{x} = \frac{1}{k}\sum_{i\in\sigma} e_i$*

- Let $\bar{x} = \arg\max(\text{MRC})$ whose support does not induce a clique, then there exist vertices u, v non adjacent noted by $u \not\sim v$ with $\bar{x}_u, \bar{x}_v > 0$; now, define $x(\nu)$ as:

$$x(\nu)_i = \begin{cases} \bar{x}_u + \nu, \text{ if } i = u \\ \bar{x}_v - \nu, \text{ if } i = v \\ \bar{x}_i, \text{ otherwise} \end{cases}$$

  i.e. $x(\nu) = \bar{x} + \nu(e^u - e^v)$ leaving $\langle e, x(\nu)\rangle = 1, x(\nu) \geq 0$ for all $\nu \ : -\bar{x}_u \leq \nu \leq \bar{x}_v$. Then, we claim that using 2nd order conditions, the support of $\bar{x}$ induces a complete l-partite graph for some l (see following lemma 1 delayed for sake of clarity) so that u, v belong to the same part on the one hand and share neighbors in remaining parts on the other hand; therefore, objective value at $x(\nu)$ is:

$$\begin{aligned} \langle \hat{A}x(\nu), x(\nu)\rangle &= \langle \hat{A}(\bar{x} + \nu(e^u - e^v)), \bar{x} + \nu(e^u - e^v)\rangle \\ &= \langle \hat{A}\bar{x}, \bar{x}\rangle + 2\nu\langle \hat{A}(e^u - e^v), \bar{x}\rangle + \nu^2(\tfrac{1}{s_u} + \tfrac{1}{s_v}) \\ &= \langle \hat{A}\bar{x}, \bar{x}\rangle + 2\nu(\tfrac{1}{s_u}\bar{x}_u - \tfrac{1}{s_v}\bar{x}_v) + \nu^2(\tfrac{1}{s_u} + \tfrac{1}{s_v}) \end{aligned}$$

  where last equality comes from complete multipartite property of $\bar{x}$ therefore, objective is a convex function of ν and its maximum is reached at either endpoint $\nu_u = -\bar{x}_u$ or $\nu_v = \bar{x}_v$; clearly, $s_u, s_v > 1$ together with $\bar{x}_u, \bar{x}_v > 0$ achieve $\langle \hat{A}x(\nu_u), x(\nu_u)\rangle > \langle \hat{A}\bar{x}, \bar{x}\rangle$ and $\langle \hat{A}x(\nu_v), x(\nu_v)\rangle > \langle \hat{A}\bar{x}, \bar{x}\rangle$ a contradiction.

- now, let the support of global maximum $\bar{x}$ be a clique σ of size k, and let us assume that all diagonal entries are equal to s along the clique

$$(\hat{A}\bar{x})_i \begin{cases} = \frac{1}{sk} + \frac{k-1}{k}, & \text{for all } i \in \sigma, \\ \leq \frac{k-1}{k}, & \text{for all } i \notin \sigma, \end{cases}$$

$$\langle \hat{A}\bar{x}, \bar{x}\rangle = 1 - \frac{s-1}{sk}$$

since at least 1 vertex from σ is not connected to i for all $i \notin \sigma$; now, let y be a solution of (MRC) such that $\langle \hat{A}\bar{x}, y\rangle$ is maximum. Since $(\hat{A}\bar{x})_i < \langle \hat{A}\bar{x}, \bar{x}\rangle$ for all $i \notin \sigma$, then y_i should be 0 for all such component or else we could improve objective value.

$$\begin{aligned} (\langle \hat{A}y, y\rangle = & \ \langle (E - \tfrac{s-1}{s} I)y, y\rangle = 1 - \tfrac{s-1}{s} \parallel y \parallel^2 \\ \langle \bar{x}, y\rangle = & \ \tfrac{1}{k}\sum_{i\in\sigma} y_i = \frac{1}{k}\langle y, e\rangle = \frac{1}{k} \\ \parallel \bar{x} \parallel^2 = & \ \tfrac{1}{k} \\ \langle \bar{x}, y\rangle^2 \leq & \ \parallel \bar{x} \parallel^2 \parallel y \parallel^2 \end{aligned}$$

since adjacency graph does not play rle outside clique. Putting all things together we arrive at $\parallel y \parallel^2 \geq \frac{1}{k}$ and $\langle \hat{A}y, y\rangle \leq \langle \hat{A}\bar{x}, \bar{x}\rangle$ with equality whenever y is colinear to $\bar{x}$ i.e $y = \bar{x}$ in our case.

The crux in above proof relies on the following lemma which could be seen as a dedicated version of theorem 1 by Gibbons et al.[11].

Lemma 1 (complete multipartite) *Let $\bar{x}$ a 2nd order condition point (2OC for short) of max clique problem for some diagonal regularization, then its support graph $G_{supp(\bar{x})}$ is complete multipartite.*

Proof 2OC from dual lagrangian requires $\langle \hat{A}y, y\rangle \leq 0, \langle e, y\rangle = 0$, known as almost negative semidefiniteness property;

- **multipartite**. Let t, u, v vertices in $G_{\text{supp}}(\bar{x})$ be such that $t \not\sim u, t \not\sim v$ and $u \sim v$. W.l.o.g. we assume $y_t > 0$ then we claim that either $y_u \geq 0$ or $y_v \geq 0$ under $\langle e, y\rangle = 0$; assume the contrary, then $\langle \hat{A}y, y\rangle = \hat{a}_{tt}y_t^2 + \hat{a}_{uu}y_u^2 + \hat{a}_{vv}y_v^2 + 2\hat{a}_{uv}y_u y_v > 0$ a contradiction. Vertices in a second OC point could then be splitted in parts having same sign w.r.t. y component.

- **completeness**. By contradiction, let t, u and v, w be vertices in two different parts $t \not\sim u, v \not\sim w$ such that $t \sim v, t \sim w$ while $u \sim v, u \not\sim w$ and $y_t, y_u \geq 0$ while $y_v, y_w \leq 0$ according to above multipartite condition (remind $y_t + y_u + y_v + y_w = 0$); then, $\langle \hat{A}y, y\rangle = \hat{a}_{tt}y_t^2 + \hat{a}_{uu}y_u^2 + \hat{a}_{vv}y_v^2 + \hat{a}_{ww}y_w^2 + 2\hat{a}_{tv}y_t y_v + 2\hat{a}_{tw}y_t y_w + 2\hat{a}_{uv}y_u y_v$; choosing, for instance $y_t = y_v = 0$ clearly leads to $\langle \hat{A}y, y\rangle = \hat{a}_{uu}y_u^2 + \hat{a}_{ww}y_w^2 > 0$ whenever either $\hat{a}_{uu} > 0$ or $\hat{a}_{ww} > 0$, a contradiction.

Remark 1 *Let us notice that $\langle \hat{A}\bar{x}, \bar{x}\rangle = 1 - \frac{s-1}{sk}$ is decreasing in s, has gradient $\nabla\langle \hat{A}\bar{x}, \bar{x}\rangle = (-1/s^2k, (s-1)/sk^2)$ and its Hessian*

$$\nabla^2\langle \hat{A}\bar{x}, \bar{x}\rangle = \begin{bmatrix} \frac{2}{s^3k} & \frac{1}{s^2k^2} \\ \frac{1}{s^2k^2} & -\frac{2(s-1)}{sk^3} \end{bmatrix}$$

is definite negative.

3 Bounding through the Legendre-Fenchel transform

The Legendre-Fenchel transform (LFT for short) of the maximum clique problem is defined by

$$\begin{aligned} u^*(y) = \; & \max_{\langle e,x\rangle=1}\{\langle y,x\rangle - \langle \hat{A}x,x\rangle\}, y \in \mathcal{R}^n \\ & = \max_{x\in\text{barycenter of faces}}\{\langle y,x\rangle + \langle x,x\rangle - 1 + \langle \hat{\bar{A}}x,x\rangle\} \end{aligned}$$

In general, LFT yields a lower enveloppe of the original function through double conjugation, i.e. $u^{**}(x) = \max_y\{\langle y,x\rangle - u^*(y)\} \leq \max_y\{\langle y,x\rangle - \langle y,x\rangle + u(x)\} = u(x)$; moreover this enveloppe is convex since letting $x^* = \arg\max_y\{\langle y,x\rangle - u(x)\}$, we have:

$$\begin{aligned} \langle y,x^*\rangle - u(x^*) \geq \; & \langle y,x'^*\rangle - u(x'^*) \\ \langle y',x'^*\rangle - u(x'^*) \geq \; & \langle y',x^*\rangle - u(x^*) \end{aligned}$$

which gives $\langle y-y', x^*-x'^*\rangle \geq 0$. In clique case, $u(x) = \langle \hat{A}x,x\rangle$; from section 2, we get in max independent set settings for constant s diagonal regularization parameter,

$$u^*(y) = \max_{\bar{x}\in\text{barycenter of faces}}\{\langle y,\bar{x}\rangle + \frac{s-1}{s}\langle \bar{x},\bar{x}\rangle\}$$

Interestingly, noting $\bar{u}^*(x)$ LFT of max independent set (equivalently max clique in complement $E-I-A$), both problems are related through:

$$\begin{aligned} \langle (A+\tfrac{1}{s}I)x,x\rangle \leq \; & 1-\bar{u}^{**}(x) \\ \langle (\bar{A}+\tfrac{1}{\bar{s}}I)x,x\rangle \leq \; & 1-u^{**}(x) \end{aligned}$$

where max clique regularization parameter s induces max independent set regularization parameter $\bar{s} = 1-1/s$ and conversely.

Then, we are in a position to rephrase the seminal paper by Amin and Hakimi[1] under previous regularization;

Claim 1 *If G is a complete graph with n vertices, then the eigenpairs of $\hat{A} = A + \frac{1}{s}I$ are $\{(n-(1-1/s),e), (-(1-1/s), ne_i - e)\}$ for $i = 1, n$.*

Letting $s > 2$, since clique size is greater than 2 in connected graphs, we derive the regularized version of Amin and Hakimi's theorem

Theorem 1 *The maximum clique size is less or equal than the number of eigenvalues less or equal than $-1/2$ for regularized adjacency matrix:*

$$\omega(G) \leq \Lambda_{\leq\frac{-1}{2}}(\hat{A}) + 1,$$

with equality holding for complete multi-partite graphs.

Proof Starting from previous claim and using separation theorem for principal submatrices of $\hat{A}$ (since it is symmetric) achieves the result in the same way as Amin and Hakimi did. Equality follows from previous claim after reordering vertices so that multi-partite blocks (diag($\frac{1}{s_i}$) are grouped along diagonal while remaining entries are all ones.

Claim 2 *E matrix of size $n \times n$ has (n, e) as an eigenpair along with 0 eigenvalues.*

Since LFT provides us the nice correspondence between max clique and max independent set regularization parameters, we arrive at $\hat{\bar{A}} = (1 - 1/s)I + \bar{A}, \hat{A} = 1/sI + A$ so that $\hat{\bar{A}} + \hat{A} = E$. Then, the minimax characterization of eigenvalues of symmetric matrices gives:

Theorem 2 *Let α_i (resp. $\bar{\alpha}_i$) be the eigenvalues of $\hat{A}$ (resp. $\hat{\bar{A}}$) ordered by non increasing values, the eigenvalues of $\hat{A} + \hat{\bar{A}}$ fulfill*

$$\Lambda_{\leq \frac{-1}{2}}(\hat{A}) + \Lambda_{\leq \frac{-1}{2}}(\hat{\bar{A}}) \leq n - \max\{\Lambda_{(\frac{-1}{2},0]}(\hat{A}), \Lambda_{(\frac{-1}{2},0]}(\hat{\bar{A}})\}n \leq \rho(\hat{A}) + \rho(\hat{\bar{A}}) \leq \sqrt{2}n$$

Proof Let λ_i be the eigenvalues of $\hat{A} + \hat{\bar{A}} = E$ in non increasing order, minmax characterization of symmetric matrices is

$$\begin{aligned} \alpha_p + \bar{\alpha}_n \leq & \ \lambda_p \leq \alpha_p + \bar{\alpha}_1 \text{ for all } p = 1, n \\ \lambda_{p+q-1} \leq & \ \alpha_p + \bar{\alpha}_q \text{ for } p + q - 1 = 1, n \end{aligned}$$

let $p = min_i\{\alpha_i \leq \frac{-1}{2}\}$, from $\lambda_n = 0 \leq \alpha_p + \bar{\alpha}_{n+1-p}$, we get $\Lambda_{<0}(\hat{A}) = \Lambda_{\leq \frac{-1}{2}}(\hat{A}) + \Lambda_{(\frac{-1}{2},0]}(\hat{A}) = n - p + 1$ and $\bar{\alpha}_{n+1-p} \geq -\alpha_p \geq \frac{1}{2}$, i.e. $\Lambda_{\leq \frac{-1}{2}}(\hat{\bar{A}}) \leq p - 1$ which we sum up to achieve one inequality involving $\Lambda_{(\frac{-1}{2},0]}(\hat{A})$; we are done by symmetry w.r.t. $\hat{\bar{A}}$.

As for second inequalities, clearly $n = \lambda_1 \leq \alpha_1 + \bar{\alpha}_1 = \rho(\hat{A}) + \rho(\hat{\bar{A}})$ on the one hand;on the other hand, $\rho(\hat{A}) = \frac{1}{s} + \rho(A), \rho(\hat{\bar{A}}) = (1 - \frac{1}{s}) + \rho(\bar{A})$ yields $\rho(\hat{A}) + \rho(\hat{\bar{A}}) = 1 + \rho(A) + \rho(\bar{A}) \leq 1 + \sqrt{2}(n-1) \leq \sqrt{2}n$ using Amin and Hakimi's result.

As Amin and Hakimi noticed, $\Lambda_{\leq \frac{-1}{2}}(\hat{A}) + 1$ is better than $\rho(\hat{A})$ for dense graphs since spectral radius increases with density; Budinich's experiments[8] incidentally report this behavior for randomly generated graphs. Finally, above result states that both upper bounds have to be used when dealing with $A, \bar{A}$ at the same time.

In practice, both upper bounds require much less effort than computing all eigenvalues since separation theorem amounts to compute Sturm sequences on tridiagonal matrices (Hessenberg form on symmetric matrices) while spectral radius is computed through Knyazev's block locally optimal conjugate gradient method[14] within a small number of loops.

Remark 2 *Incidentally, $\hat{\bar{A}} + \hat{A} = E$ implies $\nabla\langle(\hat{\bar{A}} + \hat{A})x, x\rangle = 2Ex$; it suggests that for any local optimum to max clique, max stable slope is 2 and vice versa. We will use this nice property to walk around local optimum for max clique and max stable alternately and will report quality of first 3 iterations in computational experiments (see section 4).*

4 A DC-algorithm: qualitative experiments

In this section, we report efficiency of diagonal regularization against alternating between clique and stable problem as suggested by section 3; it is shown how *good* cliques and stables are retrieved within no more than 3 iterations. M. Budinich[8] suggests "*one can conjecture that the nearer Perron's eigenvector to the best solution, the tighter the bounds*"; he bears in mind relationship between Rayleigh quotient $\langle Ax, x\rangle / \parallel x \parallel^2$ and Perron's eigenvalue. In order to computationally confirms it, we rather compare Cauchy-Schwarz inequality $\langle Ax, x\rangle \leq\parallel Ax \parallel\parallel x \parallel$ with Kantorovich inequality for a positive definite matrix (*an inverse* Cauchy-Schwarz inequality) $\langle A^{-1}x, x\rangle \parallel x \parallel^2 \leq \mu \parallel x \parallel^2 / \langle Ax, x\rangle$ where scaling factor $\mu = (\rho_1 + \rho)^2/4\rho_1\rho$ depends on min and max eigenvalues; however, in our case A matrix is not definite positive but $sI - A$ is, provided $s > \rho$. Moreover $(sI - A)^{-1} = 1/s(I + A/s + A^2/s^2 + \ldots)$ closely relates Kantorovich inequality to max clique problem; on the other hand, $(sI - A)$ definite positiveness suggests that its minimization is connected to max clique problem as well. So, we tried starting from Perron's eigenvector, Kantorovich solution to **convex** minimization of $((\rho + 1)I - A)^{-1}$ under $\langle e, x\rangle = 1$ and **convex** minimization of $((\rho+1)I - A)$ under $\langle e, x\rangle = 1$; all of them are supposed to start with a good initial guess. Contrary to Budinich's suggestion, we observe as good results for $((\rho+1)I - A)$ start as for bounded Rayleigh's quotient (either Cauchy-Schwarz or Kantorovich); furthermore, we observe bigger complete multipartite subgraphs with Kantorovich than with Perron, leaving this question opened.

First, we compare *normalized* DC-$LSL^\top$[7] (through signed Cholesky's factorization) with *standard* DC-diag (diagonal dominance enhanced through dichotomic search of minimum constant diagonal offset fulfilling convexity property); we compare as well congruence DC-$Q^\top DQ$ vs *standard* DC-diag since quadratic maximization is clearly invariant under congruence equivalence[18]. Unfortunately, both tables 2 and 3 favor DC-diag while we were expecting the converse:

- DC-$LSL^\top$ reveals a highly unstable convergence behavior in CPLEX©ilog; it possibly means that a *dedicated normalized* DC should be studied instead of a simple-minded linearization loop.
- DC-$Q^\top DQ$ is stuck in a poor starting clique and no stable set found so that walk ends up with no significant results.

Both normalization require further investigation to measure influence of linearization over stationary point finding for quadratic objective (to check whether we skip over good local optima); second, accurate simplex is turned into a rounded polyhedron so that solver could be abused by this rounding effect. Remind that we are looking for centroid of faces rather than extreme points of polyhedron; as a consequence, faces could have been twisted by rounding errors leading to unpredictable results.

Therefore, *standard* DC is kept for further experiments where columns are meant to report number of loops required in underlied algorithms to converge componentwise upto tolerance; diagonal regularization is tested against two different policies:

1. clique (stable resp.) problem takes care of all its results known so far and they are used either in a *concave strategy* 1/clqsize (1/stbsize resp.) or a *convex strategy* 1-1/clqsize (1-1/stbsize resp.)

2. clique (stable resp.) problem does take care of all results known so far and parameters are used in a complementary fashion as given by Legendre-Fenchel bounding goal, namely for a *concave strategy* (1/clqsize, 1-1/stbsize) , for a *convex strategy* (1-1/clqsize, 1/stbsize) depending on whether clique size passing through the vertex is greater than stable .

For each of these policies we also apply a diagonal regularization taking care of sibling values instead of its own values for every vertex. Since we apply first clique solving then stable solving we label parameter w.r.t. clique problem but it should be understood that for stable problem clq labels have to be changed in stb labels and vice-versa. No result in a line corresponds to a failure in CPLEX©ilog concave quadratic maximization associated to linearized DC (#DC is meant for the number of concave maximization required to reach a stationary point).

Last line of diagonal regularization parameters gives mean value of clique sizes found along the 3 loops to estimate the spread of cliques among vertices; for small sized problems we found a good covering of all the nodes through cliques within 3 loops while for 800 nodes example, the clique cover is weak due to small clique size and non overlapping cliques retrieved within 3 loops (a greater than 3 loops walk would improve this poor result).

Perron's eigenvalue is computed by block Krylov subspace algorithm tailored by A. Knyazev[14] which was designed for definite positive matrices but shows in our context, a quite stable behavior; on the contrary, Hessenberg's form leads to very sensitive eigenvalue computation which renders Perron's eigenvector less accurate compared with Knyazev's algorithm. We report under # Ritz, the number of loops required to reach an invariant vector subspace associated with Ritz values.

The Legendre-Fenchel bounding section is definitely confirmed since the best shrinking of Perron's eigenvalue is obtained with (1/clqsize,1-1/stbsize) regularization parameters.

Amin and Hakimi regularized bound and Perron bound are easily checked to fulfill the condition over edge density far to 0.5 while edge upper bound

Table 1: upper bounds comparison related to density: Bomze's regularization $1/2I$

example	clique size	density	Perron UB	$\Lambda_{\frac{-1}{2}}$ UB	edge UB
MANN_a9	16	0.927273	42.3039	**20**	43.3121
brock200_1	21	0.745427	150.071	**98**	172.586
brock800_1	23	0.649265	**520.581**	657	644.471

$(3+\sqrt{9-8(\mid \mathcal{V} \mid - \mid \mathcal{E} \mid)})/2$ is worse (a well known fact); it would be desirable to agree on examples whose density is close to 0.5 to report the meaningful feature of heuristics instead of widespread density spectrum since extreme (low or high) densities are *easy* instances.

Last, our experiments are insensitive to non connected graphs even though we assumed them connected to derive regularization features: implemented algorithm takes care of connected components one at a time, since in DIMACS max clique benchmarks, sibling max stables happen to be disconnected.

As for upper bounds, edge bound comes from connectedness of the graph around the maximum clique while spectral bounds come from section 3 for eigenvalues less than $-1/2$ and Perron's eigenvalue $\rho + 1$.

5 Concluding remarks

In this article, we refine well-known regularizations of Motzkin-Strauss's formulation of the maximum clique problem; from diagonal regularization associated with known cliques, we put further shed on the spectral properties along the search for the best clique. In the following subsections, we review our contribution.

5.1 Primitiveness

Irreducibility insures a maximum real eigenvalue but not uniqueness! If moreover, matrix is primitive then Perron's eigenvalue is strictly greater than the remaining ones and has multiplicity 1. This property is very important in Perron's eigenvector computation to escape from unstable behavior of eigenspace computation whenever multiplicity is greater than 1. Primitiveness is guaranteed as soon as trace$(A) > 0$; that's the fundamental property exploited under Bomze's diagonal regularization $\frac{1}{2}I$ and more generally for all diagonal regularizations. It allows the usage of stable eigensolver algorithms like Knyazev's.

5.2 Upper bounding

Amin and Hakimi result on upper bound was extended to the regularized standard clique case; it would be worthwhile to extend it further to the weighted case either from Perron's eigenvalue implicit formula $\rho\frac{\|\pi\|^2}{\langle e,\pi\rangle^2} \leq \frac{\omega(G,w)^4}{\omega(G,\sqrt{w})^2}$ or from combined spectral properties of both weighted adjacency and weighted complement adjacency graphs. Given $w_i \geq 0$ as weight associated with vertex i, we introduce, as most users do, $w_{ij} = \sqrt{w_i w_j}$ for edge $ij \in \mathcal{E}$, and $w_{ii} = w_i$ as diagonal element of regularized weighted matrix $\hat{W}$; regularized weighted complement is then defined by $\bar{W} = \sqrt{w}\sqrt{w}^\top - diag(w) - W$. Bounding weighted case requires to study Legendre-Fenchel transform under $\hat{\bar{W}} + \hat{W} = \sqrt{w}\sqrt{w}^\top$ as we did for standard case.

5.3 Heuristic vs structural property

Finally, we are aware of very good results obtained by Busygin's trust region technique [9]; however, from our point of view, his results are somehow surprising since his *regularization* $w_i - w_{\min}$ amounts to Motzkin-Strauss formulation in standard case. Since our study is dedicated to structural property learning, we discarded those heuristics from our scope.

5.4 Equivalence relation normalization

Our attempts to normalize quadratic programming through congruence have failed; a first guess is that nothing is lost for objective value but the facial structure has been torn up so new questions arise about canonical form of regularized problem.

For instance, let us take the congruence equivalence relation

$$\text{orbit}^\top = \{P^\top AP \mid \det(P) \neq 0\}$$

since $\langle P^\top APy, y\rangle = \langle Ax, x\rangle$ for $x = Py$, reducing initial graph modulo equivalence classes promises an actual divide-and-conquer scheme. Unfortunately, as pointed by Newman[18] the congruence equivalence relation is hard to deal with; we report a poor computational behavior whose result possibly tears up facial structure.

Newman pointed similarity equivalence relation as a more reliable alternate reduction

$$\text{orbit}^{-1} = \{P^{-1}AP \mid \det(P) \neq 0\}$$

where superscript in orbit stands for similarity transform rather than inverse operator. We could expect to retrieve all elementary divisors of the $Z-$pencil $\lambda I - A$ (not only the Perron's divisor), through the Frobenius canonical form and try to combine Perron's eigenspace to subsequent eigenspaces to improve our clique/stable walk. Our first attempts in this direction open up a numerical stability problem for big matrices; among others, Storjohann's algorithm (see[20]) fails to retrieve Frobenius normal form of big matrices unless arbitrary precision floating point computation are done; Perron's eigenspace would likely be enhanced with next eigenvalues at this price.

References

[1] A. T. Amin and S. L. Hakimi (1972) "Upper bounds of the order of a clique of a graph," *SIAM Journal on Applied Mathematics*, 22(4):569–573.

[2] F. Alizadeh (1995) "Interior point methods in semidefinite programming with applications to combinatorial optimization." *SIAM Journal on Optimization*, 5(1):13–51.

[3] I. Bomze, M. Budinich, P. Pardalos, and M. Pelillo (1999) "The maximum clique problem." *D.-Z. Du and P. M. Pardalos, editors, Handbook of Combinatorial Optimization*, volume 4. Kluwer Academic Publishers, Boston, MA.

[4] I. Bomze, M. Budinich, M. Pelillo, and C. Rossi (2002) "Annealed replication: A new heuristic for the maximum clique problem" *Discr. Appl. Math.*, 121(1–3):27–49.

[5] R. Boppana and M.M. Halldórsson (1992) "Approximating maximum independent sets by excluding subgraphs." *BIT*, 32(2), 180-196.

[6] I.M. Bomze (1997) "Evolution towards the maximum clique." *Journal of Global Optimization*, 10:143–164.

[7] A. Bojanczyk, S. Qiao, and A.O. Steinhardt (2000) "Unifying unitary and hyperbolic transformations," *Linear Algebra and its Applications*, 316, pp. 183 -196.

[8] M. Budinich (2003) "Bounds on the maximum clique of a graph." *to appear in Discr. Appl. Math. http:/www.ts.infn.it/ mbh/MCBounds.ps.Z*

[9] S. Busygin (2002) "A new trust region technique for maximum weight clique." *submitted to special issue of Discrete Applied Mathematics: Combinatorial Optimization*

[10] D. Fortin and I. Tsevendorj (2001) "The maximum clique problem: some regularizations and continuous approach",, in *Proceedings of the 12-th Baikal International Conference on Optimization Methods and their Applications* Irkutsk, Russia, pp 256-263.

[11] L. E. Gibbons, D. W. Hearn, P. M. Pardalos, and M. V. Ramana (1996) "Continuous characterizations of the maximum clique problem." *Mathematics of Operations Research*, 9:143–164.

[12] J. Hastad (1999) "Clique is hard to approximate within $|v^{1-\epsilon}|$". *Acta Mathematica,* vol 182, pp 105-142.

[13] M. Ibragimov (2001) "A method of calculating the spectral radius of a nonnegative matrix and its applications." *Economic theory*, 17:467–480.

[14] A.V. Knyazev (2001) "Toward the optimal preconditioned eigensolver: Locally optimal block preconditioned conjugate gradient method" *SIAM Journal on Scientific Computing*, 23(2):517–541.

[15] A.A. Kuznetsova and A.S. Strekalovsky (2001) "On solving the maximum clique problem." *Journal of Global Optimization*, 21(3):265–288.

[16] C. D. Meyer (1987) "Uncoupling the perron eigenvector problem." *Linear Algebra and its Appl.* vol 114/115. pp.69-94.

[17] T.S. Motzkin and E.G. Strauss (1965) "Maxima for graphs and a new proof of a theorem of Turan." *Canadian Journal of Mathematics*, 17, (4):533-540.

[18] M. Newman (1972) *Integral Matrices.* Academic Press.

[19] P. A. Parillo (2000) *Structured Semidefinite Programs and Semi-algebraic Geometry Methods in Robustness and Optimization.* PhD thesis, California Institute of Technology, Pasadena.

[20] Arne Storjohann (1998) An $O(n^3)$ algorithm for Frobenius normal form. *To appear in Proc. Int'l. Symp. on Symbolic and Algebraic Computation: ISSAC '98.*

[21] I. Tsevendorj (2001) Piecewise convex maximization problems: Global optimality conditions. *Journal of Global Optimization*, 21(1):1–14.

Table 2: **signed Cholesky vs DC**; MANN_a9: 45 vertices, 918 edges

diagonal clique regularization #clique	#stable	clique #DC $LSL^\top$	DC	clique size $LSL^\top$	DC	stable #DC $LSL^\top$	DC	stable size $LSL^\top$	DC
1/clq		208	12	4	12	549	25	3	0
		412	127	15	14	118	9	2	0
13.733	2.066	587	45	15	14	185	26	2	3
Kantorovich		631	10	12	12	117	13	2	0
		1480	14	15	9	177	43	2	3
13.4	2.066	308	58	14	16	79	20	2	0
1-1/clq		338	4	12	12	241	21	3	3
		774	120	15	16	309	37	3	2
14.355	2.066	*	80	*	14	*	5	*	3
Kantorovich		631	10	12	12	117	13	2	0
		507	10	14	12	83	44	2	3
12.755	2.066	*	116	*	16	*	37	*	2
1/clq	1-1/stb	484	4	12	12	243	82	2	3
		*	138	*	16	*	121	*	3
13.911	**2.133**	*	81	*	16	*	123	*	3
Kantorovich		631	10	12	12	213	12	2	0
		*	1	*	0	*	1	*	0
10	2	*	1	*	0	*	1	*	0
1-1/clq	1/stb	552	4	12	12	188	39	3	3
		*	120	*	16	*	130	*	2
14.622	2.066	*	56	*	14	*	44	*	0
Kantorovich		631	10	12	12	56	13	2	0
		236	1	12	0	75	1	2	0
10	2	1498	1	12	0	332	1	2	0
1/stb		486	4	12	12	264	13	3	0
		498	13	15	9	566	32	2	3
13.222	2.066	283	67	12	16	912	24	2	0
Kantorovich		631	10	12	12	56	13	2	0
		257	13	15	9	113	11	2	0
11.4	2	672	10	12	12	410	10	2	0
1-1/stb		226	4	12	12	268	61	3	3
		676	45	15	16	422	43	2	2
14.355	2.066	750	55	16	14	467	88	2	3
Kantorovich		631	10	12	12	213	12	2	0
		328	10	15	12	406	111	2	3
12.755	2.066	725	46	16	16	430	43	2	2
1/stb	1-1/clq	546	4	12	12	695	25	3	3
		538	59	16	16	182	16	2	0
13.311	2.066	533	47	16	13	150	12	2	0
Kantorovich		631	10	12	12	117	13	2	0
		171	1	15	0	54	1	2	0
10	2	171	1	15	0	54	1	2	0
1-1/stb	1/clq	325	4	12	12	599	18	3	3
		589	77	16	16	130	23	2	0
13.311	2.066	437	46	16	13	80	16	2	0
Kantorovich		631	10	12	12	117	13	2	0
		171	1	15	0	54	1	2	0
10	2	171	1	15	0	54	1	2	0

Table 3: **Congruence vs DC** ;MANN_a9: 45 vertices, 918 edges

diagonal clique regularization		clique				stable			
		#DC		size		#DC		size	
#clique	#stable	$Q^\top DQ$	DC	$Q^\top DQ$	DC	$Q^\top DQ$	DC	$Q^\top DQ$	DC
1/clq		2	12	3	12	2	25	0	0
		2	127	4	14	2	9	0	0
2.46667	2	2	45	4	14	2	26	0	3
Kantorovich		2	10	3	12	2	13	0	0
		2	14	4	9	2	43	0	3
2.46667	2	2	58	4	16	2	20	0	0
1-1/clq		2	4	3	12	2	21	0	3
		2	120	4	16	2	37	0	2
2.42222	2	2	80	4	14	2	5	0	3
Kantorovich		2	10	3	12	2	13	0	0
		2	10	4	12	2	44	0	3
2.42222	2	2	116	4	16	2	37	0	2
1/clq	1-1/stb	2	4	3	12	2	82	0	3
		2	138	4	16	2	121	0	3
2.46667	2	2	81	4	16	2	123	0	3
Kantorovich		2	10	3	12	2	12	0	0
		2	1	4	0	2	1	0	0
2.46667	2	2	1	4	0	2	1	0	0
1-1/clq	1/stb	2	4	3	12	2	39	0	3
		2	120	4	16	2	130	0	2
2.42222	2	2	56	4	14	2	44	0	0
Kantorovich		2	10	3	12	2	13	0	0
		2	1	4	0	2	1	0	0
2.42222	2	2	1	4	0	2	1	0	0
1/stb		2	4	3	12	2	13	0	0
		2	13	3	9	2	32	0	3
2.06667	2	2	67	3	16	2	24	0	0
Kantorovich		2	10	3	12	2	13	0	0
		2	13	3	9	2	11	0	0
2.06667	2	2	10	3	12	42	10	0	0
1-1/stb		2	4	3	12	2	61	0	3
		2	45	3	16	2	43	0	2
2.06667	2	2	55	3	14	2	88	0	3
Kantorovich		2	10	3	12	2	12	0	0
		2	10	3	12	2	111	0	3
2.06667	2	2	46	3	16	2	43	0	2
1/stb	1-1/clq	2	4	3	12	2	25	0	3
		2	59	3	16	2	16	0	0
2.06667	2	2	47	3	13	2	12	0	0
Kantorovich		2	10	3	12	2	13	0	0
		2	1	3	0	2	1	0	0
2.06667	2	2	1	3	0	2	1	0	0
1-1/stb	1/clq	2	4	3	12	2	18	0	3
		2	77	3	16	2	23	0	0
2.06667	2	2	46	3	13	2	16	0	0
Kantorovich		2	10	3	12	2	13	0	0
		2	1	3	0	2	1	0	0
2.06667	2	2	1	3	0	2	1	0	0

Table 4: MANN_a9: 45 vertices, 918 edges

diagonal clique regularization #clique	#stable	Perron #Ritz	value	clique #DC	size	stable #DC	size
1/clq		96	41.3039	4	12	25	0
				127	14	9	0
$(sI - A)$				61	**16**	15	0
				34	14	25	3
Kantorovich				34	14	18	0
				32	14	29	2
1-1/clq		96	41.3039	4	12	25	0
12.9556	2	8	41.6955	228	**16**	54	2
$(sI - A)$				61	**16**	15	0
14.6	2	8	41.735	51	13	11	0
Kantorovich				34	14	18	0
9.46667	2.06667	8	41.5729	287	14	35	3
1/clq	1-1/stb	96	41.3039	4	12	23	0
14.2222	2.06667	8	**40.8747**	104	15	113	3
$(sI - A)$				61	**16**	14	0
14.4889	2.06667	8	40.8813	74	**16**	20	0
Kantorovich				34	14	22	0
14	2	8	40.8754	60	14	18	0
1-1/clq	1/stb	96	41.3039	4	12	27	0
12.0667	2	9	41.6763	100	14	120	2
$(sI - A)$				61	**16**	16	0
14.6	2	8	41.735	51	13	10	0
Kantorovich				34	14	34	3
9.2	2.06667	10	41.5523	95	14	27	3
1/stb		96	41.3039	4	12	27	0
13.3111	2.06667	10	41.2928	44	**16**	34	0
$(sI - A)$				61	**16**	16	0
14.7556	2.06667	10	41.29285	46	14	22	3
Kantorovich				34	14	34	3
14.7556	2.06667	10	41.2928	35	14	32	3
1-1/stb		96	41.3039	4	12	23	0
12.0667	2	8	41.3039	36	14	99	2
$(sI - A)$				61	**16**	14	0
14.4889	2.06667	8	41.3152	41	14	100	3
Kantorovich				34	14	22	0
10.2667	2	8	41.3039	40	14	25	0
1/stb	1-1/clq	96	41.3039	4	12	25	0
12.0667	2	8	41.3039	37	14	21	2
$(sI - A)$				61	**16**	15	0
14.7556	2	8	41.3039	54	14	17	0
Kantorovich				34	14	18	0
12.0667	2	8	41.3039	37	14	21	2
1-1/stb	1/clq	96	41.3039	4	12	25	0
10.2667	2	8		54	14	17	0
$(sI - A)$				61	**16**	15	0
14.7556	2	8	41.3039	54	14	17	0
Kantorovich				34	14	18	0
10.2667	2	8	41.3039	54	14	17	0

Table 5: brock200_1: 200 vertices, 14834 edges

diagonal clique regularization #clique	#stable	Perron #Ritz	value	clique #DC	size	stable #DC	size
1/clq		32	149.071	113	16	75	5
4.58	2.18	23	148.982	133	16	107	5
$(sI-A)$				108	16	82	5
4.58	2.18	23	148.982	132	16	105	5
Kantorovich				94	14	119	5
4.875	2.225	23	**148.975**	129	15	118	5
1-1/clq		32	149.071	113	16	75	5
4.53	2.165	69	149.157	383	15	248	5
$(sI-A)$				108	16	82	5
4.53	2.165	69	149.157	383	15	248	5
Kantorovich				94	14	119	5
4.32	2.165	57	149.149	567	14	151	5
1/clq	1-1/stb	32	149.071	113	16	322	5
5.05	2.27	22	148.995	118	**18**	367	5
$(sI-A)$				108	16	320	5
5.05	2.27	22	148.995	118	**18**	367	5
Kantorovich				94	14	312	5
4.835	2.21	22	148.996	174	17	398	5
1-1/clq	1/stb	32	149.071	113	16	73	5
4.555	2.255	52	149.145	231	15	148	6
$(sI-A)$				108	16	74	5
4.555	2.255	52	149.145	231	15	148	6
Kantorovich				94	14	141	5
4.96	2.225	67	149.153	391	17	178	5
1/stb		32	149.071	113	16	73	5
4.57	2.235	23	149.05	109	17	125	6
$(sI-A)$				108	16	74	5
4.57	2.235	23	149.05	109	17	125	6
Kantorovich				94	14	141	5
4.87	2.27	23	149.047	119	16	137	6
1-1/stb		32	149.071	113	16	322	5
4.28	2.165	22	149.087	260	**18**	201	4
$(sI-A)$				108	16	320	5
4.28	2.165	22	149.087	260	**18**	201	4
Kantorovich				94	14	312	5
4.69	2.195	22	149.089	258	16	273	5
1/stb	1-1/clq	32	149.071	113	16	75	5
4.695	2.27	22	149.065	141	17	117	6
$(sI-A)$				108	16	82	5
4.695	2.27	22	149.065	147	17	117	6
Kantorovich				94	14	119	5
4.875	2.225	22	149.062	136	15	132	5
1-1/stb	1/clq	32	149.071	113	16	75	5
4.53	2.195	22	149.073	136	15	97	5
$(sI-A)$				108	16	82	5
4.53	2.195	22	149.073	137	15	96	5
Kantorovich				94	14	119	5
4.5	2.175	22	149.076	167	14	98	4

Table 6: brock800_1: 800 vertices, 207505 edges

diagonal clique regularization		Perron		clique		stable	
#clique	#stable	#Ritz	value	#DC	size	#DC	size
1/clq		6	519.581	181	15	271	7
2.73375	2.16375	90	**519.557**	216	16	221	8
$(sI-A)$				208	15	207	6
2.80375	2.13375	92	519.554	250	16	178	8
Kantorovich				181	15	271	7
2.73375	2.16375	90	**519.557**	216	16	221	8
1-1/clq		6	519.581	181	15	271	7
2.7575	2.135	112	519.605	182	15	448	8
$(sI-A)$				208	15	207	6
2.79875	2.1225	122	519.607	688	**18**	256	8
Kantorovich				181	15	271	7
2.7575	2.135	112	519.605	182	15	448	8
1/clq	1-1/stb	6	519.581	181	15	805	7
2.77	2.13125	89	519.563	161	16	597	7
$(sI-A)$				208	15	623	6
2.7025	2.10375	90	519.563	263	15	403	6
Kantorovich				181	15	805	7
2.77	2.13125	89	519.563	161	16	597	7
1-1/clq	1/stb	6	519.581	181	15	245	0
2.715	2.10375	90	519.6	294	15	254	7
$(sI-A)$				208	15	205	6
2.75125	2.06	119	519.603	605	15	379	6
Kantorovich				181	15	245	0
2.715	2.10375	90	519.6	294	15	254	7
1/stb		6	519.581	181	15	245	0
2.715	2.07375	88	519.576	214	15	271	6
$(sI-A)$				208	15	205	6
2.7875	2.07375	90	519.575	271	16	284	7
Kantorovich				181	15	245	0
2.715	2.07375	88	519.576	214	15	271	6
1-1/stb		6	519.581	181	15	805	7
2.71875	2.1125	90	519.589	532	15	398	7
$(sI-A)$				208	15	623	6
2.6875	2.1225	52	519.59	531	14	324	7
Kantorovich				181	15	805	7
2.71875	2.1125	90	519.589	532	15	398	7
1/stb	1-1/clq	6	519.581	181	15	271	7
2.68125	2.1175	90	519.578	283	14	233	7
$(sI-A)$				208	15	207	6
2.80375	2.1175	91	519.579	281	16	242	7
Kantorovich				181	15	271	7
2.68125	2.1175	90	519.578	283	14	233	7
1-1/stb	1/clq	6	519.581	181	15	271	7
2.64625	2.1825	88	519.583	188	14	205	9
$(sI-A)$				208	15	207	6
2.64875	2.1475	53	519.583	237	14	200	7
Kantorovich				181	15	271	7
2.64625	2.1825	88	519.583	188	14	205	9

Optimization and Optimal Control, pp. 121 – 139
P.M. Pardalos, I. Tseveendorj, and R. Enkhbat, Editors

Optimal synthesis of handwritten letters from their characteristic points

Masaru Kamada (kamada@cis.ibaraki.ac.jp) and Takayuki Kawahito
Department of Computer and Information Sciences, Ibaraki University, Hitachi, 316-8511, Ibaraki, Japan

Rentsen Enkhbat (renkhbat46@yahoo.com)
National University of Mongolia, P.O. Box 46/635, Ulaanbaatar, Mongolia

Abstract

A numerical local optimization of pen-down motion in writing a string of letters subject to fixed via points and pause points is presented. The square integral of jerk is minimized by following a conditional gradient method.

Keywords: Splines, handwritten characters, conditional gradient.

1 Introduction

This paper presents a numerical method for an optimization problem which arose from the field of simulated handwriting. A non-optimal approach to simulated handwriting is 'Heliscript' by Dooijes [1], in which the pen trajectory is modeled as a path on a rotating cylinder projected onto a flat plane. An optimal approach was originated by Flash and Hogan [2] who provided a sound mathematical model of pen-down handwriting motion on the basis of a minimum principle in mechanical dynamics, although their model has not yet been applied to the synthesis of continuously written strings of letters.

The handwriting model originally proposed by Flash and Hogan [2] and then extended by Kamada [3] is formulated as follows: Denote by $\boldsymbol{r}(t) = (x(t), y(t))^{\mathrm{T}}$ the position vector of the pen tip at time $t \in [0, T]$ on the plane as illustrated in figure 1. A number of via points $\boldsymbol{r}_1, \boldsymbol{r}_2, \cdots, \boldsymbol{r}_K$ on the plane which the pen visits consecutively at the unspecified times $0 < t_1 < t_2 < \cdots < t_K < T$ are assumed to exist on the plane, *i.e.*

$$\boldsymbol{r}(t_k) = \boldsymbol{r}_k, \quad k = 1, 2, 3, \cdots, K. \tag{1}$$

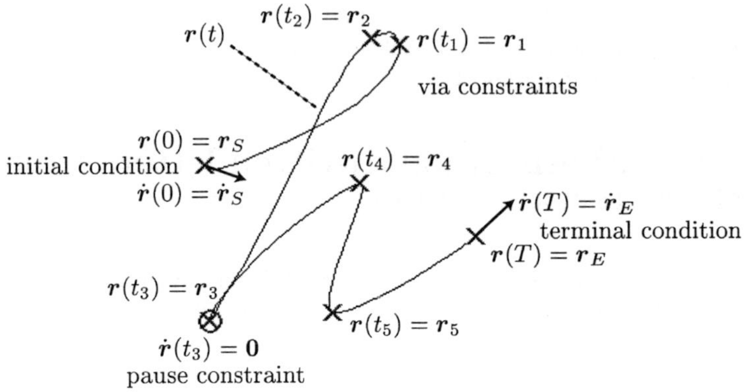

Figure 1. Characteristic points of a letter ($\times$ denotes a via point, especially $\otimes$ denotes a pause point).

Experimental knowledge by Abend et al. [4] suggests that the via points distribute around the points at which the pen has a locally minimal speed. This is compatible with the intuition that we steer the pen carefully thus slowly when writing characteristic parts of a letter. Among the via points, especially prominent are steeples, such as that at the left bottom corner of the letter 'h', at which the pen motion is not only locally slowest but pauses, *i.e.*

$$\boldsymbol{r}^{(1)}(t_{\ell_i}) = \boldsymbol{0}, \quad i = 1, 2, 3, \cdots, I. \tag{2}$$

The initial or terminal state of the pen at time $t = 0$ or $t = T$, respectively, form the boundary conditions

$$\begin{cases} \boldsymbol{r}(0) = \boldsymbol{r}_S, & \boldsymbol{r}(T) = \boldsymbol{r}_E, \\ \boldsymbol{r}^{(1)}(0) = \boldsymbol{r}'_S, & \boldsymbol{r}^{(1)}(T) = \boldsymbol{r}'_E. \end{cases} \tag{3}$$

The pen is steered so as to minimize the square integral of jerk (the rate of change in acceleration), *i.e.*

$$P[\boldsymbol{r}] = \int_0^T \left\| \boldsymbol{r}^{(3)}(t) \right\|^2 dt \longrightarrow \min. \tag{4}$$

subject to the constraints (1), (2) and the boundary conditions (3), where $||\cdot||$ denotes the Euclidean norm in the plane.

The above model yields the following problem of finding the optimal function $\boldsymbol{r}$ for expressing the handwriting motion.

PROBLEM 0. Given the constants $\boldsymbol{r}_1, \boldsymbol{r}_2, \cdots, \boldsymbol{r}_K, \ell_1, \ell_2, \cdots, \ell_I, \boldsymbol{r}_S, \boldsymbol{r}'_S, \boldsymbol{r}_E, \boldsymbol{r}'_E$, find the function $\boldsymbol{r}$ which minimizes the functional $P[\boldsymbol{r}]$ subject to (1), (2) and (3).

Since the function $\boldsymbol{r}$ depends on the variable parameters $t_1, t_2, \cdots, t_K$, we have first to find the optimal parameters and then to determine the function in practice. In the rest of this paper, however, this problem is solved in the reversal steps: (i) analytic derivation of $\boldsymbol{r}$ for specified $t_1, t_2, \cdots, t_K$; (ii) construction of a numerical method to find $t_1, t_2, \cdots, t_K$ that minimize $P[\boldsymbol{r}]$ locally. Application to synthesizing a string of letters in script style from individual letters will also be presented.

2 Preliminary result

Temporarily regarding the variable parameters $t_1, t_2, \cdots, t_K$ with Problem 0 as constants, we have the following modified problem.

PROBLEM 1. Given the constants $t_1, t_2, \cdots, t_K, \boldsymbol{r}_1, \boldsymbol{r}_2, \cdots, \boldsymbol{r}_K, \ell_1, \ell_2, \cdots, \ell_I, \boldsymbol{r}_S, \boldsymbol{r}'_S, \boldsymbol{r}_E, \boldsymbol{r}'_E$, find the function $\boldsymbol{r}$ which minimizes the functional $P[\boldsymbol{r}]$ subject to (1), (2) and (3).

The solution for this problem has been analytically derived by means of calculus of variations in [3] as follows:

PROPOSITION 1. The solution for Problem 1 is expressed as

$$\boldsymbol{r}(t) = \sum_{m=0}^{5} \boldsymbol{a}_m t^m + \sum_{k=1}^{K} \boldsymbol{b}_k (t_k - t)_+^5 + \sum_{i=1}^{I} \boldsymbol{c}_i (t_{\ell_i} - t)_+^4, \tag{5}$$

where $(\cdot - t)$ denotes the truncated function

$$(\cdot - t)_+ := \begin{cases} (\cdot - t) & \text{if} \cdot > t \\ 0 & \text{if} \cdot \leq t, \end{cases} \tag{6}$$

and the coefficients $\boldsymbol{a}_m = (a_{mx}, a_{my})^{\mathrm{T}}$, $\boldsymbol{b}_k = (b_{kx}, b_{ky})^{\mathrm{T}}$, and $\boldsymbol{c}_i = (c_{ix}, c_{iy})^{\mathrm{T}}$ are uniquely determined by (1), (2), (3), and the natural boundary conditions:

$$\boldsymbol{r}^{(3)}(0) = \boldsymbol{r}^{(3)}(T) = \boldsymbol{0}. \tag{7}$$

The first and second terms of the solution (5) represent a quintic spline defined as a piecewise quintic polynomial which is four-times continuously differentiable [5]. The first term makes $\boldsymbol{r}(t)$ be a quintic polynomial for $t \in [t_K, T]$. The second term $\boldsymbol{b}_k(t_k - t)_+^5$ is four-times continuously differentiable and superimposes an additional polynomial $\boldsymbol{b}_k(t_k - t)^5$ only where $t < t_k$ to make a different polynomial for $t \in [t_{k-1}, t_k)$, $(k = 1, 2, \cdots, K$; and we have let $t_0 = 0$ and $t_{K+1} = T$ for the convenience of notation). The point t_k where the spline switches from one polynomial to another is called a knot [5].

The third term $\boldsymbol{c}_{\ell_i}(t_{\ell_i} - t)_+^4$ of (5) violates the definition of quintic spline since it contributes a quartic polynomial component that is only three-times continuously differentiable. Besides, it is widely known that numerical evaluation of $\boldsymbol{r}(t)$ according to (5) suffers from cancellation of significant digits as t_k and t get far away from each other.

A good remedy for both the quartic term and the numerical difficulty is extended B-splines [5]. The extended quintic B-spline M_j having the knots $\tau_j \leq \tau_{j+1} \leq \cdots \leq \tau_{j+6}$ is defined [5] as

$$M_j(t) = [\tau_j, \tau_{j+1}, \cdots, \tau_{j+6}](\cdot - t)_+^5, \tag{8}$$

where $[\tau_j, \tau_{j+1}, \cdots, \tau_{j+m}]$ denotes the operator of divided difference

$$[\tau_j, \tau_{j+1}, \cdots, \tau_{j+i}]g(\cdot) = \begin{cases} \frac{[\tau_{j+1},\cdots,\tau_{j+i}]g(\cdot) - [\tau_j,\cdots,\tau_{j+i-1}]g(\cdot)}{\tau_{j+i}-\tau_j} & \text{if } \tau_j < \tau_{j+i} \\ \left.\frac{d^i g(\tau)}{dt^i}\right|_{\tau=\tau_j} & \text{if } \tau_j = \tau_{j+i} \end{cases}$$

$$i = 0, 1, 2, \cdots. \tag{9}$$

B-splines having distinct knots such as $\tau_j < \tau_{j+1} < \cdots < \tau_{j+6}$ are called ordinary B-splines with simple knots. Extended B-splines are allowed to have a few knots at the same position, which are called multiple knots.

Numerical evaluation of the B-splines is fast and accurate mainly due to the local support property [5]

$$M_j(t) = 0 \quad \text{for} \quad t \notin (\tau_j, \tau_{j+6}). \tag{10}$$

Moreover, the quartic term $\boldsymbol{c}_{\ell_i}(t_{\ell_i} - t)_+^4$ in (5) that violated the definition of quintic splines can be embedded in the extended quintic B-splines having a double knot (or two knots at the same position as $\tau_j = \tau_{j+1}$) at t_{ℓ_i} as presented by the following proposition [3].

PROPOSITION 2. The representation (5) of $\boldsymbol{r}$ by the truncated functions is equivalent to a different representation in terms of the B-splines

$$\boldsymbol{r}(t) = \sum_{j=-5}^{K+I} \boldsymbol{d}_j M_j(t) \tag{11}$$

in $[0, T]$ in the sense that the coefficients for each of these representations uniquely determine those required by the other representation in order to express the same function in $[0, T]$. Here, the knots $\{\tau_j\}_{j=1}^{K+I}$ are set as

$$\tau_j = \begin{cases} t_j, & 1 \leq j \leq \ell_1 \\ t_{j-i}, & \ell_i + i \leq j \leq \ell_{i+1} + i, \quad (i = 1, 2, \cdots, I-1) \\ t_{j-I}, & \ell_I + I \leq j \leq K + I \end{cases} \tag{12}$$

so that we have a simple knot at t_k if $k \notin \{\ell_1, \cdots, \ell_I\}$ and a double knot at t_k if $k \in \{\ell_1, \cdots, \ell_I\}$. The other knots $\tau_{-5}, \cdots, \tau_0$, $\tau_{K+I+1}, \cdots, \tau_{K+I+6}$ are arbitrary points such that $\tau_{-5} < \cdots < \tau_0 = 0$ and $T = \tau_{K+I+1} < \cdots < \tau_{K+I+6}$.

Combining Propositions 1 and 2, we have the following theorem.

THEOREM 1. *The solution* $\boldsymbol{r}$ *for Problem 1 is an extended quintic spline represented by (11) as a linear combination of the extended quintic B-splines* M_j, $(j = -5, -4, \cdots, K+I)$ *on the basis of the knots* $\tau_{-5}, \tau_{-4}, \cdots, \tau_{K+I+6}$ *set by (12) so that we have a simple knot at simple via point* t_k *such that* $k \notin \{\ell_1, \cdots, \ell_I\}$ *and a double knot (or two knots at the same position) at pause point* t_k *such that* $k \in \{\ell_1, \cdots, \ell_I\}$. *The other knots are arbitrary as long as they satisfy* $\tau_{-5} < \tau_{-4} < \cdots < \tau_{-1} < \tau_0 = 0$ *and* $T = \tau_{K+I+1} < \tau_{K+I+2} < \cdots < \tau_{K+I+6}$. *The coefficients* $\boldsymbol{d}_{-5}, \boldsymbol{d}_{-4}, \cdots, \boldsymbol{d}_{K+I}$ *are uniquely determined by the conditions (1), (2), (3), and (7) of which matrix form is*

$$DB = X, \tag{13}$$

where

$$D = (\boldsymbol{d}_{-5} \quad \boldsymbol{d}_{-4} \quad \cdots \quad \boldsymbol{d}_{K+I}), \tag{14}$$

$$X = (\boldsymbol{r}_S \quad \boldsymbol{r}'_S \quad \mathbf{0} \mid \boldsymbol{r}_1 \quad \boldsymbol{r}_2 \quad \cdots \quad \boldsymbol{r}_K \mid \underbrace{\mathbf{0} \ \cdots \ \mathbf{0}}_{I} \mid \boldsymbol{r}_E \quad \boldsymbol{r}'_E \quad \mathbf{0}), \tag{15}$$

and

$$B = \left(\begin{array}{cccc} M_{-5}(0) & M_{-4}(0) & \cdots & M_{K+I}(0) \\ M^{(1)}_{-5}(0) & M^{(1)}_{-4}(0) & \cdots & M^{(1)}_{K+I}(0) \\ M^{(3)}_{-5}(0) & M^{(3)}_{-4}(0) & \cdots & M^{(3)}_{K+I}(0) \\ \hline M_{-5}(t_1) & M_{-4}(t_1) & \cdots & M_{K+I}(t_1) \\ M_{-5}(t_2) & M_{-4}(t_2) & \cdots & M_{K+I}(t_2) \\ \vdots & \vdots & \ddots & \vdots \\ M_{-5}(t_K) & M_{-4}(t_K) & \cdots & M_{K+I}(t_K) \\ \hline M^{(1)}_{-5}(t_{\ell_1}) & M^{(1)}_{-4}(t_{\ell_1}) & \cdots & M^{(1)}_{K+I}(t_{\ell_1}) \\ M^{(1)}_{-5}(t_{\ell_2}) & M^{(1)}_{-4}(t_{\ell_2}) & \cdots & M^{(1)}_{K+I}(t_{\ell_2}) \\ \vdots & \vdots & \ddots & \vdots \\ M^{(1)}_{-5}(t_{\ell_I}) & M^{(1)}_{-4}(t_{\ell_I}) & \cdots & M^{(1)}_{K+I}(t_{\ell_I}) \\ \hline M_{-5}(T) & M_{-4}(T) & \cdots & M_{K+I}(T) \\ M^{(1)}_{-5}(T) & M^{(1)}_{-4}(T) & \cdots & M^{(1)}_{K+I}(T) \\ M^{(3)}_{-5}(T) & M^{(3)}_{-4}(T) & \cdots & M^{(3)}_{K+I}(T) \end{array}\right). \tag{16}$$

REMARK 1. Theorem 1 becomes merely a special case of the famous theorem on the smoothest interpolation problem by de Boor [6] and Schoenberg [7] if there are no pause constraints (2), whereas the pause constraints have been experimentally shown in [3] to be crucial for reproduction of steeples. If we fail to require a pause constraint (and do not place double knots consequently) at a steeple such as the left bottom corner of 'h' in figure 1, the steeple is likely to be deformed into a smooth curve or a loop.

REMARK 2. Extended B-splines having multiple knots appeared at the same time as ordinary B-splines in [5] where extended splines which interpolate the function values as well as the derivatives at multiple knots are constructed. However, the optimality of these functions in the context of Theorem 1 is the original result of [3].

The following formulae are prepared for the technical derivation in the next section. Introducing the notation

$$L_j^{m,n}(t) = \begin{cases} [\tau_j, \cdots, \tau_{j+n}](\cdot - t)_+^{m-1} & \text{if } \tau_j < t < \tau_{j+m}, \\ 0 & \text{elsewhere,} \end{cases}$$
$$n = 1, 2, \cdots, m; \quad m \geq 1, \tag{17}$$

we have the following recurrence formulae for inductive evaluation of M_j,

$$M_j(t) = L_j^{6,6}(t) \tag{18}$$

$$L_j^{m,n}(t) = \frac{L_{j+1}^{m,n-1}(t) - L_j^{m,n-1}(t)}{\tau_{j+n} - \tau_j} \quad \text{for } n = m, m-1, \cdots, 2 \tag{19}$$

$$L_j^{m,1}(t) = \begin{cases} (m-1)(\tau_j - t)_+^{m-2} & \text{if } \tau_j = \tau_{j+1} \\ \frac{(\tau_{j+1}-t)_+^{m-1} - (\tau_j - t)_+^{m-1}}{\tau_{j+1} - \tau_j} & \text{if } \tau_j < \tau_{j+1} \end{cases} \tag{20}$$

from (8), (9) and (10) and the fact that $\tau_{j+n} > \tau_j$, $(n \geq 2)$ in the present case where we have only simple and double knots. By using the same notation, a well-known recurrence formula [5] concerning derivatives of the B-splines with respect to the variable t can be written as follows:

$$\left(\frac{d}{dt}\right)^{\nu} L_j^{m,m}(t) = \frac{m-1}{\tau_{j+m} - \tau_j} \left\{ \left(\frac{d}{dt}\right)^{\nu-1} L_j^{m-1,m-1}(t) - \left(\frac{d}{dt}\right)^{\nu-1} L_{j+1}^{m-1,m-1}(t) \right\},$$
$$\nu = 1, 2, \cdots, m-1; m \geq 2. \tag{21}$$

3 Numerical optimization method

Problem 0 is reduced to the following problem by substituting the analytic solution presented in Theorem 1 for the function $\boldsymbol{r}$ being sought.

PROBLEM 2. Given the constants $\boldsymbol{r}_1$, $\boldsymbol{r}_2$, $\cdots$, $\boldsymbol{r}_K$, ℓ_1, ℓ_2, $\cdots$, ℓ_I, $\boldsymbol{r}_S$, $\boldsymbol{r}'_S$, $\boldsymbol{r}_E$, $\boldsymbol{r}'_E$, find $\boldsymbol{t} = (t_1, t_2, \cdots, t_K)$ in the feasible set

$$F = \{\boldsymbol{t} = (t_1, t_2, \cdots, t_K) \,|t_k < t_{k+1}, \ \ k = 0, 1, 2, \cdots, K\}, \ \ (t_0 = 0, t_{K+1} = T) \tag{22}$$

which minimizes the objective function

$$P(\boldsymbol{t}) = \mathrm{tr}(DGD^{\mathrm{T}}) \tag{23}$$

subject to (13), where

$$G = \begin{pmatrix} \int_0^T M_{-5}^{(3)}(t)M_{-5}^{(3)}(t)dt & \cdots & \int_0^T M_{-5}^{(3)}(t)M_{K+I}^{(3)}(t)dt \\ \int_0^T M_{-4}^{(3)}(t)M_{-5}^{(3)}(t)dt & \cdots & \int_0^T M_{-4}^{(3)}(t)M_{K+I}^{(3)}(t)dt \\ \vdots & & \vdots \\ \int_0^T M_{K+I}^{(3)}(t)M_{-5}^{(3)}(t)dt & \cdots & \int_0^T M_{K+I}^{(3)}(t)M_{K+I}^{(3)}(t)dt \end{pmatrix}. \tag{24}$$

THEOREM 2. *Problems 0 and 2 are equivalent.*

(Proof) Theorem 1 implies that if $\boldsymbol{r}$ is the solution for Problem 0, it is represented in the form of (11) and satisfies (13) which is equivalent to (1), (2), (3) and (7). Substitute (11) for $\boldsymbol{r}$ in (4). Then we have the functional rearranged as

$$\begin{aligned} P[\boldsymbol{r}] &= \int_0^T \left\| \sum_{j=-5}^{K+I} \boldsymbol{d}_j M_j^{(3)}(t) \right\|^2 dt \\ &= \operatorname{tr}\left(\sum_{i=-5}^{K+I} \sum_{j=-5}^{K+I} \boldsymbol{d}_i \int_0^T M_i^{(3)}(t) M_j^{(3)}(t) dt \boldsymbol{d}_j^{\mathrm{T}} \right) \\ &= \operatorname{tr}(DGD^{\mathrm{T}}) \\ &= P(\boldsymbol{t}), \end{aligned} \tag{25}$$

which is the objective function of Problem 2. So the solution $\boldsymbol{t}$ for Problem 2 is the same as the one that should have been already obtained by solving Problem 0.

Conversely, if the solution $\boldsymbol{t}$ for Problem 2 is obtained, the solution $\boldsymbol{r}$ for Problem 0 is determined by (13). ■

REMARK 3. The problem of optimizing the times $0 < t_1 < t_2 < \cdots < t_K < T$ for the interpolation of given spatial points is called the data parameterization problem in the general theory of splines. Flash and Hogan solved, in their experimental confirmation process of their model [2], a special case of the problem, where a single intermediate via point is allowed, by reducing the problem to a higher order algebraic equation. If we attacked the present case in this course, it would be required to construct and solve complicated simultaneous higher order algebraic equations because of the many via points involved. The case of cubic splines having simple knots was solved by Marin [8]. He provided an analytic solution for the one-dimensional data and an iterative optimization procedure for the multi-dimensional data. Unfortunately, those results are not applicable to the present case of extended quintic splines represented by B-splines having simple and double knots.

Problem 2 can be solved numerically by a constrained optimization method. The only trouble is how to compute the gradient

$$\text{grad}P = \left(\frac{\partial P}{\partial t_1}, \frac{\partial P}{\partial t_2}, \cdots, \frac{\partial P}{\partial t_K} \right). \tag{26}$$

However, the gradient is much complicated in the present case since the variables t_1, t_2, $\cdots$, t_K are tied to the knots τ_1, τ_2, $\cdots$, τ_{K+I} by (12) and because the knots determine the shape of B-splines. In order to compute $\frac{\partial P}{\partial t_k}$, we need the following five lemmas.

The first lemma writes down $\frac{\partial P}{\partial t_k}$ in terms of B and G and their partial differentials.

LEMMA 1.

$$\begin{aligned} \frac{\partial P}{\partial t_k} &= \text{tr}\left(-(XB^{-1}\frac{\partial B}{\partial t_k}B^{-1})G(XB^{-1})^{\mathrm{T}} + (XB^{-1})\frac{\partial G}{\partial t_k}(XB^{-1})^{\mathrm{T}} \right. \\ &\qquad \left. -(XB^{-1})G\left(XB^{-1}\frac{\partial B}{\partial t_k}B^{-1}\right)^{\mathrm{T}} \right), \quad k = 1, 2, \cdots, K. \end{aligned} \tag{27}$$

(See Appendix for a proof.)

Matrix G used in (27) is defined as (24) by means of integration, which causes troubles in its numerical evaluation. The following lemma resolves the integration to rewrite G as an arithmetic function of the B-splines and their derivatives.

LEMMA 2. The elements of matrix G can be rearranged as follows:

$$\begin{aligned} &\int_0^T M_i^{(3)}(t)M_j^{(3)}(t)dt \\ &= \left[M_i^{(3)}(t)M_j^{(2)}(t)\right]_0^T - \left[M_i^{(4)}(t)M_j^{(1)}(t)\right]_0^T + \left[M_i^{(5)}(t)M_j(t)\right]_0^T \\ &\quad -\Bigg\{ \sum_{\substack{0 \le p \le 6 \\ \text{and} \\ 0 < \tau_{i+p} < \tau_{i+p+1} \le T}} \left[M_i^{(5)}(\tau_{i+p}+0) - M_i^{(5)}(\tau_{i+p}-0)\right] M_j(\tau_{i+p}) \\ &\qquad - \sum_{\substack{0 \le p \le 6 \\ \text{and} \\ 0 < \tau_{i+p} = \tau_{i+p+1} < T}} \left[M_i^{(4)}(\tau_{i+p}+0) - M_j^{(4)}(\tau_{i+p}-0)\right] M_j^{(1)}(\tau_{i+p}) \Bigg\}. \end{aligned} \tag{28}$$

(See Appendix for a proof.)

The B-splines and their derivatives $M_j^{(\nu)}(t)$, $(\nu = 0, 1, 2, \cdots, 5)$ used in (28) to represent the elements of G can be recursively evaluated as $M_j^{(\nu)}(t) = (\frac{d}{dt})^\nu L_j^{6,6}(t)$ by (18), (19), (20) and (21). The same is true of B represented by (16). Its inverse B^{-1}

can be computed by Gaussian elimination since it follows from Theorem 1 that B is invertible.

In addition to B^{-1} and G, we have to evaluate $\frac{\partial B}{\partial t_k}$ and $\frac{\partial G}{\partial t_k}$ composing $\frac{\partial P}{\partial t_k}$ in (27). Then we need the following lemma which is obtained by differentiating the both sides of (21) with respect to t_k.

LEMMA 3.

$$\frac{\partial}{\partial t_k}\left\{\left(\frac{d}{dt}\right)^{\nu} L_j^{m,m}(t)\right\}$$
$$=\begin{cases} \frac{m-1}{(\tau_{j+m}-\tau_j)^2}\left\{\left(\frac{d}{dt}\right)^{\nu-1} L_j^{m-1,m-1}(t)-\left(\frac{d}{dt}\right)^{\nu-1} L_{j+1}^{m-1,m-1}(t)\right\} \\ \quad +\frac{m-1}{\tau_{j+m}-\tau_j}\left\{\frac{\partial}{\partial t_k}\left(\frac{d}{dt}\right)^{\nu-1} L_j^{m-1,m-1}(t)-\frac{\partial}{\partial t_k}\left(\frac{d}{dt}\right)^{\nu-1} L_{j+1}^{m-1,m-1}(t)\right\} \\ \qquad\qquad \text{if } t_k=\tau_j, \\ -\frac{m-1}{(\tau_{j+m}-\tau_j)^2}\left\{\left(\frac{d}{dt}\right)^{\nu-1} L_j^{m-1,m-1}(t)-\left(\frac{d}{dt}\right)^{\nu-1} L_{j+1}^{m-1,m-1}(t)\right\} \\ \quad +\frac{m-1}{\tau_{j+m}-\tau_j}\left\{\frac{\partial}{\partial t_k}\left(\frac{d}{dt}\right)^{\nu-1} L_j^{m-1,m-1}(t)-\frac{\partial}{\partial t_k}\left(\frac{d}{dt}\right)^{\nu-1} L_{j+1}^{m-1,m-1}(t)\right\} \\ \qquad\qquad \text{if } t_k=\tau_{j+m}, \\ \frac{m-1}{\tau_{j+m}-\tau_j}\left\{\frac{\partial}{\partial t_k}\left(\frac{d}{dt}\right)^{\nu-1} L_j^{m-1,m-1}(t)-\frac{\partial}{\partial t_k}\left(\frac{d}{dt}\right)^{\nu-1} L_{j+1}^{m-1,m-1}(t)\right\} \\ \qquad\qquad \text{otherwise,} \end{cases}$$
$$\nu=m-1,m-2,\cdots,2,1;\quad m\geq 2. \tag{29}$$

Recursive application of (29) to $M_j^{(\nu)}(t_p)=(\frac{d}{dt})^{\nu}L_j^{6,6}(t_p)$, $(p=0,1,2,\cdots,K+1)$ for the purpose of evaluating $\frac{\partial B}{\partial t_k}$ eventually requires $\frac{\partial}{\partial t_k}L_j^{6-\nu,6-\nu}(t_p)$. Differentiating the both sides of (19) and (20), with attention to which of τ_j, τ_{j+1}, τ_{j+n} and t_p may happen to be identical with the differentiating variable t_k, we have the following (30) and (31), respectively.

LEMMA 4.

$$\frac{\partial}{\partial t_k}L_j^{m,n}(t_p) = \begin{cases} \frac{1}{(\tau_{j+n}-\tau_j)^2}\left(L_{j+1}^{m,n-1}(t_p)-L_j^{m,n-1}(t_p)\right) \\ \quad +\frac{1}{\tau_{j+n}-\tau_j}\left(\frac{\partial}{\partial t_k}L_{j+1}^{m,n-1}(t_p)-\frac{\partial}{\partial t_k}L_j^{m,n-1}(t_p)\right) & \text{if } t_k=\tau_j, \\ \frac{-1}{(\tau_{j+n}-\tau_j)^2}\left(L_{j+1}^{m,n-1}(t_p)-L_j^{m,n-1}(t_p)\right) \\ \quad +\frac{1}{\tau_{j+n}-\tau_j}\left(\frac{\partial}{\partial t_k}L_{j+1}^{m,n-1}(t_p)-\frac{\partial}{\partial t_k}L_j^{m,n-1}(t_p)\right) & \text{if } t_k=\tau_{j+n}, \\ \frac{1}{\tau_{j+n}-\tau_j}\left(\frac{\partial}{\partial t_k}L_{j+1}^{m,n-1}(t_p)-\frac{\partial}{\partial t_k}L_j^{m,n-1}(t_p)\right) & \text{otherwise,} \end{cases}$$
$$n=2,3,\cdots,m, \tag{30}$$

$$\frac{\partial}{\partial t_k} L_j^{m,1}(t_p) = \begin{cases} (m-1)(m-2)(t_k - t_p)^{m-3} & \text{if } t_p < t_k = \tau_j = \tau_{j+1}, \\ -(m-1)(m-2)(\tau_j - t_p)^{m-3} & \text{if } t_p = t_k < \tau_j = \tau_{j+1}, \\ -(m-2)(\tau_{j+1} - \tau_j)^{m-3} & \text{if } t_p = t_k = \tau_j < \tau_{j+1}, \\ \frac{1}{\tau_{j+1}-\tau_j}(-(m-1)(\tau_{j+1} - t_p)^{m-2} & \\ \qquad +(m-1)(\tau_j - t_p)^{m-2}) & \text{if } t_p = t_k < \tau_j < \tau_{j+1}, \\ \frac{1}{(\tau_{j+1}-\tau_j)^2}((\tau_{j+1} - t_p)^{m-1} - (\tau_j - t_p)^{m-1}) & \\ \qquad -\frac{1}{\tau_{j+1}-\tau_j}(m-1)(\tau_j - t_p)^{m-2} & \text{if } t_p < \tau_j = t_k < \tau_{j+1}, \\ \frac{-1}{(\tau_{j+1}-\tau_j)^2}((\tau_{j+1} - t_p)^{m-1} - (\tau_j - t_p)^{m-1}) & \\ \qquad +\frac{1}{\tau_{j+1}-\tau_j}(m-1)(\tau_{j+1} - t_p)^{m-2} & \text{if } t_p < \tau_j < \tau_{j+1} = t_k, \\ (m-2)(\tau_{j+1} - \tau_j)^{m-3} & \text{if } t_p = \tau_j < \tau_{j+1} = t_k, \\ 0 & \text{otherwise,} \end{cases} \tag{31}$$

$$p = 0, 1, 2, \cdots, K+1; \quad k = 1, 2, \cdots, K.$$

Note that we can check the identity of the variables τ_j, τ_{j+1}, τ_{j+n}, t_p and t_k by comparing their values as in (30) and (31) since $t_1 < t_2 < \cdots < t_K$ are distinct, $\{\tau_1, \tau_2, \cdots, \tau_{K+I}\} = \{t_1, t_2, \cdots, t_K\}$ and $\{\tau_{-5}, \cdots, \tau_0, \tau_{K+I+1}, \cdots, \tau_{K+I+6}\} \cap \{t_1, t_2, \cdots, t_K\} = \phi$.

The above formulae are also applicable to the evaluation of $\frac{\partial G}{\partial t_k}$. Partial differentials of the components $M_j^{(\nu)}(0)$, $M_j^{(\nu)}(T)$ and $M_i^{(\nu)}(\tau_{i+p})$ used in (28) to represent elements of G can be reduced to $\frac{\partial}{\partial t_k} L_j^{6-\nu,6-\nu}(0)$, $\frac{\partial}{\partial t_k} L_j^{6-\nu,6-\nu}(T)$ and $\frac{\partial}{\partial t_k} L_i^{6-\nu,6-\nu}(\tau_{i+p})$ by (29). Then they can be evaluated by (30) and (31) because $\{0, \tau_1, \tau_2, \cdots, \tau_{K+I}, T\} = \{t_0, t_1, t_2, \cdots, t_K, t_{K+1}\}$. Partial differentials of the other components $M_i^{(4)}(\tau_{i+p} \pm 0)$ and $M_i^{(5)}(\tau_{i+p} \pm 0)$ in (28) can also be reduced by (29) to $\frac{\partial}{\partial t_k} L_i^{2,2}(\tau_{i+p} \pm 0)$ and $\frac{\partial}{\partial t_k} L_i^{1,1}(\tau_{i+p} \pm 0)$, respectively. Then their evaluation requires different formulae given in the following lemma, which are obtained by a straight forward calculation based on (19) and (20).

LEMMA 5.

$$\frac{\partial}{\partial t_k} L_i^{2,2}(\tau_q + 0) = \begin{cases} \frac{1}{(\tau_{i+2}-\tau_i)^2} & \text{if } t_k = \tau_i \le \tau_{i+1} = \tau_q < \tau_{i+2}, \\ \frac{-1}{(\tau_{i+2}-\tau_i)^2} & \text{if } \tau_i \le \tau_{i+1} = \tau_q < \tau_{i+2} = t_k, \\ 0 & \text{otherwise.} \end{cases} \tag{32}$$

$$\frac{\partial}{\partial t_k} L_i^{2,2}(\tau_q - 0) = \begin{cases} \frac{1}{(\tau_{i+2}-\tau_i)^2} & \text{if } t_k = \tau_i < \tau_{i+1} = \tau_q \le \tau_{i+2}, \\ \frac{-1}{(\tau_{i+2}-\tau_i)^2} & \text{if } \tau_i < \tau_{i+1} = \tau_q \le \tau_{i+2} = t_k, \\ 0 & \text{otherwise.} \end{cases} \tag{33}$$

$$\frac{\partial}{\partial t_k} L_i^{1,1}(\tau_q + 0) = \begin{cases} \frac{1}{(\tau_{i+1}-\tau_i)^2} & \text{if } t_k = \tau_q = \tau_i < \tau_{i+1}, \\ \frac{-1}{(\tau_{i+1}-\tau_i)^2} & \text{if } \tau_q = \tau_i < \tau_{i+1} = t_k, \\ 0 & \text{otherwise.} \end{cases} \tag{34}$$

$$\frac{\partial}{\partial t_k} L_i^{1,1}(\tau_q - 0) = \begin{cases} \frac{1}{(\tau_{i+1}-\tau_i)^2} & \text{if } t_k = \tau_i < \tau_q = \tau_{i+1}, \\ \frac{1}{(\tau_{i+1}-\tau_i)^2} & \text{if } \tau_i < \tau_q = \tau_{i+1} = t_k, \\ 0 & \text{otherwise.} \end{cases} \tag{35}$$

Combining the formulae listed in Lemmas 1 – 5, we can compute grad$P$ as asserted in the following theorem.

THEOREM 3. *The gradient* grad$P$ *of the objective function* $P(\boldsymbol{t})$ *can be evaluated arithmetically by the formulae in Lemmas 1 – 5.*

(Proof.) The partial differentials $\frac{\partial P}{\partial t_k}$, $(k = 1, 2, \cdots, K)$ constituting the gradient can be computed by (27) of Lemma 1 from B, G, $\frac{\partial B}{\partial t_k}$, $\frac{\partial G}{\partial t_k}$, and constant X therein. Recall that B is always invertible from Theorem 1.

The B-splines and their higher order derivatives composing matrix B can be evaluated by the recurrence formulae (19) and (20) on the basis of (18).

The element of G is represented by an arithmetic function of the B-splines and their higher order derivatives, and $M_i^{(4)}(\tau_q \pm 0)$ in the case $\tau_q = \tau_{q+1}$ and $M_i^{(5)}(\tau_q \pm 0)$ in the case $\tau_q < \tau_{q+1}$ as (28) of Lemma 2. The last two quantities can be arithmetically translated by (21) into $L_i^{2,2}(\tau_q \pm 0)$ and $L_i^{1,1}(\tau_q \pm 0)$, respectively, on the basis of (18).

To compute the partial differential $\frac{\partial B}{\partial t_k}$, we first use the recurrence formula (29) in Lemma 3 on the basis of (18) to translate the partial differentials of derivatives of B-splines into those of $L_j^{m,m}(t_p)$. Then another set of recurrence formulae (30) and (31) in Lemma 4 is applicable to compute $\frac{\partial}{\partial t_k} L_j^{m,m}(t_p)$.

Matrix $\frac{\partial G}{\partial t_k}$ can be evaluated in the same manner as $\frac{\partial B}{\partial t_k}$ except for $L_i^{2,2}(\tau_q \pm 0)$ and $L_i^{1,1}(\tau_q \pm 0)$. We use (32) – (35) in Lemma 5 to differentiate $L_i^{2,2}(\tau_q \pm 0)$ and $L_i^{1,1}(\tau_q \pm 0)$ with respect to t_k. ■

Now we have obtained a means to compute the gradient. In order for the convenience in practice, we shall modify Problem 2 without changing the practical meanings as follows:

PROBLEM 2'. Given the constants ε, $\boldsymbol{r}_1$, $\boldsymbol{r}_2$, $\cdots$, $\boldsymbol{r}_K$, ℓ_1, ℓ_2, $\cdots$, ℓ_I, $\boldsymbol{r}_S$, $\boldsymbol{r}'_S$, $\boldsymbol{r}_E$, $\boldsymbol{r}'_E$, find $\boldsymbol{t} = (t_1, t_2, \cdots, t_K)$ in the feasible set

$$F' = \{\boldsymbol{t} = (t_1, t_2, \cdots, t_K) \,|\, t_k + \varepsilon \le t_{k+1},\ k = 0, 1, 2, \cdots, K\}, \quad (t_0 = 0, t_{K+1} = T) \tag{36}$$

which *locally* minimizes the objective function $P(\boldsymbol{t})$ subject to (13).

REMARK 4. The original open feasible set F is modified to be a little smaller closed set F' by introducing the parameter $\varepsilon > 0$. This parameter means the limit beyond which t_k must not be close to its neighbors t_{k-1} and t_{k+1}. Since ε can be chosen arbitrarily small, there is virtually no difference between F and F' in practice.

REMARK 5. Problem 2' is a hunt for a *local* minimum instead of global ones, which is compelling with this kind of problems.

Incorporating the means to compute the gradient in the established conditional gradient method [9], we have the following algorithm.

ALGORITHM.

Step 0. Choose $\boldsymbol{t}^{(0)} = (t_1^{(0)}, t_2^{(0)}, \cdots, t_K^{(0)}) \in F'$ as the initial values. Choose an accuracy as $\theta > 0$. Set $i := 0$.

Step 1. Compute $\mathrm{grad}P = (\frac{\partial P}{\partial t_1}, \frac{\partial P}{\partial t_2}, \cdots, \frac{\partial P}{\partial t_K})$ at $\boldsymbol{t} = \boldsymbol{t}^{(i)}$ by (18), (19), (20), (21) and the formulae listed in Lemmas 1–5.

Step 2. Find a solution $\bar{\boldsymbol{t}}^{(i)}$ of the linear programming problem

$$\sum_{k=1}^{K} \left(\frac{\partial P(\mathbf{t})}{\partial t_k} \bigg|_{\mathbf{t}=\boldsymbol{t}^{(i)}} \right) (\bar{t}_k^{(i)} - t_k^{(i)}) \;\rightarrow\; \min, \;\bar{\boldsymbol{t}}^{(i)} = (\bar{t}_1^{(i)}, \bar{t}_2^{(i)}, \cdots, \bar{t}_K^{(i)}) \in F'.$$

Step 3. Compute h_i as

$$h_i = \left| \sum_{k=1}^{K} \left(\frac{\partial P(\mathbf{t})}{\partial t_k} \bigg|_{\mathbf{t}=\boldsymbol{t}^{(i)}} \right) (\bar{t}_k^{(i)} - t_k^{(i)}) \right|.$$

Step 4. If $h_i \leq \theta$ then terminate and $\boldsymbol{t}^{(i)}$ is the solution. Otherwise take a point $\boldsymbol{t}^{(i+1)} = (t_1^{(i+1)}, t_2^{(i+1)}, \cdots, t_K^{(i+1)}) \in F'$ that is constructed as

$$\boldsymbol{t}^{(i+1)} = \boldsymbol{t}^{(i)} + \alpha_i(\bar{\boldsymbol{t}}^{(i)} - \boldsymbol{t}^{(i)}), \;\alpha_i \in (0, 1]$$

such that $P(\boldsymbol{t}^{(i+1)}) < P(\boldsymbol{t}^{(i)})$. Set $i := i + 1$ and return to **Step 1**.

Convergence of the algorithm is ensured by the following theorem.

THEOREM 4. *The sequence* $\mathbf{t}^{(i)}$, $(i = 0, 1, 2, \cdots)$ *generated by the algorithm converges to the set of local minimizers* S *of Problem 2', i.e.*

$$\inf_{s \in S} \|\boldsymbol{t}^{(i)} - s\| \rightarrow 0 \text{ as } i \rightarrow \infty.$$

(Proof) A proof is immediate from Theorem 3.5 of [9] since the feasible set F' is closed, bounded and convex. ■

4 Application

Letters written by a volunteer on an electromagnetic tablet is shown in figure 2. Their characteristic points were extracted as via points (marked by $\times$) and pause points (marked by $\otimes$) where the pen speed was locally minimal and especially zero, respectively. The times when to visit those points were taken from the data. While the best pause points can apparently be chosen in this way, the via points may not be chosen at the best positions. Although mischosen via points would cause disastrous results in synthesis, the points in figure 2 were not so bad.

Letters reproduced as an extended quintic spline interpolation from the characteristic points, the times, and the boundary conditions are plotted in figure 3. Overall shapes as well as minute steeples are well reproduced. That is the result made possible solely by Theorem 1. Optimization of the times seems to have already been made by the human brain and hand.

On the other hand, synthesis of a string of letters needs the algorithm developed in this paper. Since the model of writing a letter should also be true of writing a string of letters in one continuous stroke, the pen motion to write the string in script style may be synthesized as the optimal function for the concatenation of characteristic points taken from individual letters. Good concatenated points can be built up by appending those for succeeding letters one by one on the right hand side with some spacing. Boundary conditions of the intermediate letters should be discarded because there are no boundaries between the connected letters. A good initial guess for the times to visit the concatenated points might be the accumulation of the time intervals between adjacent points. Several strings of letters synthesized in that way are plotted as light curves in figure 4. These light curves deviate from what we usually write, probably because the initial guess for the times is not right. The dark curves in figure 4 represent synthesized strings with the times optimized by the present algorithm. They look more similar to our handwriting.

This technique can be used for enhancing the impression of computer-printed documents such as personal invitation letters or greeting cards. Extension of the present technique to other calligraphic writing such as Chinese brush-writing requires a three-dimensional model because the pen-up motion between the strokes on the paper surface is as relevant as the pen-down motion.

Acknowledgment

The anonymous reviewer gave many valuable comments that greatly improved the presentation of this paper. The possible extension to include the pen-up motion just mentioned was suggested by the same reviewer.

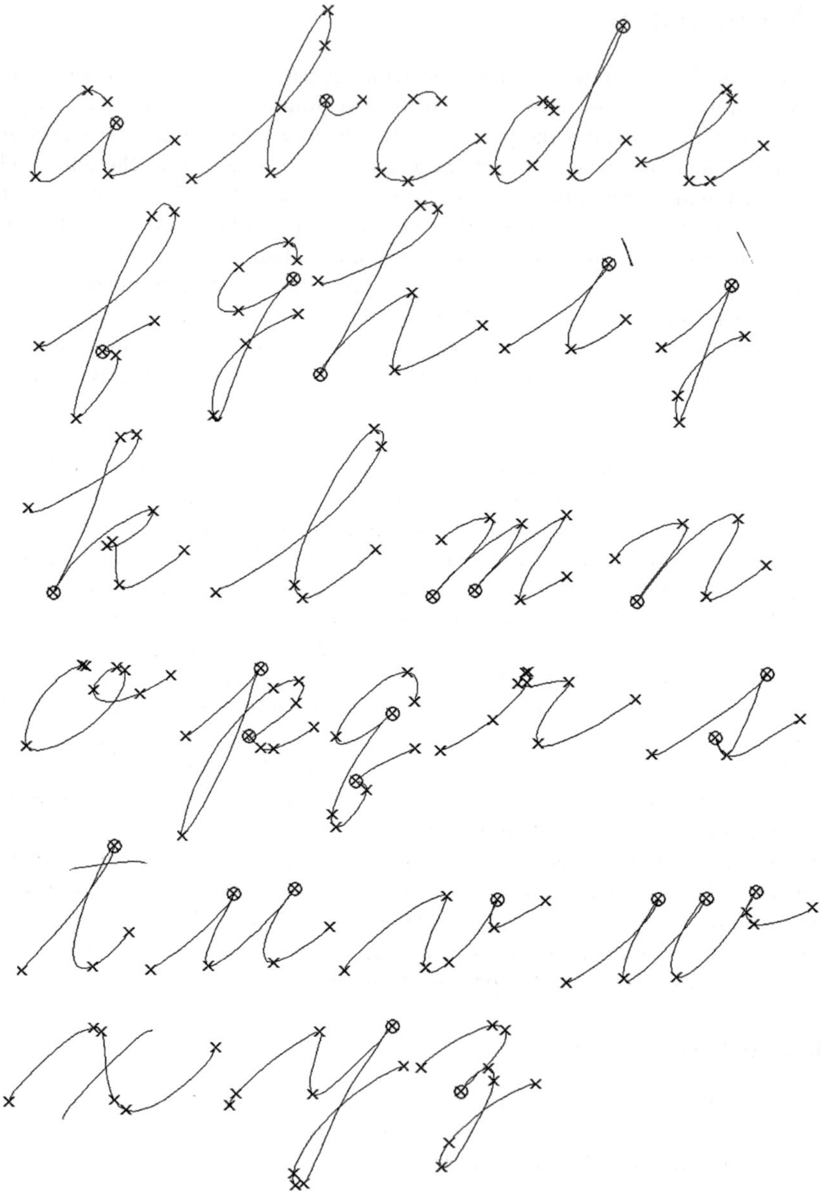

Figure 2. Letters written by a volunteer ($\times$ denotes a via point, especially $\otimes$ denote a pause point).

Figure 3. Reproduction of letters in figure 2.

Figure 4. Strings of letters synthesized on the basis of individual letters in figure 2 with the initial guess of the times to visit the concatenated characteristic points (light curves) and with the optimized times (dark curves).

References

[1] E. H. Dooijes (1989), "Rendition of quasi-calligraphic script defined by pen trajectory," In J. André and R. D. Hersch (eds.) *Proc. Int'l. Conf. on Raster Imaging and Digital Typography,* Cambridge Univ., 521-260.

[2] T. Flash and N. Hogan (1985), "The coordination of arm movements: an experimentally confirmed mathematical model," *Journal of Neurosciences,* vol.5, 1688-1703.

[3] M. Kamada (2000), "On the optimality of splines for expressing the handwriting process," (invited paper), *Proceedings of the X European Signal Processing Conference,* CD-ROM, FriAmSS1-4.

[4] W. Abend, E. Bizzi and P. Morasso (1982), "Human arm trajectory formation," *Brain,* vol.105, 331-348.

[5] H. B. Curry and I. J. Schoenberg (1966), "On Pólya frequency functions IV: The fundamental spline functions and their limits," *Journal d'Analyse Mathematique,* vol.17, 71-107.

[6] C. de Boor (1963), "Best approximation properties of spline functions of odd degree," *Journal of Mathematics and Mechanics,* vol.12, 747-750.

[7] I. J. Schoenberg (1964), "On interpolation by spline functions and its minimal properties," *Internat. Schriftenreihe Numer. Math.,* vol.5, 109-129.

[8] S. P. Marin (1984), "An approach to data parameterization in parametric cubic spline interpolation problems," *J. Approx. Theory,* vol.41, 64-86.

[9] O. V. Vasiliev (1996), *Optimization Methods,* Advanced series in mathematical science and engineering, World Federation, Atlanta.

Appendix

Proof of Lemma 1. From (23), we have

$$\frac{\partial P}{\partial t_k} = \frac{\partial P}{\partial t_k}\mathrm{tr}(DGD^{\mathrm{T}}) = \mathrm{tr}\left(\frac{\partial D}{\partial t_k}GD^{\mathrm{T}} + D\frac{\partial G}{\partial t_k}D^{\mathrm{T}} + DG\left(\frac{\partial D}{\partial t_k}\right)^{\mathrm{T}}\right). \tag{A.1}$$

Differentiating the both sides of (13), we have

$$\frac{\partial D}{\partial t_k}B + D\frac{\partial B}{\partial t_k} = O,$$

which implies

$$\frac{\partial D}{\partial t_k} = -D\frac{\partial B}{\partial t_k}B^{-1} = -XB^{-1}\frac{\partial B}{\partial t_k}B^{-1}. \tag{A.2}$$

Equations (A.1) and (A.2) give (27). ■

Proof of Lemma 2. Three fold partial integration of an element of G is

$$\int_0^T M_i^{(3)}(t)M_j^{(3)}(t)dt$$
$$= \left[M_i^{(3)}(t)M_j^{(2)}(t)\right]_0^T - \left[M_i^{(4)}(t)M_j^{(1)}(t)\right]_0^T + \left[M_i^{(5)}(t)M_j(t)\right]_0^T - \int_0^T M_i^{(6)}(t)M_j(t)dt. \tag{A.3}$$

The integrand of the last term in (A.3) includes the sixth-order derivative $M_i^{(6)}(t)$ of the quintic B-spline $M_i(t)$. The fourth-order derivative $M_i^{(4)}(t)$ of $M_i(t) = L_i^{6,6}(t)$ is translated by (21) into a weighted sum of $L_i^{2,2}(t)$, which is further translated by (19) into a weighted sum of $L_i^{2,1}(t)$. Equation (20) implies that $L_i^{2,1}(t)$ is a continuous piecewise linear function which switches at τ_i and τ_{i+1} in the case of simple knots $\tau_i < \tau_{i+1}$ and that $L_i^{2,1}(t)$ is a step function which is discontinuous at τ_i in the case of double knot $\tau_i = \tau_{i+1}$. Consequently, $M_i^{(4)}(t)$ is a piecewise linear function which is switching continuously at simple knots and discontinuously at double knots. Then the fifth-order derivative $M_i^{(5)}(t)$ is a sum of a piecewise constant function, which is discontinuous at simple knots, and delta functions δ at double knots. Finally, the sixth-order derivative $M_i^{(6)}(t)$ is a sum of the delta functions at simple knots and derivatives of delta function at double knots as follows:

$$\begin{aligned} M_i^{(6)}(t) \;=\; & \sum_{\substack{0 \le p \le 6 \\ \text{and} \\ \tau_{i+p} < \tau_{i+p+1}}} \left[M_i^{(5)}(\tau_{i+p}+0) - M_i^{(5)}(\tau_{i+p}-0)\right]\delta(\tau_{i+p}-t) \\ & + \sum_{\substack{0 \le p \le 6 \\ \text{and} \\ \tau_{i+p} = \tau_{i+p+1}}} \left[M_i^{(4)}(\tau_{i+p+1}+0) - M_i^{(4)}(\tau_{i+p}-0)\right]\delta^{(1)}(\tau_{i+p}-t). \end{aligned} \tag{A.4}$$

The coefficients for δ and $\delta^{(1)}$ in (A.4) are the difference of the function values just before and after the corresponding discontinuities. Then we can resolve the integral as follows:

$$\int_0^T M_i^{(6)}(t)M_j(t)dt \;=\; \sum_{\substack{0 \le p \le 6 \\ \text{and} \\ 0 < \tau_{i+p} < \tau_{i+p+1} \le T}} \left[M_i^{(5)}(\tau_{i+p}+0) - M_i^{(5)}(\tau_{i+p}-0)\right] M_j(\tau_{i+p})$$

$$- \sum_{\substack{0 \le p \le 6 \\ \text{and} \\ 0 < \tau_{i+p} = \tau_{i+p+1} < T}} \left[M_i^{(4)}(\tau_{i+p+1} + 0) - M_i^{(4)}(\tau_{i+p} - 0) \right] M_j^{(1)}(\tau_{i+p}) \tag{A.5}$$

by the property

$$\int_0^T f(t)\delta(\tau - t)dt = f(\tau) \quad \text{and} \quad \int_0^T f(t)\delta^{(1)}(\tau - t)dt = -f^{(1)}(\tau)$$

of the delta function for $0 < \tau < T$ and twice differentiable f. Equations (A.3) and (A.5) give (28). ■

Optimization and Optimal Control, pp. 141 – 153
P.M. Pardalos, I. Tseveendorj, and R. Enkhbat, Editors

The Role of Optimization in Economics [1]

Donald W. Katzner (dkatzner@econs.umass.edu)
University of Massachusetts
Amherst, USA

Keywords: optimization in economics

As a discipline, Economics is primarily concerned with the allocation to alternative employments of scarce resources that have various possible uses. How, it asks, are the finite amounts of land, labor, and capital that are available to a society apportioned to individual and government activities, and to the production of the various products that benefit individuals and society at large? How do those allocations modify over time? And how might allocations that are thought to be inappropriate or unsatisfactory be changed? Because the quantities of resources are limited to the extent that all needs and desires cannot be met, it is clear that choices have to be made. And therefore, much of economics is concerned with the analysis and consequences of the making of choices or, in the jargon of the discipline, the making of decisions.

Of the many analytical tools that could be brought to bear on the decision problem, the most prominent has been optimization. For almost 200 years, economists have been concerned to find, in a substantial number of contexts, the optimum solution values of vectors of decision variables, thereby discovering the attainable extrema of certain appropriately defined objective functions. The logic behind that approach is derived from the idea that the "best" alternative from any collection of available options ought to be pursued, and hence the necessity arises of making the optimal choice decision under the circumstances presented. In all of this, economists are conscious of the fact that most economic choice decisions are inevitably constrained decisions. The finiteness of resources imposes constraints at all analytical levels, from individual consumers and firms, to industries and markets, to the economy as a whole. Such constraints may be either implicit or explicit, thus affecting the nature of the mathematics required.

[1]Thanks are due to Rob Kusner and Douglas Vickers for their help in the preparation of this paper.

Indeed, both unconstrained or, respectively, constrained optimization techniques in relation to first- and second-order maximization or minimization conditions have been employed in the past. Although the appearance of optimization in economics has not, as will be indicated below, gone unchallenged, it nevertheless retains a central place in the discipline to this day.

The exact moment at which economists became aware, either explicitly or implicitly, of the idea of optimization and its relevance to Economics is not clear. But it is possible to pinpoint what may be the first appearance of optimization as a mathematical issue in the Economics literature. That initial taste, and it was, indeed, a very small sip, was a fleeting one.

For in this instance, mathematics relating only peripherally to the optimization problem seems to have crept into Economics surreptitiously through the back door and was thereafter quickly expelled. The story, according to Groenewegen [6, pp. vii-viii, xxvi-xxvii] and Robertson [16, pp. 526-527], is as follows: In 1771, an Italian economist by the name of Pietro Verri published a book whose title translated into English is *Reflections on Political Economy.* The book immediately met with some success, and its initial edition was reprinted several times with various and differing annotations into 1772. One of the reprinted versions (the sixth) contained some mathematical commentary attributed to Verri's friend and mathematician, Paolo Frisi. But a subsequent, and what was considered to be the definitive edition of the book that was published in 1781 deleted, among other things, Frisi's commentary.

Two parts of Frisi's commentary are of interest here. First, early in his book, Verri [21, p. 19] argued, [2] without using mathematical symbolism, that "... the price of things will be in direct proportion to the number of buyers and in inverse proportion to the number of sellers." Verri then went on to suggest that the ends of the state could best be served by keeping the price as low as possible [21, p.21]. In his commentary, Frisi approximated Verri's proportionality statement by the equation $P = C/V$, where P represented price, C the number of buyers, and V the number of sellers. To examine the properties of Verri's "minimum" price value, should the state be able to maintain it, Frisi then expressed that equation in differential form and equated the price differential dP to zero. This led him to the conclusion that the minimum (or, for that matter, any value of P) could be maintained by offsetting the changes in one of C or V by changes in the other so as to keep the ratio required for achieving the minimum (or, respectively, another value of P) intact. [3]

Later on in his book, Verri discusses the problem of obtaining maximum production with minimum labor [21, p. 68]. Frisi's commentary here asserts that this

---

[2]It should be noted that the following references to and quotation from Verri's work are based on the translation of the 1781 edition, which is not the same as the sixth version on which Frisi based his commentary. However, it is safe to assume that the material in question is similar in spirit, if not identical in wording, to that to which Frisi was responding.

[3]That is, $dP = [VdC - CdV]/V^2 = 0$, from which it follows that $dC/dV = C/V$. See Robertson [16, p. 526].

problem, when reduced to analytical terms, is an isoperimetric problem – that is, it is mathematically equivalent to the problem of finding the maximum area that can be enclosed within a perimeter of fixed length. Although Frisi provides some mathematical discussion relating to the problem as set in the context posed by Verri, he does not solve it. Nevertheless, this commentary, together with that described above, seem to be the earliest mathematical references to optimization in Economics.

The broader historical context concerning the development of the mathematics of optimization at the time Verri's book and Frisi's commentary were published is worth noting. First, the use of the calculus to determine the values at which functions have maxima and minima had been worked out by Fermat 100 years earlier in the middle of the 17th century. [4] Second, the isoperimetric problem, which was related to the Principle of Least Action (the doctrine dating to antiquity that nature acts in the shortest, most efficient way) and Fermat's Principle of Least Time (that light only travels along paths that take the least time), had recently been solved by Lagrange in the 1750's. In so doing, Lagrange created a general technique for dealing with such matters to which Euler subsequently gave the name "calculus of variations."[5] Because of the likelihood that there was much discussion of Lagrange's "new" method among mathematicians at the time, it is not surprising that Frisi chose to point out in his commentary the implicit presence of the isoperimetric problem in Verri's book. Third, around or shortly after the time that the first edition of Verri's book was being reprinted with annotations, Lagrange was developing what later came to be known as his multiplier method for determining the critical values of functions that have maximum and minimum values under constraints.[6] But, as will be pointed out momentarily, the mathematics of constrained optimization did not enter the Economics literature for another 100 years.

It is possible that Frisi's use of mathematics to express indirectly the idea of optimization — short-lived in print though it was — inspired other, more direct, mathematical discussions of optimization in Economics.[7] Be that as it may, in 1815, Georg von Buquoy, a German economist, gave an analysis of profit maximization, with respect to the depth to which, when planting crops, fields are plowed, that was remarkably prescient of what was to come. Robertson [16, p. 527]provided a translation:[8]

> "Let the law according to which the value of the crop $[R]$ depends upon the depth of plowing $[x]$ be given by the equation $R = f(x)$, and let the law according to which costs $[C]$ depend upon the depth of plowing be expressed by the equation $C = g(x)$. The net revenue can generally be expressed at any depth by $[R - C =]\ f(x) - g(x)$, from which that value

[4]See, for example, Kline [14, pp. 347-348].
[5]*Ibid.*, pp. 580-583.
[6]Grattan-Guinness [5, pp. 326-327].
[7]See Groenewegen [6, p. xxvii].
[8]The mathematical notation employed below has been modified slightly from the original.

> [of x] can be found at which the net revenue is a maximum. And this maximum will be at that value of x at which the first derived function of $f(x) - g(x)$ disappears and at which simultaneously the second derived function becomes negative."

Although it is expressed in terms of plowing depth, this, of course, is the marginal-cost-equals-marginal-revenue statement of the necessary first-order condition for "profit" maximization, along with a statement of the second-order condition sufficient to ensure a maximum at the critical point.

It was the French economist Agustin Cournot [3] who, in 1838, placed the mathematics of optimization in many of the economic contexts in which it appears today. Profit was the difference between revenue and costs, the latter both expressed as functions of output, and upon equating its derivative to zero, Cournot derived the marginal-cost-equals-marginal-revenue standard that attends virtually all of today's introductory and intermediate text books. He analyzed profit maximization under perfect competition (Ch. 8) and monopoly (Ch. 5). He developed the analysis of a market with two competing firms (duopoly) who are simultaneously maximizing their own profit under the assumption that the competing firm's output remains fixed (Ch. 7). And he even explored the meaning of revenue maximization along a demand curve in terms of first- and second-order maximization conditions (Ch. 4). In all of this Cournot gives extensive accounts of the derivations and implications of first- and second-order conditions. It is with Cournot that optimization and the mathematics associated with optimization became an integral part of economic analysis.

However, up to the middle of the 19th century, the constraints imposed by the finiteness of resources remained implicit in mathematical discussions of optimization in economics. Cournot's presentations in particular did not require explicit account be taken of them. The first appearance of explicit constraints in relation to optimization occurred in 1854 with the verbal statement by Hermann H. Gossen [4, pp. 108-109], another German economist, of the first-order condition resulting from constrained maximization of utility (pleasure) subject to the (linear) budget constraint:

> "Man obtains the maximum of life pleasure if he allocates all his earned money ... between the various pleasures and determines the ... [proportion of income for a specific pleasure] in such a manner that the last atom of money spent for each pleasure offers the same amount of pleasure."

Then in 1871 the English economist, W. Stanley Jevons [8, pp. 99-100] wrote down the first-order mathematical equations for a slightly different maximization problem and, in Switzerland in 1874, Léon Walras [22, p. 127] spelled out both first- and second-order conditions for, among other things, that same problem.[9]. Thus 100

[9] The page references in this Jevons citation are to the fifth (1957) edition of his *The Theory of Political Economy*; those to Walras' *Elements of Pure Economics* are to the English translation of the 1926 edition

years after Frisi, constrained optimization finally entered the Economics literature in a formal, mathematical way. (In what follows, the tern "optimization" will refer to optimization in both constrained and unconstrained formats.)

Regardless, during the more than 160 years since Cournot, the use of optimization and the mathematics of optimization in economics exploded.

Today optimization and the mathematics of optimization appear almost everywhere across the discipline. At least three broad categories of applications may be identified. First, optimization has become the cornerstone of models that purport to explain the behavior of individual consumers and firms, and sometimes even collections of consumers and collections of firms. In this category fall such applications (some with explicit constraints and some without) as utility maximization, expenditure minimization, cost minimization, profit maximization, revenue maximization, maximization of the rate of return on invested capital, and the determination of various solutions, characterized in terms of optimization criteria, of a wide variety of games.

Second, optimization has been employed to define an ideal that a collective or a society might want to attain.[10] Now societies are distinct from individuals. They are made up of individuals and have an aggregative quality that individuals lack. Societies have priorities and motivations that are generally different from those of individuals, and their behaviors fall in different spheres of activity. Explanations of the behaviors of individuals and societies, therefore, must, of necessity, deviate from each other. The optimization ideals that a society might pursue include Pareto optimality (a circumstance in which resources are fully employed and in which it is impossible to make one person better off without making someone else worse off), welfare maximization, maximization of the value of national output, optimal taxation, and optimal economic growth.

Third, optimization is used as a convenience. For example, it may be possible to frame propositions more economically by drawing on a "maximum principle."[11] Thus a representative, maximizing consumer might serve as the basis for formulating market demand or aggregate consumption functions. And it is often advantageous to establish the existence of "equilibrium" in various forms — such as, say, the existence of equilibrium in a perfectly competitive economy — by placing that equilibrium in the context of an optimization problem. In all three categories the assumptions and conditions under which optimization is pursued vary enormously. Thus optimization can be applied in a static (timeless) or dynamic framework; it can determine the choice of a vector, a function, or a path; it may involve a differentiable or only a continuous objective function; and it may result in interior or boundary solutions.

The extensive appearance of optimization in Economics raises two central questions that will be addressed in the remainder of this paper. On the one hand, why has

[10] Of course, models intended to explain behavior may also be taken to define ideal behavior. But their primary purpose remains explanation.

[11] Samuelson [17, p. 251].

optimization become so central to economic theorizing? On the other, is optimization likely to remain so in the future? To answer these questions requires separate consideration of each of the three categories of optimization applications described above. The former question will be discussed first; the latter will be taken up at the end of this essay.

In reference to the question of why optimization has become fundamental in economic analysis, the last of the above categories of applications of optimization may be disposed of rather easily. For the applications that fall in this category are all applications of economical or mathematical expediency. That is, optimization is applied in such cases because it provides a simple, direct, or otherwise appropriate method of demonstrating a desired result. In terms of one of the examples cited above, to prove the existence of equilibrium in a competitive economy generally requires the use of a fixed point theorem. But under certain conditions, that proof can be reduced to the application of a maximization problem, namely, the maximization of a "community utility" function subject to the economy-wide "transformation" function,[12] and the existence of the equilibrium falls out of that problem upon invoking the Weierstrass Theorem that a continuous function defined on a compact set always achieves a maximum (and minimum) value on that set. Not only is the proof of existence greatly simplified by going this route, but the proof itself can be illustrated geometrically in an intuitive and suggestive way. A more general proof that equates the competitive equilibrium to a Nash equilibrium in a noncooperative game has been given by Arrow and Debreu [1]. The latter, of course, is defined in relation to a specific optimization problem [1, p. 273].

Consider, now, the first of the above categories in which optimization figures centrally in models that are constructed to explain the behavior of individual economic agents (consumers or firms). These models all attempt to clarify what is happening in the real world. That is, they purport to give relatively accurate and seemingly satisfactory reasons why certain behavior has been or might be observed.[13] This, in turn, implies that there must be acceptable accordances between any particular model itself and the reality that is to be explained. For it is the presence of such accordances that permits the investigator to conclude that the model is a credible and appropriate reflection of what is actually going on. One way to establish the presence of acceptable accordances is by subjecting the observable parts of the model to empirical tests.

[12]See Katzner [9, Sect. 8.4]. The transformation function indicates the maximum amount of a consumption or investment commodity that can be produced by the economy, given the levels of production of all other consumer and investment goods and the available factor supplies. It is implicit in this definition that all production takes place as efficiently as possible.

[13]But it does not follow that any such model is capable of immediate concrete application to specific real-world situations. To confuse a model of a person or a firm with, respectively, an actual person or an actual firm, is to commit what Machlup called the "fallacy of misplaced concreteness." According to Machlup [15, p. 9], the fallacy occurs when theoretical symbols are used "... as though they had a direct, observable, concrete meaning." What a model does do, of course, is to establish principles and modes of causation that apply only in very broad and general ways.

Not all parts of all models in Economics relate to variables that are observable. Some things cannot be seen. Thus Economists expend considerable energy deriving properties in relation to observable variables that follow from assumptions made with respect to unobservable variables. It is the former properties that can be subjected to tests against data that have been recorded in reality.

Optimization fits into explanatory models in that it is usually reflective of what is thought to be the driving force behind the behavior of agents in them. One common motive that is often seen as important in directing behavior in reality is self interest. Self interest focuses an agent's attention on the achievement of a goal that, in the eyes of the agent, lies in the interest of that agent. Rational behavior is the manifestation of general goal-oriented (not necessarily self-interested) behavior in a model.

In the case of the Economist's theory of consumer behavior, for example, it is assumed that the consumer has fixed preferences among vectors of quantities of commodities, and that these preferences are represented by a utility function (which is, in this case, the objective function) with appropriate properties defined over the space of all nonnegative commodity vectors. The consumer is then assumed to be rational in the particular sense that, in making purchases, he always chooses that basket that maximizes utility subject to the limitations imposed by the prices of commodities (assumed to be given) and the amount of money he has to spend (also assumed given). It is the optimization, then, that translates the consumer's preferences into buying action. Once translated, the characteristics of the consumer's buying behavior, which are observable and which emerge from the properties of his fixed preferences or utility function through the maximization, can be determined and subjected to empirical tests. Clearly, optimization lies at the very core of this explanation of consumer behavior. It is no less than the representation in the model of what is thought to be the fundamental propellent that determines the consumer's choice decisions and brings about his purchases. Optimization plays a similar role in most explanatory models in Economics.

But there are problems relating to the use of optimization in these models. To illustrate them, attention will focus on the model of consumer behavior described above. Also, although they are distinct concepts and may be interpreted in different ways, it will be convenient in what follows to conflate the meaning of the terms "self interest," "rationality," and "optimization" so that the latter two both refer to constrained utility maximization and are taken to be the representation of self interest in the model.

The first difficulty arises with respect to the establishment of acceptable accordances between the model and reality through empirical testing of the observable parts of the model. To conduct such tests, it is necessary to observe the buying behavior of the individual at different times when the prices he faces and/or the amount of money he has to spend have changed. The rate of response of observed quantities purchased to changes in any price or alterations of funds available may be characterized in terms of appropriate partial derivatives, for example, $\partial q/\partial p$ or $\partial q/\partial m$, where q, p, and m, refer, respectively, to the quantity of the commodity purchased, its price, and the money available to spend. Or alternatively, the proportional change in quantity purchased, $\partial q/q$, may be expressed as a ratio of the proportional change in price, $\partial p/p$, to provide what is referred to as the "price-elasticity of demand," $(\partial q/\partial p)(p/q)$. A corresponding expression may be derived for the "funds-available" or "income elasticity of demand." For any test to be successful, the behavior so observed, described, in part, in the language of partial derivatives or elasticities, has to be consistent with the corresponding characteristics predicted by the model, given the fixed preferences of the consumer in it. But the fixed preferences of the model are a depiction of predilections in the mind of the consumer, and the economist who is testing the model has no control over them. Thus any collection of purchase data that contradicts those predicted characteristics, and which, therefore, would cast doubt on the relevance of the model, can always be explained away as a consequence of changing preferences in the consumer's mind that occurred in the interval between the observation of one data point and another, and that could not be controlled.

It follows that the model cannot be falsified and remains viable as a potential explanation of consumer behavior regardless of whether the observations are consistent with it or not, and hence the ability to establish acceptable accordances through empirical testing breaks down.

Where, then, do the acceptable accordances come from? Elsewhere [11] it has been suggested that these are derived in part from a judgment by the investigator that the assumption content of the model is consistent with the dominant features of the cultural heritage of the particular consumer in question. The argument may be briefly summarized as follows: Mental processes are heavily influenced by the cultural milieu in which they take place [13]. In any culture, some cultural traits are dominant over others. Since the purchases of a consumer are determined through mental processes, it is both reasonable and realistic to abstract from what are thought to be less important cultural characteristics and focus on the dominant traits of the cultural heritage of that individual in making assumptions about what motivates those purchases. In the case of the Western world, the relevant dominant trait is self interest. That trait, according to Weber [23], was institutionalized as a central tenet of Western culture during the 17th century through the dogmas of certain religious texts. Self interest as a motive remains strong and dominant today — especially in America. It is therefore thought to be appropriate and accurate to explain consumer purchasing behavior in Western societies in terms of constrained utility maximization that reflects self interest.

It follows from this argument, of course, that in societies where optimization is irrelevant or inappropriate as a representation of the existing dominant cultural traits, it should not figure into explanations of consumer buying behavior. This, it has been suggested [11], is the case with Japan. According to Bellah [2], the dominant traits of Japanese culture (these were institutionalized during the Tokugawa era from 1600 to 1868) include loyalty, the necessity to fulfill obligation, and frugality. But self interest and other motives that might be represented in terms of an optimization problem are not among them. By previous argument, then, the model of constrained utility maximization can not be used successfully to explain consumer behavior in Japan. The same is true of most explanatory models that employ optimization techniques when attempting to explain the economic behavior of Japanese agents in other contexts and in general. Additional examples of societies in which deviations from self-interested behavior seem to occur have been found by Henrich *et al* [7].

For the most part, modern explanatory Economics was developed by Western economists to explain Western economic behavior. It is not surprising, then, that there are not many explanatory models in the Economics literature that represent the driving force behind human behavior by something other than an optimization.[14] One could still argue, however, that all economic behavior (both Western and non-Western) is explainable in terms of some form of optimization as long as the objective function is appropriately modified from its Western specification to fit the situation at hand. Indeed, a mathematical theorem of this sort has been proved by Katzner [12] within the framework of the theory of choice. But applying optimization techniques as representative of the motivating propellant of human behavior when optimization cannot represent the actual force that brings about that behavior is to produce inaccurate and incorrect explanation. And this can be dangerous not only because it leads to misunderstanding, but also because economists often use their explanatory models as the basis for policy recommendations. Hence the use of optimization models where unwarranted can actually lead to considerable harm.

Returning to the original question, then, one may conclude that optimization has become so important in economic explanations of individual behavior not because it has been empirically established that it is universally the engine that drives that behavior but, rather, because it is a consequence of the peculiarities of Western culture and the fact that models using optimization were created primarily to explain behavior in that environment.

Another difficulty, actually an entire class of difficulties, relating to the application of optimization in explanatory economic models has more to do with the form in which optimization is applied than with the application of optimization itself. For, in the absence of the ability to verify empirically that any particular assumption of optimization is appropriate, and the consequent reliance on the judgment of the investigator as to whether there are the requisite accordances between a model and the reality it is intended to explain, certain aspects of some applications of optimization in Economics have come under heavy criticism. These criticisms frequently have to do with whether or not the individual or individuals whose behavior is in question are actually able to know or obtain or enough information to be able to optimize in the manner spelled out in the model. Two examples of such criticisms will be detailed here, each drawing attention to a perceived weakness in the utility maximization model of consumer behavior.

[14] A few examples have been provided by Katzner [12].

Simon [20] has suggested that to be able to maximize utility as described above, the consumer has to have complete knowledge of the commodities that might make up each purchase and complete knowledge of his preferences for all possible purchases. For only then can the true maximizing choice be determined. But in most situations, the number of options is so great and the amount of information so vast, that such knowledge is beyond the capacity of the consumer to secure and manage. Faced with this difficulty, consumers do the best that they can by choosing a vector that "satisfices" or that seems to be "good enough." Thus, to restore its optimizing character, the model of consumer behavior needs to be modified by introducing certain limits on the numbers of goods and the numbers of different possible quantities of goods that can be purchased, and that have heretofore been absent.

Shackle [18,19] extends this lack of information to cover much greater territory. Among other things, and remaining in the context of the utility-maximization model, he is concerned about the time that elapses between the moment a decision to purchase is made and the consumption of that purchase which necessarily occurs later. Since the future is unknown and unknowable, the individual cannot know what his circumstances will be and what he himself will be like at the time of consumption, and hence he cannot know what his preferences will be at that moment. Something bought today may turn out to be despised or useless when the time comes to consume it. And restricting the numbers of goods that can be purchased and their possible quantities is not enough to provide an adequate explanation of purchasing behavior. Nor is the introduction of probabilities in the context of mathematical expectation, the theory of games, or some other formal structure, very satisfactory since the individual does not have enough information at his disposal to determine those probabilities in a meaningful way either. Rather, a complete restructuring of the decision model is required in which nonprobabilistic uncertainty and the consequent ignorance in which the individual decides plays a major role. Shackle [18] proposed such model in which a concept called "potential surprise" replaces probability and in which optimization occurs in reference to the maximization of what are referred to as "attractiveness" and "decision index" functions.[15]

Turning to the second category of optimizations, every element of which characterizes an ideal that might in some way be considered desirable by a collective or a society, it is clear that the optimizations involved do not themselves need to represent something present in reality. Each model, of course, has to have appropriate associations with the real world with respect to part or all of its structural components. But the optimization itself appears only as the representation of an end towards which policy makers, say, are striving or that investigators are studying. Because societies are comprised of individuals, such ends or ideals are necessarily amalgams of individual thought forms. Those amalgams are constrained by institutions and, in the process of formation, are buffeted by pressures from various sources. The latter include, in part, cultural habits and political forces. Cultural imperatives are clearly significant

[15] An extensive discussion of Shackle's approach and his decision model appears in Katzner [10].

because, in addition to what has been said of them earlier, they largely determine what is considered to be important by the individuals in the society whose economy is at issue, and are therefore crucial inputs to the crucible out of which society's ends flow. In American society, for example, where self interest dominates, and where success as reflected in accumulated wealth and consumption is highly respected, the emergence of the maximization of national income, economic growth, and wealth, as desirable ends is quite natural.

With respect to the study of ends by investigators, any end related to society's ideals or not could be employed as the basis for a standard of comparison against which an analyst might contrast actual or potential situations. But, of course, the choice of that end, if it is to reflect the values of the society under investigation, has to be sensitive to that society's cultural heritage. Regardless, in light of the role that optimization plays in explanatory models, and taking into account both the prevalence in Western societies of cultural predilections that give rise to the employment of optimization, and the fact that much of current Economics was developed in the context of and to apply to Western societies, the widespread use of optimization to define desirable ends is not surprising. Thus, in addition to having been the main impetus for the extensive utilization of optimization in Economics in explanations of individual behavior, Western cultural forces have also been responsible to a considerable degree for the frequent appearance of optimization in the economic characterization of collective or societal ideals.

With regard to the future of optimization in Economics, it is clear that as long as optimization remains a convenient mathematical tool in, to cite a previously discussed example, establishing the existence of equilibrium in economic models, optimization will continue to be used in such contexts. Optimization is also likely to remain important in explaining behavior where self interest continues as the primary force generating actual economic behavior. Under such conditions, one would further expect the use of optimization in characterizing ideal situations to flourish as well. Likewise, with respect to cultures in which self interest or other motives that could be represented by optimization continue to have no influence in determining behavior in general, optimization ought not and probably will not figure significantly in explanatory models and in analyzing ideal circumstances. But, of course, cultural shifts over time that cause the institutionalization or deinstitutionalization of motives that can be represented by optimization will alter the prevalence and usefulness of the application of optimization in Economics.

References

[1] Arrow, K.J., and G. Debreu, (1954),"Existence of an Equilibrium for a Competitive Economy," *Econometrica* 22, pp. 265-290.

[2] Bellah, R.N., (1957), *Tokugawa Religion* (Glencoe: Free Press)

[3] Cournot, A., (1960) *Researches into the Mathematical Principles of the Theory of Wealth* New York: Kelley.

[4] Gossen, H.H., (1983) *The Laws of Human Relations and the Rules of Human Action Derived Therefrom*, R.C. Blitz, trans. (Cambridge: MIT Press).

[5] Grattan-Guinness, I.,(1997) *The Norton History of the Mathematical Sciences* (New York: Norton).

[6] Groenewegen, P.D., (1993) "Introduction," *Reflections on Political Economy* by P. Verri, trans. B. McGilvray and P.D. Groenewegen (Fairfield: Kelley).

[7] Henrich, J., (2002) *et al*, "In Search of Homo Economicus: Behavioral Experiments in 15 Small-Scale Societies," *American Economic Review* 91, no.2 , pp. 73-78.

[8] Jevons, W.S.,(1965) *The Theory of Political Economy* (New York: Kelley).

[9] Katzner, D.W., (1988) *Walrasian Microeconomics: An Introduction to the Economic Theory of Market Behavior* (Reading: Addison-Wesley).

[10] ——, (1998) *Time, Ignorance, and Uncertainty in Economic Models* (Ann Arbor: University of Michigan Press).

[11] ——, (1999) "Western Economics and the Economy of Japan," *Journal of Post Keynesian Economics* 21, pp. 503-521.

[12] ——, (2000) "Culture and the Explanation of Choice Behavior," *Theory and Decision* 48, pp. 241-262.

[13] ——, (2003) "What Are the Questions?" *Journal of Post Keynesian Economics* (forthcoming).

[14] Kline, M., (1972) *Mathematical Thought from Ancient to Modern Times* (New York: Oxford University Press, 1972).

[15] Machlup, F.,(1967) "Theories of the Firm: Marginalist, Behavioral, Managerial," *American Economic Review* 57, pp. 1-33.

[16] Robertson, R.M., (1949) "Mathematical Economics before Cournot," *Journal of Political Economy* 57, pp. 523-536.

[17] Samuelson, P.A., (1972) "Maximum Principles in Analytical Economics," *American Economic Review* 67, pp. 249-262.

[18] Shackle, G.L.S., (1969) *Decision, Order and Time in Human Affairs*, 2nd. ed. (Cambridge: Cambridge University Press).

[19] ——, (1972)*Epistemics and Economics* (Cambridge: Cambridge University Press.

[20] Simon, H.A., (1957) *Administrative Behavior*, 2nd. ed. (New York: Macmillan).

[21] Verri, P., (1993)*Reflections on Political Economy*, trans. B. McGilvray and P.D. Groenewegen (Fairfield: Kelley).

[22] Walras, L., (1954) *Elements of Pure Economics*, W. Jaffé, trans. (Homewood: Irwin).

[23] Weber, M., (1958) *The Protestant Ethic and the Spirit of Capitalism*, trans. T. Parsons (New York: Scribner).

Optimization and Optimal Control, pp. 155 - 164
P.M. Pardalos, I. Tseveendorj, and R. Enkhbat, Editors

A Global Optimization Approach to Solving Equilibrium Programming Problems

O.Khamisov (khamisov@isem.sei.irk.ru)
Institute of Energy Systems, Irkutsk, Russia

Abstract.

Some types of equilibrium programming problem can be reduced to an implicit nonconvex mathematical programming problem. In general the reduced problem can have many locally optimal solutions. Only globally optimal solutions correspond to equilibrium points. In this paper we suggest a global optimization scheme based on concept of the nonlinear support function for solving equilibrium programming problems.

Keywords: Global optimum, support function, Nash equilibrium, d.c. optimization.

In this paper we consider the equilibrium programming problem in the form suggested in [1]. Find $y^* \in \Omega \subset E^n$:

$$y^* \in Argmin\{\Phi(x, y^*) : g(x) \leq 0, x \in \Omega\} \qquad (EP)$$

or prove that such point does not exist (E^n is the nth dimensional Euclidean space). Problem (EP) is in fact the problem of finding a fixed point of an extremal type mapping. There can be other equivalent formulations of equilibrium programming problem, see for example [7]. Many interesting applications such as the Nash equilibrium problem [11], complementarity problem [7], variational inequality problem [2], bilevel programming problem [4], inverse mathematical programming problem [5], and others can be represented in the form (EP). From the practical point of view it is very important to have methods and algorithms for solving (EP). Mainly the present algorithms use local and convex optimization techniques like subgradients, linearization, projections and separating planes ([3], [8]). Different convexity type assumptions must be satisfied in order to guarantee the convergence of the algorithms. In our paper we suggest a global optimization approach based on the concept of nonlinear support function [10].

1 Reduction to a global optimization problem

In this section we use ideas from [11] to show how problem (EP) can be reduced to a global optimization problem and describe the main approach for solving the reduced problem. Let us introduce the auxiliary function

$$\Psi(x,y) = \Phi(x,y) - \Phi(y,y) \tag{1.1}$$

and set

$$D = \{x \in \Omega : g(x) \leq 0\}.$$

We will assume that function $\Phi(x,y)$ is continuous and the set D is compact. Then problem (EP) is equivalent to the following one. Find $y^* \in D$:

$$\Psi(x,y^*) \geq 0, \forall x \in D \tag{1.2}$$

or prove that such point does not exist. Inequality (1.2) holds iff the following inequality is true

$$\max_{y \in D} \min_{x \in D} \Psi(x,y) \geq 0. \tag{1.3}$$

If we introduce the function

$$P(y) = \min_{x \in D} \Psi(x,y) \tag{1.4}$$

then in order to check inequality (1.3) it is sufficient to solve the problem

$$P(y) \to \max, \tag{1.5}$$

$$y \in D. \tag{1.6}$$

In general function $P(y)$ can be multiextremal. Let y^* be a globally optimal solution of problem (1.5)-(1.6). If $P(y^*) < 0$ then problem (EP) does not have equilibrium solutions, otherwise y^* is equilibrium solution. From (1.1) and (1.4) it follows that $P(y) \leq 0, \forall y \in D$ and if y^* is an equilibrium solution then

$$P(y^*) = 0. \tag{1.7}$$

Hence, condition (1.7) can be used as necessary and sufficient global optimality condition for problem (1.5)-(1.6) when the equilibrium point exists. Due to (1.4) problem (1.5)-(1.6) is an implicit optimization problem. In order to calculate the value of the objective function $P(y)$ we have to solve the auxiliary problem (1.4). The min-representation of function $P(y)$ can be used in creation of a numerical iterative approach to solving problem (1.5)-(1.6).
Let $\overline{y} \in D$ be an arbitrary point. Solve problem (1.4). Let $\overline{x}$ be the corresponding optimal value of problem (1.4). If $P(\overline{y}) = 0$ then $\overline{y}$ is equilibrium point and we are done. Otherwise, from (1.4) we have

$$P(\overline{y}) = \Psi(\overline{x},\overline{y}), \tag{1.8}$$

$$P(y) \leq \Psi(\overline{x}, y), \forall y \in D. \tag{1.9}$$

We say that the function $\Psi(\overline{x}, y)$ is a support majorant function for function $P(y)$. Hence, we can suggest the following approach for solving problem (1.5)-(1.6) which coincides with the support planes method in convex programming [9] and Pijavskii method in global optimization [12].
Step 0. Take $y^0 \in D$ and set $k = 0$.
Step 1. Solve the problem

$$\Psi(x, y^k) \to \min, \tag{1.10}$$

$$x \in D. \tag{1.11}$$

Let x^k be a globally optimal solution of problem (1.10)-(1.11).
Step 2. If $\Psi(x^k, y^k) = 0$ then y^k is equilibrium solution, goto Step 6.
Step 3. Solve the problem

$$\min_{0 \leq j \leq k} \{\Psi(x^j, y)\} \to \max, \tag{1.12}$$

$$y \in D. \tag{1.13}$$

Let y^{k+1} be a globally optimal solution of problem (1.12)-(1.13), Ψ_k its optimal value, i.e. $\Psi_k = \min_{0 \leq j \leq k}\{\Psi(x^j, y^{k+1})\}$.
Step 4. If $\Psi_k < 0$ then the equilibrium point does not exist, goto Step 6.
Step 5. Set $k = k + 1$ and goto Step 1.
Step 6. Stop.
If the number of iterations is finite then either we have found an equilibrium point or have proved that such point does not exist. If the number of iterations is infinite then direct application of Theorem 2 from [5] gives the following result.
Theorem. *If functions $\Psi(x^j, y)$ form a family of equicontinuous functions then every accumulation point of sequence $\{y^k\}$ is a global maximum of function $P(y)$ over D.*
Two problems (1.10)-(1.11) and (1.12)-(1.13) must be solved during each iteration. In the particular case when the function $\Psi(x, y)$ is convex in x and concave in y then the procedure described by Step.0-Step.6 finds a saddle point of $\Psi(x, y)$ over D. If set D is a convex set then both problems (1.10)-(1.11) are convex programming problems. In general these problems can be hard global optimization problems. ¿From (1.1) and (1.10) it follows that instead of problem (1.10)-(1.11) we can solve the problem

$$\Phi(x, y^k) \to \min, \tag{1.14}$$

$$x \in D. \tag{1.15}$$

So, the properties of problem (1.14)-(1.15) depend on properties of function $\Phi(x, y)$ with respect to the first argument x. Properties of the second problem (1.12)-(1.13) depend on properties of function $\Phi(x, y)$ with respect to both arguments (x, y). In the next section we consider problem (EP) in which function $\Phi(x, y)$ can be represented as a difference of two convex functions.

2 Equilibrium problem with d.c. function $\Phi(x,y)$

We will assume that the function $\Phi(x,y)$ is a d.c. function [6], i.e. it can be represented as a difference of two convex functions

$$\Phi(x,y) = G(x,y) - H(x,y), \tag{2.1}$$

where $G(x,y)$ and $H(x,y)$ are convex functions determined on $E^n \times E^n$ and set D is a convex compact set. These assumptions are not very restrictive since the set of d.c. functions is dense in the set of continuous functions and in practice almost all nonconvex functions can be represented in the form (2.1). Under these assumptions problem (1.14)-(1.15) is a problem of minimization of a d.c. function over a convex compact set

$$G(x,y^k) - H(x,y^k) \to \min, \tag{2.2}$$

$$x \in D. \tag{2.3}$$

Different global optimization methods exist for solving problem (2.2)-(2.3) [6]. We will not discuss in this paper what method should be used for solving this problem. At the present state it is of interest to apply global optimization methods even for general equilibrium problems with small number of variables. From (2.1) we obtain the following form of the second auxiliary problem (1.12)-(1.13)

$$\min_{0\le j\le k}\{G(x^j,y) - H(x^j,y)\} - [G(y,y) - H(y,y)] \to \max, \tag{2.4}$$

$$y \in D. \tag{2.5}$$

In [6] it is shown that the minimum over a finite family of d.c. functions is again a d.c. function. In particular

$$\min_{0\le j\le k}\{G(x^j,y) - H(x^j,y)\} = \sum_{j=1}^{k} G(x^j,y) - \max_{0\le j\le k}\{\sum_{i=1,i\neq j}^{k} G(x^i,y) + H(x^i,y)\}.$$

Hence, problem (2.4)-(2.5) can be rewritten in the d.c. form

$$[\sum_{j=1}^{k} G(x^j,y) + H(y,y)] - [\max_{0\le j\le k}\{\sum_{i=1,i\neq j}^{k} G(x^i,y) + H(x^i,y)\} + G(y,y)] \to \max, \tag{2.6}$$

$$y \in D, \tag{2.7}$$

where the functions

$$\sum_{j=1}^{k} G(x^j,y) + H(y,y)$$

and

$$\max_{0\le j\le k}\{\sum_{i=1,i\neq j}^{k} G(x^i,y) + H(x^i,y)\} + G(y,y)$$

are convex. So, the iterative scheme described by Step 0. - Step 6. in the previous section can be used for solving problem (EP) with d.c. function $\Psi(x,y)$ and at each iteration two d.c. optimization problems (2.2)-(2.3) and (2.6)-(2.7) must be solved.

3 Finding the Nash equilibrium

In this section we apply the global optimization approach of Section 1 for finding the Nash equilibrium in n-person noncooperative game with d.c. loss functions. Let $X^i \subset E^{n_i}$ be the strategy set of and $f_i(x^1, ..., x^i, ..., x^n)$ be the loss function of player $i, i = 1, ..., n$. Let us recall the well-known definition of the Nash equilibrium. A multistrategy

$$\overline{x} = (\overline{x}^1, ..., \overline{x}^n) \in X = X^1 \times X^2 \times ... \times X^n$$

is called the Nash equilibrium iff

$$f_i(\overline{x}^1, ..., \overline{x}^i, ..., \overline{x}^n) \leq f_i(\overline{x}^1, ..., x^i, ...\overline{x}^n), \forall x^i \in X^i, i = 1, .., n.$$

The problem of finding the Nash equilibrium can be converted into problem (EP) if we set

$$\Phi(x, y) = \sum_{i=1}^{n} f_i(y^1, ..., x^i, ..., y^n), D = X, n = n_1 + ... + n_n.$$

<u>Example 1.</u> Let $n = 2, n_1 = n_2 = 1$,

$$f_1(x_1, x_2) = \sin(x_1 + x_2), x_1 \in X^1 = [0, 18],$$

$$f_2(x_1, x_2) = x_1 \cos(x_2), x_2 \in X^2 = [0, 18]$$

In this example the corresponding function $P(y)$ in (1.5) has explicit form

$$P(y_1, y_2) = -1 - y_1 - \sin(y_1 + y_2) - y_1 \cos(y_2).$$

The Nash equilibrium is given by $y_1^{eq} = \frac{\pi}{2}, y_2^{eq} = \pi$. However, there are other equilibrium points what can be seen from Picture 1 where function $P(y_1, y_2)$ is shown.

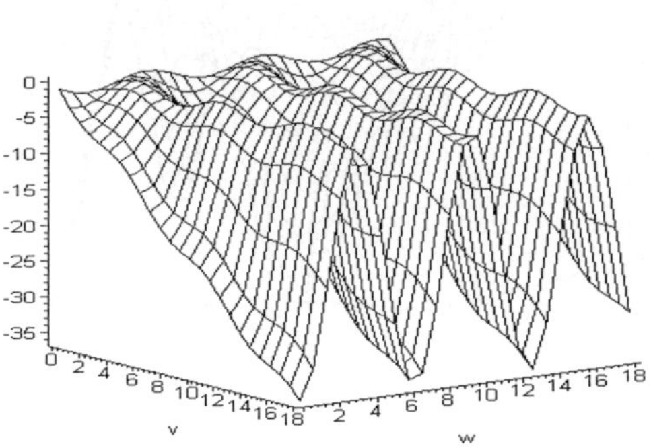

Picture 1.

Example 2. Consider the two person non-cooperative game with the following loss functions and sets of strategies

$$f_1(x_1, x_2) = x_2 \sin(x_1 + x_2) + 0.1(x_2 - x_1)^2, x_1 \in X^1 = [0, 10],$$

$$f_2(x_1, x_2) = \sin(x_1 x_2) - x_1 \cos(x_2) + 0.1(x_1 - x_2)^2, x_2 \in X^2 = [0, 10].$$

Both functions are d.c. functions since

$$f_1(x_1, x_2) = (f_1(x_1, x_2) + 10(x_1^2 + x_2^2)) - 10(x_1^2 + x_2^2),$$

$$f_2(x_1, x_2) = (f_2(x_1, x_2) + 100(x_1^2 + x_2^2)) - 100(x_1^2 + x_2^2).$$

In this example we could not draw the picture of the corresponding function $P(y_1, y_2)$. This example was solved numerically using the iterative procedure described in Section 1. The starting point was $y_1^0 = 5, y_2^0 = 5$. Then the corresponding global optimal solution of problem (2.2)-(2.3) is $x_1^0 = 5.957, x_2^0 = 6.014$. The picture of the first support majorant function (see (1.8)-(1.9)) $\Psi(x^0, y)$ is shown in Picture 2. The horizontal plane corresponds to zero level. So, regions with possible equilibrium points are those with nonnegative values of the first majorant function.

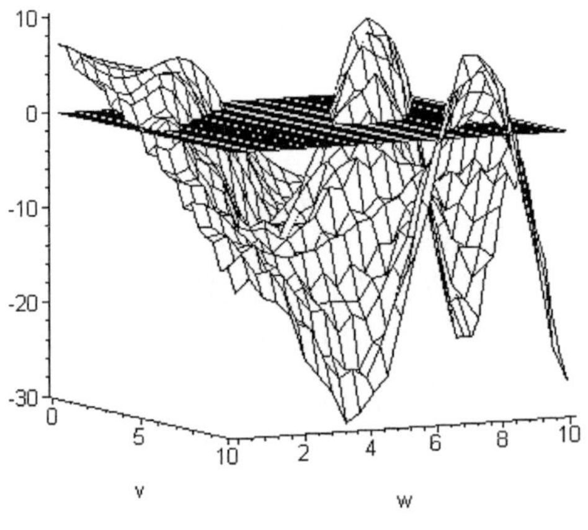

Picture 2.

In ten iterations an approximate solution was obtained. The accuracy was set to 0.01. Computational results by iterations are given in Table 1, where k is the number of iteration, y and x correspond to y^k and x^k respectively, f gives values of the loss functions, column $\Psi(x,y)$ contains values $\Psi(x^k,y^k) = P(y^k)$ (as soon as this value is less than the given accuracy computations stop), Ψ_k is optimal value obtained at Step 3 (if an equilibrium point exists then this value tends to zero).

Table 1.

k	y	x	f	$\Psi(x,y)$	Ψ_k
0	$y_1 = 5$ $y_2 = 5$	$x_1 = 5.957$ $x_2 = 6.014$	$f_1 = -4.905$ $f_2 = -5.692$	-6.326	9.376
1	$y_1 = 3.248$ $y_2 = 7.597$	$x_1 = 3.506$ $x_2 = 5.504$	$f_1 = -5.880$ $f_2 = -2.628$	-3.512	6.561
2	$y_1 = 4.609$ $y_2 = 6.477$	$x_1 = 4.577$ $x_2 = 6.427$	$f_1 = -6.105$ $f_2 = -5.206$	-0.036	3.786
3	$y_1 = 1.351$ $y_2 = 3.302$	$x_1 = 1.519$ $x_2 = -0.514$	$f_1 = -2.965$ $f_2 = -1.469$	-2.263	2.373
4	$y_1 = 5.069$ $y_2 = 5.992$	$x_1 = 5.036$ $x_2 = 5.947$	$f_1 = -5.897$ $f_2 = -5.663$	-0.033	1.524
5	$y_1 = 4.192$ $y_2 = 7.037$	$x_1 = 4.044$ $x_2 = 5.732$	$f_1 = -6.116$ $f_2 = -4.227$	-1.118	1.196
6	$y_1 = 5.469$ $y_2 = 5.512$	$x_1 = 5.485$ $x_2 = 6.550$	$f_1 = -5.512$ $f_2 = -6.112$	-1.236	1.020
7	$y_1 = 4.332$ $y_2 = 6.802$	$x_1 = 4.268$ $x_2 = 6.751$	$f_1 = -6.141$ $f_2 = -4.107$	-0.040	0.920
8	$y_1 = 4.824$ $y_2 = 6.202$	$x_1 = 4.838$ $x_2 = 6.194$	$f_1 = -6.010$ $f_2 = -5.617$	-0.020	0.895
9	$y_1 = 2.666$ $y_2 = 8.665$	$x_1 = 8.515$ $x_2 = -0.358$	$f_1 = -8.665$ $f_2 = -2.399$	-11.116	0.554
10	$y_1 = 5.266$ $y_2 = 5.768$	$x_1 = 5.264$ $x_2 = 5.762$	$f_1 = -5.724$ $f_2 = -5.421$	-0.003	0.498

In order to give an impression of how the approach works we give pictures corresponding to thetwo last iterations. Pictures 3 and 4 show the loss functions and current approximations to the equilibrium point of the first and second players at iteration 9, pictures 5 and 6 show the same for last iteration 10.

Finally, it is interesting to note that one of the equilibrium points is the zero point $(0,0)$ with zero values of the loss functions. At the approximate equilibrium point $y^{10} = (5.266, 5.768)$ values of the loss functions are essentially less. So, this fact shows the efficiency of application of global optimization technique.

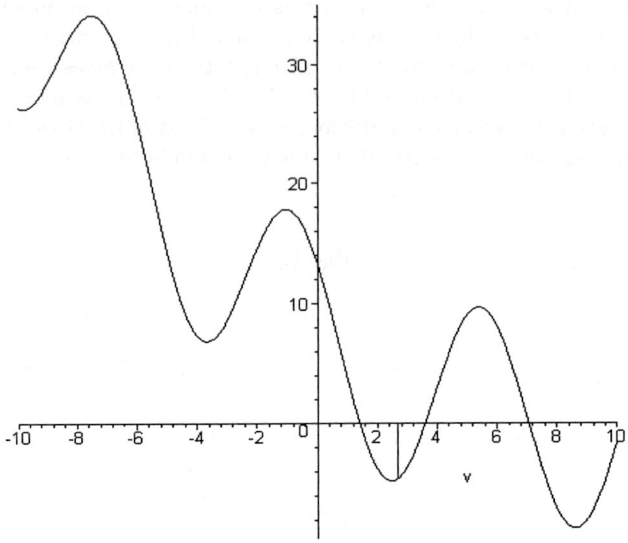

Picture 3.

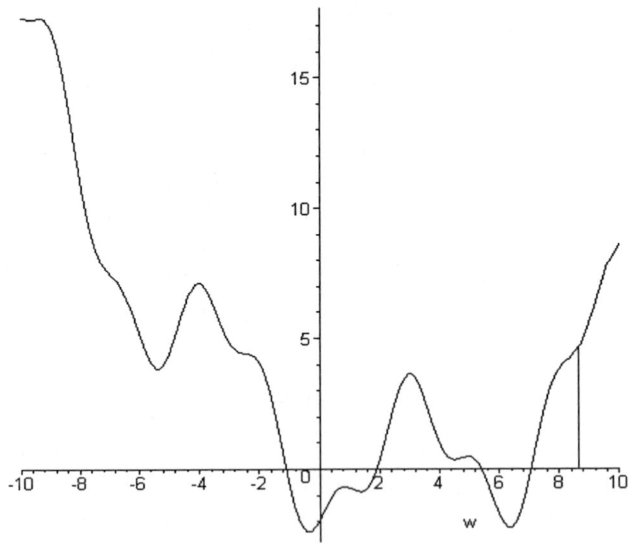

Picture 4.

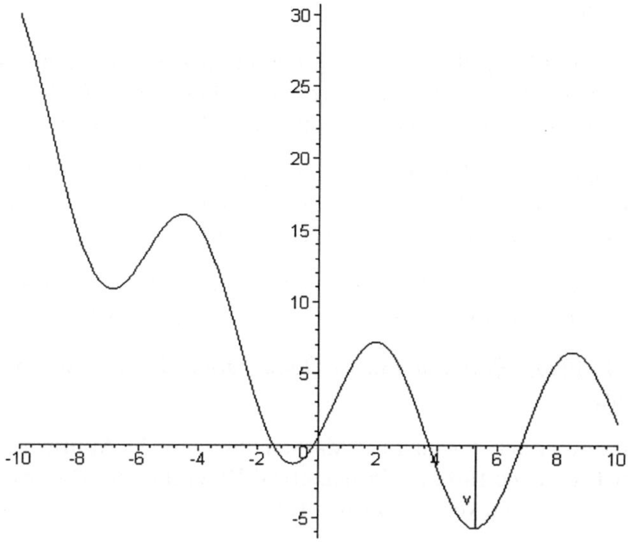

Picture 5.

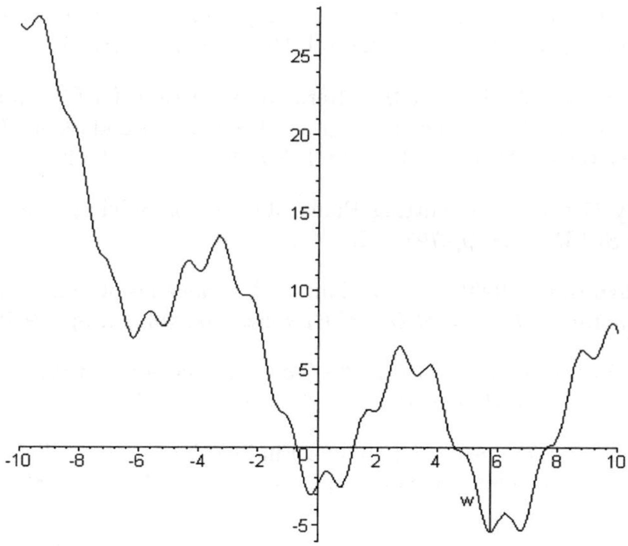

Picture 6.

References

[1] A.S. Antipin (1997), "Equilibrium programming problems: prox-regularization and prox-methods". *Recent Advances in Optimization. Lecture Notes in Economics and Mathematical Systems*, Springer, Berlin, pp.1-18.

[2] A.S. Antipin (2000), "Methods for solving variational ineaquality problems with connected variables", *Journal of Computational Mathematics and Mathematical Physics*, vol. 40, N9, pp.1291-1307 (in Russian)

[3] A.S. Antipin, "Gradient Approach of Computing Fixed Points of Equilibrium Problems", *Journal of Global Optimization* 24 (3):285-309.

[4] J.F. Bard (1996), *Practical Bilevel Optimization*, Kluwer Academic Publishers, Dordrecht.

[5] V.P. Bulatov (1998) "Numerical Methods Solving the Multiextremal Problems Connected with the Inverse Mathematical Programming Problems", *Journal of Global Optimization*, vol. 12, pp.405-413.

[6] R. Horst, H. Tuy (1993) *Global Optimization. Deterministic Approaches*, Springer Verlag, Berlin

[7] A.N. Iusem and W. Sosa (2001), "New Existence Results for Equilibrium Problems", Instituto de Mathematica Pura e Aplicada, Estrada Dona Castorina 110, Jardim Botânico, CEP, 22460-320 RJ, Rio de Janeiro, Brazil.

[8] A.N. Iusem and W. Sosa (2001), "Iterative Algorithms for for Equilibrium Problems", Instituto de Mathematica Pura e Aplicada, Estrada Dona Castorina 110, Jardim Botânico, CEP, 22460-320 RJ, Rio de Janeiro, Brazil.

[9] J. Kelley (1960), "The cutting Plane Method for Solving Convex Programs", *Journal SIAM*, vol.8, pp.703-712.

[10] O.V. Khamisov, (1999), "On Optimization Properties of Functions with a Concave Minorant", *Journal of Global Optimization*, vol. 14, pp. 79-101.

[11] H. Nikaido and K. Isoda (1955), "A Note on Noncoopearative Convex Games", *Pacific Journal of Mathematics*, vol. 5, suppl. I, pp. 807-815.

[12] S. Pijavskii (1967), "An Algorithm for Finding the Absolute Minimum of a Function", Theory of Optimal Solutions, vol. 2, Kiev, IK AN USSR, pp.13-24 (in Russian).

Optimization and Optimal Control, pp. 165 – 174
P.M. Pardalos, I. Tseveendorj, and R. Enkhbat, Editors

Comparison of Convex Relaxations for Monomials of Odd Degree

Leo Liberti (l.liberti@ic.ac.uk)
Centre for Process Systems Engineering,
Imperial College of Science, Technology and Medicine,
London, UK.

Abstract

Branch-and-Bound is one of the most widely used technique for performing global optimization of Nonlinear Programs (NLPs) and Mixed-Integer Nonlinear Programs (MINLPs). In most implementations of this technique the lower bound is obtained by solving a convex relaxation of the original problem. As shown in [9],[10],[11] the convex relaxation can be formed by replacing the nonconvex terms with their convex relaxations. Tight convex underestimators are already available for many types of nonconvex terms, including bilinear and trilinear products, linear fractional terms, and concave and convex univariate functions. However, terms which are piecewise concave and convex are not explicitly catered for. A frequently occurring example of such a term is the monomial of odd degree. We propose a novel convex nonlinear envelope for odd power terms of the form x^{2k+1} ($k \in \mathbb{N}$), where $x \in [a, b]$ and $a < 0 < b$; the derived relaxation is continuous and differentiable everywhere in $[a, b]$. We then make a comparison with an existing convex relaxation for odd power monomials where we show that the novel convex relaxation gains better results in a Branch-and-Bound implementation.

Keywords: convex relaxation, monomial, branch-and-bound, global optimization.

1 Introduction

In this paper, we address NLPs in the following formulation:

$$\left.\begin{array}{lcr} \min_x & f(x) & \\ \alpha & \leq g(x) \leq & \beta \\ a & \leq x \leq & b \end{array}\right\} \qquad (1)$$

where x are the problem variables, α, β are the lower and upper bounds of the constraints, a, b are the lower and upper bounds of the variables. f is the (possibly nonlinear) objective function and g is a set of (possibly nonlinear) constraints.

The most effective deterministic methods for the global solution of MINLPs in the general formulation (1) are spatial Branch-and-Bound algorithms. The whole search space is iteratively partitioned into progressively smaller regions. For each region, lower and upper bounds to the objective functions are computed, and the lowest upper bound is stored as the current best solution to the problem. A region is fathomed and discarded when the bounds are close enough or when the objective function lower bound for the region is greater than the current best solution. The algorithm terminates when there are no more regions to examine, at which point the current best solution is the solution of the problem.

One of the most critical steps in the algorithm is the generation of the lower bound: a good lower bound means rapid fathoming of regions, and as a consequence a lot less regions to examine during the algorithm run. There are many methods to provide a lower bound to the objective function. Lagrangian relaxation, either applied directly [4],[1] or as part of an altogether different algorithm (e.g. Generalized Benders Decomposition), relies on a reformulation of the problem so that it becomes box-constrained:

$$\left.\begin{array}{ll} \max_{u,v} & \inf\{f(x) + u(g-\beta) + v(\alpha - g) \mid a \leq x \leq b\} \\ & u, v \geq 0 \end{array}\right\} \qquad (2)$$

Another method to provide lower bounds (esp. with respect to bilinear terms) is the RLT (Reformulation-Linearization Technique), proposed in [8]. By this method, new constraints are obtained by multiplying existing linear constraints by problem variables and other problem constraints and substituting the resulting bilinearities so that the new constraint becomes linear. On solving this linear problem to optimality, we find a lower bound to the original problem. One of the most common methods for providing lower bounds, however, is that of forming and locally solving a convex relaxation of the original problem [7],[2],[11]. The fact that the relaxation is convex implies that the local solution is also the global one, and hence a guaranteed underestimator of the objective function value.

The convex relaxation of any MINLP can be formed by isolating the nonconvex terms and replacing them with the respective convex relaxations. Tight convex underestimators are already available for many types of nonconvex term, including bilinear and trilinear products, linear fractional terms, and concave and convex univariate functions. However, terms which are piecewise concave and convex are not explicitly catered for. A frequently occurring example of such a term is x^{2k+1}, where $k \in \mathbb{N}$ and the range of x includes zero. In this article, we propose two convex relaxations for such terms: one nonlinear and one linear. We then briefly compare these relaxations to another possible relaxation of the monomial of odd degree, based on reformulating it to a bilinear product, and we show that our proposed relaxation is tighter. Finally, we present some numerical results that highlight this point.

2 Novel Convex Relaxation

Consider the monomial x^{2k+1} in the range $x \in [a, b]$ where $a < 0 < b$. Let c, d be the x-coordinates of the points C, D where the tangents from points A and B respectively meet the curve (see fig. 1 below). The shape of the convex underestimator of x^{2k+1} depends on the relative magnitude of b and c. In particular, if $c < b$ (as in fig. 1), a convex underestimator can be formed from the tangent from $x = a$ to $x = c$ followed by the curve x^{2k+1} from $x = c$ to $x = b$. On the other hand, if $c > b$ (cf. fig. 2), a convex underestimator is the straight line passing through A and B (similarly for the concave overestimator of x^{2k+1} in the range $x \in [a, b]$). If $d > a$, the overestimator is given by the upper tangent from B to D followed by the curve x^{2k+1} from D to A, as in fig. 1. If $d < a$ the overestimator is the straight line from A and B. Note that conditions $c > b$ and $d < a$ cannot both hold simultaneously.

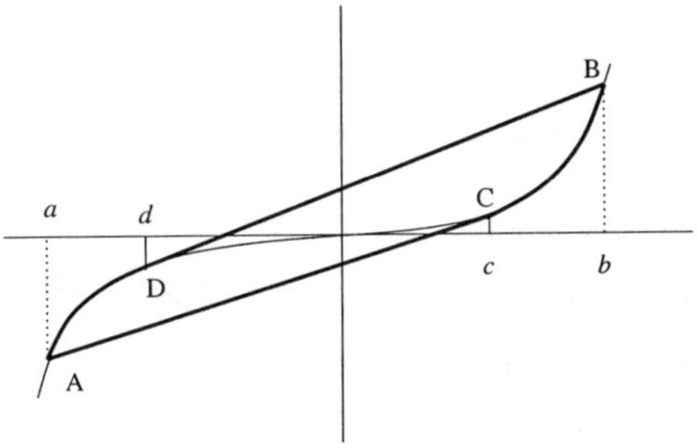

Figure 1: Tightest (nonlinear) convex envelope of x^{2k+1}.

In order to build the convex envelope of x^{2k+1} we therefore require the equations of the tangents that pass through points A, C and B, D. We do this by equating the slope of the line $\overline{AC}$ to the gradient of x^{2k+1} at $x = c$:

$$\frac{c^{2k+1} - a^{2k+1}}{c - a} = (2k+1)c^{2k} \tag{3}$$

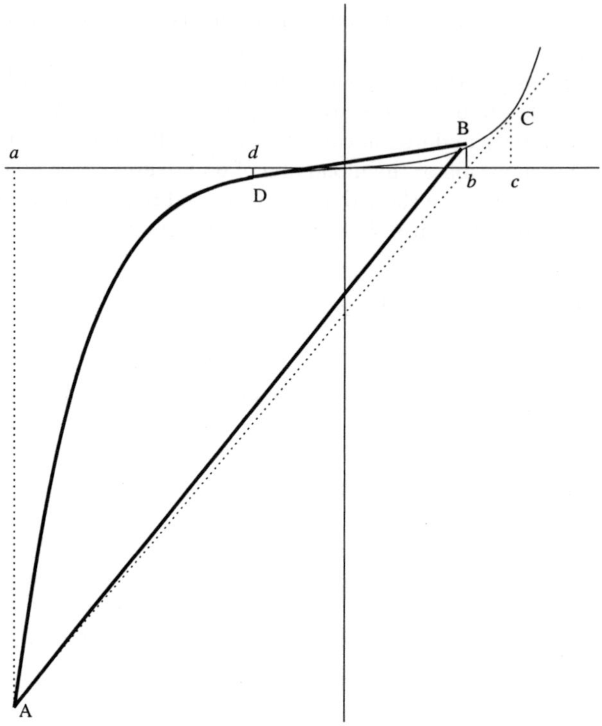

Figure 2: The case when $c > b$.

Hence c is a root of the polynomial:

$$P^k(x,a) \equiv (2k)x^{2k+1} - a(2k+1)x^{2k} + a^{2k+1} \tag{4}$$

It can be shown by induction on k that:

$$P^k(x,a) = a^{2k-1}(x-a)^2 Q^k(\frac{x}{a}) \tag{5}$$

where the polynomial $Q^k(x)$ is defined as:

$$Q^k(x) \equiv 1 + \sum_{i=2}^{2k} i x^{i-1}. \tag{6}$$

Thus, the roots of $P^k(x,a)$ can be obtained from the roots[1] of $Q^k(x)$. However, generally, polynomials of degree higher than 4 cannot be solved by radicals. In fact

[1]Although $P^k(x,a)$ also has $x = a$ as a root, this is not of practical interest.

the Galois group of $Q^3(x) \equiv 6x^5+5x^4+4x^3+3x^2+2x+1$ is isomorphic to S_5 which is not solvable. For details on Galois theory and the solvability of polynomials, see [12]. Notice further that $Q^k(x)$ does not depend on the range of x; moreover, $Q^k(x)$ has exactly one real root, r_k, for any $k \geq 1$, and this lies in $[-1+1/2k, -0.5]$ [5]. Hence, the roots of $Q^k(x)$ for different k can be computed *a priori* to arbitrary precision using simple numerical schemes (e.g. bisection). A table of these roots is presented in table 1 for $k \leq 10$.

k	r_k	k	r_k
1	-0.5000000000	6	-0.7721416355
2	-0.6058295862	7	-0.7921778546
3	-0.6703320476	8	-0.8086048979
4	-0.7145377272	9	-0.8223534102
5	-0.7470540749	10	-0.8340533676

Table 1: Numerical values of the roots of $Q^k(x)$ for $k = 1,..,10$ (to 10 significant digits).

If the roots shown in the second column of table 1 are denoted by r_k, then the tangent points c and d in fig. 1 are simply $c = r_k a$ and $d = r_k b$. The lower and upper tangent lines are given respectively by:

$$a^{2k+1} + \frac{c^{2k+1} - a^{2k+1}}{c-a}(x-a) \tag{7}$$

$$b^{2k+1} + \frac{d^{2k+1} - b^{2k+1}}{d-b}(x-b) \tag{8}$$

Hence, the convex envelope for $z = x^{2k+1}$ when $x \in [a,b]$ and $a < 0 < b$:

$$l_k(x) \leq \quad z \quad \leq u_k(x) \tag{9}$$

is as follows:

- If $c < b$, then:

$$l_k(x) = \begin{cases} a^{2k+1}\left(1 + R_k\left(\frac{x}{a} - 1\right)\right) & \text{if } x < c \\ x^{2k+1} & \text{if } x \geq c \end{cases} \tag{10}$$

otherwise:

$$l_k(x) = a^{2k+1} + \frac{b^{2k+1} - a^{2k+1}}{b-a}(x-a) \tag{11}$$

- If $d > a$, then:

$$u_k(x) = \begin{cases} x^{2k+1} & \text{if } x \leq d \\ b^{2k+1}\left(1 + R_k\left(\frac{x}{b} - 1\right)\right) & \text{if } x > d \end{cases} \tag{12}$$

otherwise:

$$u_k(x) = a^{2k+1} + \frac{b^{2k+1} - a^{2k+1}}{b - a}(x - a) \tag{13}$$

where we have used the constant $R_k \equiv \frac{r_k^{2k+1}-1}{r_k-1}$.

We can relax the nonlinear convex envelope to a linear one by dropping the "follow the curve" requirements on either side of the tangency points, and using the lower and upper tangent as convex underestimator and concave overestimator respectively, as follows:

$$a^{2k+1}\left(1 + R_k\left(\frac{x}{a} - 1\right)\right) \quad \leq z \leq \quad b^{2k+1}\left(1 + R_k\left(\frac{x}{b} - 1\right)\right) \tag{14}$$

We can tighten the envelope further by also considering the tangents to the curve at the endpoints A, B:

$$(2k+1)b^{2k}x - 2kb^{2k+1} \quad \leq z \leq \quad (2k+1)a^{2k}x - 2ka^{2k+1}. \tag{15}$$

A similar idea for producing the convex envelope of x^3 was discussed in [7] (p. 159), but the treatment was limited to one demonstrative example.

3 Alternative Convex Relaxation

One common way to derive a convex relaxation for the monomial of odd degree x^{2k+1} is to reformulate it to wx and add a constraint $w = x^{2k}$. The bilinear term $z = wx$ can then be relaxed via the McCormick's convex under- and over-estimators [7]

$$\begin{aligned} z &\geq w^L x + x^L w - w^L x^L \\ z &\geq w^U x + x^U w - w^U x^U \\ z &\leq w^U x + x^L w - w^U x^L \\ z &\leq w^L x + x^U w - w^L x^U. \end{aligned}$$

where $w^L \leq w \leq w^U, x^L \leq x \leq x^U$; whereas the term x^{2k} is convex and hence an envelope is given by the function itself and the secant. Thus the reformulation is:

$$\begin{aligned} z &= wx \\ w &= x^{2k} \\ a \leq\ & x \leq b \\ 0 \leq\ & w \leq w^U = \max\{a^{2k}, b^{2k}\} \end{aligned}$$

k	Iterations (novel rel.)	Iterations (alternative rel.)	k	Iterations (novel rel.)	Iterations (alternative rel.)
1	11	43	10	15	59
2	7	63	9	11	59
3	19	55	8	19	59
4	15	59	11	19	59
5	11	63	12	19	9
6	11	55	13	15	9
7	19	55	14	15	59

Table 2: Numerical results from the simple test problem.

and the convex relaxation is:

$$\begin{array}{rcl} aw \leq & z & \leq bw \\ w^U x + bw - w^U b \leq & z & \leq w^U x + aw - w^U a \\ x^{2k} \leq & w & \leq a^{2k} + \dfrac{b^{2k} - a^{2k}}{b-a}(x-a) \\ a \leq & x & \leq b \\ 0 \leq & w & \leq w^U \end{array}$$

After some algebraic manipulation, we can eliminate w to obtain the following non-linear convex envelope for z:

$$\frac{w^U a}{a-b}(x-b) \quad \leq z \leq \quad \frac{w^U b}{b-a}(x-a) \tag{16}$$

$$bx^{2k} + w^U(x-b) \quad \leq z \leq \quad ax^{2k} + w^U(x-a) \tag{17}$$

$$a\left(a^{2k} + \frac{b^{2k}-a^{2k}}{b-a}(x-a)\right) \quad \leq z \leq \quad b\left(a^{2k} + \frac{b^{2k}-a^{2k}}{b-a}(x-a)\right) \tag{18}$$

4 Numerical Results

We have generated some numerical results that illustrate the difference between the two convex envelopes described above for x^{2k+1}. We solved the problem

$$\left.\begin{array}{rl} \min_{x,y} & x - y \\ y & = x^{2k+1} \\ -2 & \leq x, y \leq 2 \end{array}\right\} \tag{19}$$

to global optimality using the spatial Branch-and-Bound algorithm described in [11] within the $oo\mathcal{OPS}$ implementation [6], both with the novel convex (linear) relaxation

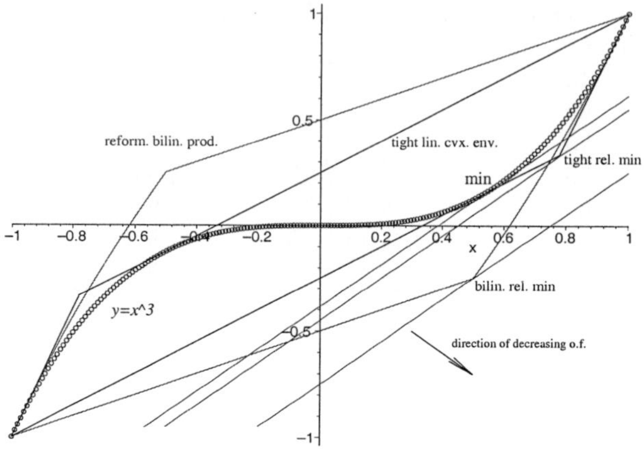

Figure 3: Graphical description of simple test problem (in 2D).

and with the alternative convex relaxation. Table 2 lists the number of iterations taken by the algorithm when k varies. The first column lists the results relative to the novel convex relaxation (section 2), the second those relative to the alternative convex relaxation (section 3).

The results clearly substantiate the theory: the novel convex relaxation gives better performance in comparison to the alternative relaxation based on reformulation to bilinear product. In the case of x^3 we can see why this happens in fig. 3, 4. The direction of minimization of the objective function $x - y$ is such that the minimum over the original feasible region is very near the minimum over the novel convex relaxation. However, the minimum over the alternative convex relaxation is further away. Hence the performance gain.

5 Conclusion

We have proposed a tight novel convex envelope for monomials of odd degree when the range of the defining variable includes zero, i.e. when they are piecewise convex and concave. We have then compared it with another possible relaxation (based on reformulation to bilinear product) and shown that the former performs better than the latter when tested in a Branch-and-Bound algorithm.

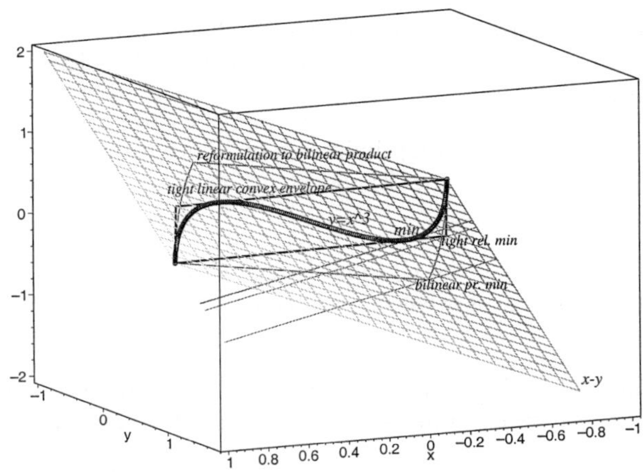

Figure 4: Graphical description of simple test problem (in 3D).

References

[1] Adhya, N., M. Tawarmalani, and N. V. Sahinidis (1999) 'A Lagrangian Approach to the Pooling Problem'. *Industrial and Engineering Chemistry Research* **38**, 1956–1972.

[2] Adjiman, C. S., S. Dallwig, C. A. Floudas, and A. Neumaier (1998) 'A global optimization method, αBB, for general twice-differentiable constrained NLPs: I. Theoretical advances'. *Computers & Chemical Engineering* **22**(9), 1137–1158.

[3] Bazaraa, M. S., H. D. Sherali, and C. Shetty (1993) *Nonlinear Programming: Theory and Algorithms.* Chichester: Wiley, second edition.

[4] Ben-Tal, A., G. Eiger, and V. Gershovitz (1994) 'Global Minimization by Reducing the Duality Gap'. *Mathematical Programming* **63**, 193–212.

[5] Liberti, L. and C. C. Pantelides (2003) To appear, 'Tightest Convex Envelopes of Monomials of Odd Degree'. *Journal of Global Optimization*, 25(2): 157-168.

[6] Liberti, L., P. Tsiakis, B. Keeping, and C. C. Pantelides (2001) '$oo\mathcal{OPS}$'. Centre for Process Systems Engineering, Chemical Engineering Department, Imperial College, London, UK, v.1.24.

[7] McCormick, G. P.(1976) 'Computability of Global Solutions to Factorable Nonconvex Programs: Part I — Convex Underestimating Problems'. *Mathematical Programming* **10**, 146–175.

[8] Sherali, H. D.(1998) 'Global Optimization of Nonconvex Polynomial Programming Problems having Rational Exponents'. *Journal of Global Optimization* **12**, 267–283.

[9] Smith, E. M. (1996) 'On the Optimal Design of Continuous Processes'. Ph.D. thesis, Imperial College of Science, Technology and Medicine, University of London.

[10] Smith, E. M. and C. C. Pantelides (1997) 'Global Optimisation of Nonconvex MINLPs'. *Computers and Chemical Engineering* **21**, S791–S796.

[11] Smith, E. M. and C. C. Pantelides (1999) 'A symbolic reformulation/spatial branch-and-bound algorithm for the global optimisation of nonconvex MINLPs'. *Computers and Chemical Engineering* **23**, 457–478.

[12] Stewart, I.(1989) *Galois Theory.* London: Chapman & Hall, second edition.

Optimization and Optimal Control, pp. 175 – 183
P.M. Pardalos, I. Tseveendorj, and R. Enkhbat, Editors

Structural Stability of Vector Optimization Problems [1]

Paulo Mbunga (mbunga@mathematik.hu-berlin.de)
Institut für Mathematik
Humboldt-Universität zu Berlin
Campus Adlershof, Rudower Chaussee 25, Haus II, 4. Etage D-12489 Berlin

Abstract

We study global stability properties for vector optimization problems of the type:

$$\mathcal{VOP}(f,H,G): \qquad \min\left\{f(x) = (f_1(x),\ldots,f_l(x)) \mid x \in M[H,G]\right\},$$

where

$$M[H,G] := \{x \in \mathbb{R}^n \mid h_i(x) = 0, \quad g_j(x) \le 0, \quad i \in I, j \in J\}$$

with

$$I := \{1,\ldots,m\}, \quad J := \{1,\ldots,s\}, \quad L := \{1,\ldots,l\}.$$

We extend Guddat/Jongen's [5] concept of structural stability of scalar nonlinear optimization problems to vector optimization problems. Under the assumption that $M[H,G]$ is compact we prove the necessary condition for the structural stability of a vector optimization problem, i.e. the scalar problem

$$\mathcal{P}^{\max}(f,H,G): \qquad \min\left\{\max_{l\in L} f_l(x) \mid x \in M[H,G]\right\}$$

has to be structurally stable.
Keywords: vector optimization, structural stability, strong stability.

[1]This work was supported by the Graduiertenkolleg Geometrie und Nichlineare Analysis of the Humboldt-Universität.

1 Introduction

We consider the vector optimization problem

$\mathcal{VOP}(f,H,G)$: $$\min\{f(x)=(f_1(x),\dots,f_l(x)) \mid x\in M[H,G]\},$$

where

$$M[H,G]:=\{x\in\mathbb{R}^n \mid h_i(x)=0,\quad g_j(x)\le 0,\quad i\in I, j\in J\}$$

with

$$I:=\{1,\dots,m\},\quad J:=\{1,\dots,s\},\quad L:=\{1,\dots,l\}.$$

Unless otherwise statet, we will assume that the functions f_k, g_j, $k\in L, j\in J$ belong to $C^2(\mathbb{R}^n,\mathbb{R})$. The lower level set corresponding to $u\in\mathbb{R}^{|L|}$ will be denoted as follows:

(1) $$\mathcal{L}^u(f,H,G):=\{x\in M[H,G] \mid f(x)\underset{D}{\le} u\},\quad u\in\mathbb{R}^{|L|}.$$

The partial ordering $\underset{D}{\le}$ is induced by a convex cone D with

(2) $$x\underset{D}{\le} y \quad :\Longleftrightarrow \quad y-x\in D.$$

Definition 1 *The vector optimization problems $\mathcal{VOP}(f,H,G)$ and $\mathcal{VOP}(\tilde f,\tilde H,\tilde G)$ are called equivalent if there exist continuous mappings $\phi:\mathbb{R}^{|L|}\times\mathbb{R}^n\longrightarrow\mathbb{R}^n$ and $\psi:\mathbb{R}^{|L|}\longrightarrow\mathbb{R}^{|L|}$ with the properties $P1-P3$:*

P1 For every $u\in\mathbb{R}^{|L|}$ the mapping $\phi_u:\mathbb{R}^n\longrightarrow\mathbb{R}^n$ is a homeomorphism from $\mathbb{R}^n$ onto itself, where $\phi_u:=\phi(u,x)$.

P2 The mapping ψ is a homeomorphism from $\mathbb{R}^{|L|}$ onto itself and ψ is D-monotonically increasing; i.e.

(3) $$q_1\underset{D}{\le} q_2 \text{ implies } \psi(q_1)\underset{D}{\le}\psi(q_2) \text{ for } q_1,q_2\in\mathbb{R}^{|L|}.$$

P3 $\phi_u[\mathcal{L}^u(f,H,G)]=\mathcal{L}^{\psi(u)}(\tilde f,\tilde H,\tilde G)$ for all $u\in\mathbb{R}^{|L|}$.

Definition 2 *The vector optimization problems $\mathcal{VOP}(f,H,G)$ is called* structurally stable if there exists a C^2_S-neighborhood $\mathcal{O}$ of (f,H,G) with the property that $\mathcal{VOP}(f,H,G)$ and $\mathcal{VOP}(\tilde f,\tilde H,\tilde G)$ are equivalent for all $(\tilde f,\tilde H,\tilde G)\in\mathcal{O}$.

The C^2_S-topology above for the product $\prod_{i=1}^r C^2(\mathbb{R}^n,\mathbb{R})$ will be the product topology generated by the strong (or Whitney-) C^2-topology C^2_S on each factor $C^2(\mathbb{R}^n,\mathbb{R})$ (cf. [6],[9]). A typical base-neighbourhood of a function $\rho \in C^2(\mathbb{R}^n,\mathbb{R})$ is the set $\rho + \mathcal{W}_\epsilon$, where $\mathcal{W}_\epsilon$ is defined as follows with the aid of a continuous positive $\epsilon : \mathbb{R}^n \longrightarrow \mathbb{R}$,

$$(4) \quad \mathcal{W}_\epsilon = \{\eta \in C^2(\mathbb{R}^n,\mathbb{R}) \mid |\eta(x)| + \sum_i |\frac{\partial \eta}{\partial x_i}(x)| + + \sum_{i,j} |\frac{\partial^2 \eta}{\partial x_i \partial x_j}(x)| < \epsilon(x), \forall x \in \mathbb{R}^n\}.$$

We have the chance of a more local control for data corresponding to the scalar problem $\mathcal{P}(r,H,G)$, $r \in C^2(\mathbb{R}^n,\mathbb{R})$ by putting for any subset U of $\mathbb{R}^n$:

$$(5) \quad norm[(r,H,G),U] = \sup_{x\in U} \max_{\eta\in\{r,h_i,i\in I,g_j,j\in J\}} \left\{ |\eta(x)| + \sum_i |\frac{\partial \eta}{\partial x_i}(x)| + \sum_{i,j} |\frac{\partial^2 \eta}{\partial x_i \partial x_j}(x)| \right\}.$$

Remark 1 *If $|L| = 1$ and $D = \mathbb{R}_+^{|L|}$, then we have the structural stability for the scalar nonlinear problem introduced in [5].*

We consider the following scalar optimization problem

$$\mathcal{P}^{\max}(f,H,G): \qquad \min\left\{\max_{l\in L} f_l(x) \mid x \in M[H,G]\right\}.$$

For the corresponding lower level set we shall use the notation

$$\mathcal{L}^t_{\max}(f,H,G) := \left\{x \in M[H,G] \mid \max_{l\in L} f_l(x) \le t\right\}, \quad t \in R.$$

We use the notations given in Weber[14] and denote the canonical projection of the space $\mathbb{R}^{n+1}$ on the first n components or on the last component by $pr_{(1,\dots,n)}$ and pr_{n+1}, respectively; i.e.

$$\check{x} = \left(pr_{(1,\dots,n)}(\check{x}), pr_{n+1}(\check{x})\right) = (x, x_{n+1}) \in \mathbb{R}^n \times \mathbb{R}.$$

Let us now look at the following equivalent differentiable optimization problem which in particular reveals an objective function of class C^2.

$$\mathcal{P}^{\vee}(\check{f},\check{H},\check{G}): \qquad \min\{\check{f}(\check{x}) := pr_{n+1}(\check{x}) \mid x \in M[\check{H},\check{G}]\},$$

where

$$M[\check{H},\check{G}] := \left\{(x,x_{n+1}) \in \mathbb{R}^{n+1} \mid x \in M[H,G], f_l(x) - x_{n+1} \le 0 \quad (l \in L)\right\},$$

namely with $\check{h}_i(\check{x}) := h_i(x)(i \in I)$ and

$$\check{g}_j(\check{x}) := \begin{cases} g_j(x), \text{ for } j \in J \\ f_l(x) - x_{n+1}, \text{ for } j = s + l, \quad l \in L. \end{cases}$$

$\mathcal{P}^{\vee}(\check{f}, \check{H}, \check{G})$ is strongly related to scalarization methods for vector optimization problems. As pointed out in Weidner[16], this scalarization method includes the known procedures of Brosowski and Conci [3], Helbig[8], Wierzbicki[17] and other papers cited in [7]. Roughly speaking, each minimal point of $\mathcal{P}^{\vee}(\check{f}, \check{H}, \check{G})$ is weakly efficient and vice versa each weakly efficient and hence each efficient point solves the scalar problem $\mathcal{P}^{\vee}(\check{f}, \check{H}, \check{G})$.

We note that the lower level sets $\mathcal{L}^t_{\max}(f, H, G)$ of $\mathcal{P}^{\max}(f, H, G)$ are just the projection in the sense of $pr_{(1,\dots,n)}$ of the corresponding level sets $\mathcal{L}^{\vee t}(\check{f}, \check{H}, \check{G})$ due to $\mathcal{P}^{\vee}(\check{f}, \check{H}, \check{G})$. Both sets are homotopy-equivalent (cf. Jongen et al.[9], Lemma 4.2.1) and $M[\check{H}, \check{G}]$ is just the part of the epigraph $E(\max_{l \in L} f_l)$ over the feasible set $M[H, G]$.

For $x \in \mathbb{R}^n$ and $\rho > 0$ let $B(x, \rho)$ denote the Euclidean ball in $\mathbb{R}^n$, centered at x with radius ρ. Following M. Kojima [11] we define:

Definition 3 *Let $x^u \in M[H, G]$ be a Kuhn-Tucker point for $\mathcal{P}(r, H, G)$, $r \in C^2(\mathbb{R}^n, \mathbb{R})$. Then x^u is called* strongly stable if for some $\bar{\delta} > 0$ and each $\delta \in (0, \bar{\delta}]$ there exists an $\alpha > 0$ such that whenever $(\tilde{r}, \tilde{H}, \tilde{G})$ satisfies $norm[(r - \tilde{r}, H - \tilde{H}, G - \tilde{G}), B(x^u, \delta)] \leq \alpha$, the ball $B(x^u, \delta)$ contains a Kuhn-Tucker point for $\mathcal{P}(\tilde{r}, \tilde{H}, \tilde{G})$ which is unique in $B(x^u, \delta)$.

Lemma 1 (Kojima [11]) *Let $\bar{x}^u \in M[H, G]$ be a Kuhn-Tucker point for $\mathcal{P}(r, H, G)$, $r \in C^2(\mathbb{R}^n, \mathbb{R})$.*

1. *If LI is satisfied at $\bar{x}^u$, then $\bar{x}^u$ is a strongly stable Kuhn-Tucker point if and only if the matrix $D^2 L_{[\lambda,\mu]}(\bar{x}^u)$ has nonvanishing determinants with a common sign on the subspace $W(\bar{x}^u, \bar{J})$ for all $\bar{J}$ with $J_+(\bar{x}^u) \subset \bar{J} \subset J_o(\bar{x}^u)$, where*

$$J_+(\bar{x}^u) = \{j \in J_o(\bar{x}^u) | \mu_j > 0\}$$
$$W(\bar{x}^u, \bar{J}) = \{\xi \in \mathbb{R}^n | Dh_i(\bar{x}^u)\xi = 0, i \in I, Dg_j(\bar{x}^u)\xi = 0, j \in \bar{J}\}$$
$$L_{[\lambda,u]}(x) := r(x) + \sum_{i \in I} \lambda_i h_i(x) + \sum_{j \in J_o(\bar{x}^u)} u_j g_j(x).$$

2. *Let MFCQ be satisfied at $\bar{x}^u$, but LI not. Then, $\bar{x}^u$ is a strongly stable Kuhn-Tucker point if and only if for every (λ, μ) the matrix $D^2 L_{[\lambda,\mu]}(\bar{x}^u)$ is positive definite on the subspace $W(\bar{x}^u, J_+(\bar{x}^u))$.*

Firstly we note that the structural stability of the problem $\mathcal{P}^{\max}(f, H, G)$ was characterized in Weber[14].

Theorem 1 (Main result) *Let* $D = \mathbb{R}_+^{|L|}$ *and the vector optimization problem* $\mathcal{VOP}(f, H, G)$ *with compact feasible set* $M[H, G]$ *be* structurally stable. Then the problem $\mathcal{P}^{\max}(f, H, G)$ is *structurally stable.*

2 Lemmas and preliminary results

Lemma 2 *Let* $t \in \mathbb{R}$, $D = \mathbb{R}_+^{|L|}$ *and* $u := (t, \ldots, t) = te$, $e = (1, \ldots, 1) \in \mathbb{R}^{|L|}$. *Then the following relations between lower level sets are fulfilled:*

$$\mathcal{L}^t_{\max}(f, H, G) = pr_{(1,\ldots,n)}\mathcal{L}^{\vee t}(\check{f}, \check{H}, \check{G}) = \mathcal{L}^u(f, H, G). \tag{6}$$

Proof. The first equality is obvious. We show the second equality.
$\mathcal{L}^u(f, H, G) \subseteqq pr_{(1,\ldots,n)}\mathcal{L}^{\vee t}(\check{f}, \check{H}, \check{G})$: Let $x \in \mathcal{L}^u(f, H, G)$. Then $x \in M[H, G]$, $f_l(x) \leq t$, $l \in \{1, \ldots, L\}$. This implies that $(x, t) \in M(\check{H}, \check{G})$ and so $x \in pr_{(1,\ldots,n)}\mathcal{L}^{\vee t}(\check{f}, \check{H}, \check{G})$.
$\mathcal{L}^u(f, H, G) \supseteqq pr_{(1,\ldots,n)}\mathcal{L}^{\vee t}(\check{f}, \check{H}, \check{G})$: Let $x \in pr_{(1,\ldots,n)}\mathcal{L}^{\vee t}(\check{f}, \check{H}, \check{G})$. Then, there exists a $z \in \mathbb{R}$ with $(x, z) \in \mathcal{L}^{\vee t}(\check{f}, \check{H}, \check{G})$. This implies that $x \in M[H, G]$, $f_l(x) \leq z$, $l \in L$ and $z \leq t$ and so $f_l(x) \leq t$, $l \in L$. That means $x \in \mathcal{L}^u(f, H, G)$. □

Lemma 3 *Let D be a pointed, closed and convex cone. Then there exists a* $\bar{u} \in \mathbb{R}^{|L|}$ *with:*

$$u_1 \leq_D \bar{u}, \quad u_2 \leq_D \bar{u} \tag{7}$$

for all $u_1, u_2 \in \mathbb{R}^{|L|}$.

Proof. We follow the idea by Tanino in [12] (Proof of the Theorem 3.3.1). Pick

$$d \in intD$$

and define

$$y(\alpha) = \alpha d \qquad \alpha \in \mathbb{R}.$$

We claim that, for any $y \in \mathbb{R}^{|L|}$, there exists $\alpha > 0$ such that

$$y \in y(\alpha) - D.$$

If it is not true, then applying the standard separation theorem to the ray $\{y - \alpha d : \alpha > 0\}$ and the convex cone $-D$, we deduce the existence of $\nu \in \mathbb{R}^p \setminus \{0\}$ such that

$$\langle \nu, y - \alpha d \rangle \geq 0 \qquad \text{for all } \alpha > 0$$

and

$$\langle \nu, d' \rangle \geq 0 \qquad \text{for all } d' \in D.$$

Since $d \in intD$, $\langle \nu, d \rangle > 0$ from the latter inequality and so

$$\langle \nu, y - \alpha d \rangle < 0$$

for sufficiently large α, which is a contradiction.
We can therefore choose $\alpha_1, \alpha_2 > 0$ with

$$u_1 \in \alpha_1 d - D \qquad \text{and } u_2 \in \alpha_2 d - D.$$

Take $\alpha_{\max} := \max\{\alpha_1, \alpha_2\}$. Since $\alpha_i d \underset{D}{\leq} \alpha_{\max} d, i = 1, 2$, we can observe that the inequalities (7) are satisfied by $\bar{u} := \alpha_{\max} d$. □

3 Proof of the necessity part

Let us now assume that the problem $\mathcal{VOP}(f, H, G)$ is *structurally stable* according to the Definition 2. Then from the results given by Weber in [10] (see also [14] and [5]) we have to verify the following conditions:

($C1$) The Mangasarian-Fromovitz constraint qualification is satisfied at every point of $M[H, G]$.

($C2$) Every Kuhn-Tucker point of $\mathcal{P}^{\max}(f, H, G)$ is strongly stable.

($C3$) Different Kuhn-Tucker points have different ($f-$) values.

Proof of (C1): We show that the structural stability of $\mathcal{VOP}(f, H, G)$ implies (C1). Consider a C^2_s-neighborhood $\mathcal{O}$ of (f, H, G) with the property: for every $(\tilde{f}, \tilde{H}, \tilde{G}) \in \mathcal{O}$ the set $M[\tilde{H}, \tilde{G}]$ is compact and moreover $\mathcal{VOP}(f, H, G)$ is equivalent with $\mathcal{VOP}(\tilde{f}, \tilde{H}, \tilde{G})$. Choose an arbitrary element $(\tilde{f}, \tilde{H}, \tilde{G}) \in \mathcal{O}$. Let ϕ and ψ be mappings which establish the equivalence according Definition 1. Take $\alpha, \tilde{\alpha} \in \mathbb{R}^{|L|}$ with

$$\alpha := \left(\alpha_1, \ldots, \alpha_{|L|}\right), \quad \alpha_i := \max_{x \in M[H,G]} f_i(x) \tag{8}$$

$$\tilde{\alpha} := \left(\tilde{\alpha}_1, \ldots, \tilde{\alpha}_{|L|}\right), \quad \tilde{\alpha}_i := \max_{x \in M[\tilde{H},\tilde{G}]} \tilde{f}_i(x) \tag{9}$$

and consider the vectors α and $\psi^{-1}(\tilde{\alpha})$. Lemma 7 implies the existence of a $\bar{u} \in \mathbb{R}^{|L|}$ such that $\alpha \underset{D}{\leq} \bar{u}$ and $\psi^{-1}(\tilde{\alpha}) \underset{D}{\leq} \bar{u}$. We note that $\mathcal{L}^{\bar{u}}(f, H, G) = M[H, G]$. Since ψ is monotonically increasing, we have $\tilde{\alpha} \underset{D}{\leq} \psi(\bar{u})$. Thus $\mathcal{L}^{\psi(\bar{u})}(\tilde{f}, \tilde{H}, \tilde{G}) = M[\tilde{H}, \tilde{G}]$. From $\phi_{\bar{u}}(\mathcal{L}^{\bar{u}}(f, H, G)) = \mathcal{L}^{\psi(\bar{u})}(\tilde{f}, \tilde{H}, \tilde{G})$, we have $\phi_{\bar{u}}(M[H, G]) = M[\tilde{H}, \tilde{G}]$. Hence, $\phi_{\bar{u}}$ establishes a homeomorphism between $M[H, G]$ and $M[\tilde{H}, \tilde{G}]$. So, we have shown:

for every $(\tilde{H}, \tilde{G})$ in some C^2_S-neighborhood of (H, G) the sets $M[H, G]$ and $M[\tilde{H}, \tilde{G}]$ are homeomorphic. Then, with the help of the Stability Theorem from [5] we may equivalently say that MFCQ is satisfied at every point $x \in M[H, G]$. Hence, the condition (C1) is valid.

Proof of (C2): Let us assume that there is a stationary point which is not strongly stable, i.e. condition (C2) is not valid. In order to arrive at a contradiction, the perturbation result from Weber[14] plays a central part.

Lemma 4 (Perturbation Lemma, Weber[14]) *Let $M[H, G]$ be compact and suppose that (C1) holds. Suppose further, that $\mathcal{P}^{\max}(V, H, G)$ has a Kuhn-Tucker point $\bar{x}^u$ which is not strongly stable. Then, there exist a number $k \in \mathbb{N}$ and functions $(V^1, \tilde{H}, \tilde{G})$, $(V^2, \tilde{H}, \tilde{G})$, both arbitrarily C^2_S-near (V, H, G), with the following properties:*

- *(i) Problem $\mathcal{P}^{\max}(V^1, \tilde{H}, \tilde{G})$ has k Kuhn-Tucker points, exactly one of which is not strongly stable.*
- *(ii) Problem $\mathcal{P}^{\max}(V^2, \tilde{H}, \tilde{G})$ has at least $k+1$ Kuhn-Tucker points, all of them being strongly stable.*
- *(iii) Conditions $C1$ and $C3$ hold for problems $\mathcal{P}^{\max}(V^i, \tilde{H}, \tilde{G}), i = 1, 2$.*

For $r \in C^2(\mathbb{R}^n, \mathbb{R})$ we denote the set $\{x \in M[H, G] \mid a \leq r(x) \leq b \in\}$ by $\mathcal{L}^a_b(r, H, G)$ for $a \leq b$, $a, b \in \mathbb{R}$.

The next Lemma is proved by Guddat and Jongen in [5].

Lemma 5 *Let the feasible set $M[H, G]$ corresponding to the problem $\mathcal{P}(r, H, G)$ be compact and suppose that Condition $C1$ is fulfilled. Let $a, b \in \mathbb{R}$ be given with $a < b$.*

- *(i) If $\mathcal{L}^a_b(r, H, G)$ contains no Kuhn-Tucker point, then $\mathcal{L}^a(r, H, G)$ and $\mathcal{L}^b(r, H, G)$ are homeomorphic.*
- *(ii) Suppose that $\mathcal{L}^a_b(r, H, G)$ contains exactly one Kuhn Tucker point $\bar{x}^u$ with $a < r(\bar{x}^u) < b$, and suppose that $\bar{x}^u$ is strongly stable. Then the sets $\mathcal{L}^a(r, H, G)$ and $\mathcal{L}^b(r, H, G)$ are not homeomorphic.*

Lemma 6 *Let $M[H, G]$ be compact and suppose that Condition $C1$ is fulfilled. Let $V^1, V^2 \in C^2(\mathbb{R}^n, \mathbb{R})^{|L|}$ be functions satisfying Condition $C3$ w.r.t. $\mathcal{P}^{\max}(V^i, H, G), i = 1, 2$, with the following additional properties:*

- *(i) Problem $\mathcal{P}^{\max}(V^1, H, G)$ has k Kuhn-Tucker points, exactly one of which is not strongly stable.*
- *(ii) Problem $\mathcal{P}^{\max}(V^2, H, G)$ has at least $k+1$ Kuhn-Tucker points, all of them being strongly stable.*

Then, the vector optimization problems $\mathcal{VOP}(V^i, H, G), i = 1, 2$ are not equivalent.

Proof. We refer to the n-dimensional problems $\mathcal{P}^{\max}(V^i, H, G), i = 1, 2$ due to the corresponding $(n+1)$-dimensional problems $\mathcal{P}^{\vee}(\check{f}^i, \check{H}, \check{G}), i = 1, 2$. For the two latter problems, we study the topological behaviour of the lower level sets, when the level parameter t increasingly traverses the line of real numbers. Taking account of Condition $C3$ formulated for both problems and the Lemma 5, it follows that the homeomorphy-type of the lower level set $\mathcal{L}^{\vee t}(\check{f}^2, \check{H}, \check{G})$ changes at least $k+1$, and at least $k-1$, but (only) at most k such changes for $\mathcal{L}^{\vee t}(\check{f}^1, \check{H}, \check{G})$. The pairwise homotopy equivalence between the lower level sets in $\mathbb{R}^n$ and $\mathbb{R}^{n+1}$ gives rise to the analogous description in dimension n of the topological behavior of $\mathcal{L}^t_{\max}(V^i, H, G), i = 1, 2$. From Lemma 2 (6) it follows that $\mathcal{L}^{te}(V^i, H, G), i = 1, 2$ have two different numbers for the changes of the topological type in $\mathbb{R}^n$, as we move along the line $u(t) = te, e \in \mathbb{R}^{|L|}, t \in \mathbb{R}$. If the problems $\mathcal{VOP}(V^i, H, G), i = 1, 2$ would be equivalent, then these two numbers have to coincide. □

We now apply the perturbational result for our indirect proof of Condition $C2$. If this condition does not hold, we can apply Lemma 4 and Lemma 6 in order to obtain two non-equivalent problems $\mathcal{VOP}(f^i, \tilde{H}, \tilde{G}), i = 1, 2$ with each $(f^i, \tilde{H}, \tilde{G})$ arbitrarily C^2_S-near to (f, H, G). This is in contradiction with structural stability.

Proof of (C3): We know the validity of $C1$ and $C2$. If $\mathcal{P}^{\max}(f, H, G)$ does not satisfy Condition $C3$, whenever the equation $\max_{l \in L} f(\bar{x}^u_1) = \max_{l \in L} f(\bar{x}^u_2)$ holds for two different Kuhn-Tucker points of $\mathcal{P}^{\max}(f, H, G)$ then an addition of one small positive and one small negative constant to all functions $f_k \quad (k \in L)$ locally at $\bar{x}^u_1$ and $\bar{x}^u_2$, respectively, leads to a problem $\mathcal{P}^{\max}(\bar{f}, H, G)$ satisfying Condition C3. Let k the number of Kuhn-Tucker points for $\mathcal{P}^{\max}(\bar{f}, H, G)$. Then, reasoning as in the proof of Lemma 6, the homeomorphy-type of $\mathcal{L}^t_{\max}(\bar{f}, H, G)$ for increasing t, changes exactly k-times. The corresponding number of changes w.r.t. $\mathcal{P}^{\max}(f, H, G)$ is less than k. So, as in Lemma 6, $\mathcal{VOP}(f, H, G)$ and $\mathcal{VOP}(\bar{f}, H, G)$ are not equivalent. This leads again to a contradiction with the condition on structural stability of $\mathcal{VOP}(f, H, G)$.

References

[1] Bank, B.; Guddat, J.; Klatte, D.; Kummer, B. and Tammer, K.: Non-Linear Parametric Optimization. Akademie-Verlag. Berlin, 1982.

[2] Guddat, J., Jongen, H. Th., Rueckmann, J.: On stability and stationary points in nonlinear optimization, J. Aust. Math. Soc., Ser. B 28 (1986), 36 – 56.

[3] Brosowski, B. and Conci, A.: On vector optimization and Parametric Programming. Segundas Jornadas Latino Americanas de Matematica Aplicada, 2(1983), 483 – 495.

[4] Guddat, J.; Guerra Vasquez, F. and Jongen, H. Th.: Parametric Optimization, Singularities, Pathfollowing and Jumps. John Wiley, Chichester 1990.

[5] Guddat, Juergen and Jongen, H. Th.: Structural Stability in Nonlinear Optimization. In: Optimization 18 (1987)5, 617 – 631.

[6] Hirsch, M. W.: Differential Topology, Springer-Verlag, Berlin, Heidelberg, New York, 1976.

[7] Helbig, S.: On Quatitative and Qualitative Stability of a Scalarization Method, I. In Parametric Optimization and Related Topics III, 255 – 270, Peter Lang Verlag, Franfurt am Main 1993.

[8] Helbig, S.: Approximation of the Efficient Point Set by Perturbation of the Ordering Cone, Zeitschrift für Operations Research 35(1991), 197 – 220.

[9] Jongen, H. Th.; Jonker, P., Twilt, F.: Nonlinear Optimization in $\mathbb{R}^n$, I. Morse Theory, Chebishev approximation, Peter Lang Verlag, Frankfurt, 1983.

[10] Jongen, H. Th and Weber, G.-W. : Nonlinear Optimization: Characterization of Structural Stability. In: J. Global Optimization 1 (1991), 47 – 64.

[11] Kojima, M.: Strongly stable stationary solutions in nonlinear programs. In: S.M. ROBINSON (ed.): Analysis and computation of fixed points. Academic Press, New York, 1980.

[12] Sawaragi Y., Nakayama H., Tanino T.: Theory of Multiobjective Optimization. Academic Press, Inc., Orlando 1985.

[13] The Luc, D.: Theory of Vector Optimization. Springer-Verlag, Berlin, 1989.

[14] Weber, G.: Minimization of a max-type function: Characterization of structural stability. In Parametric Optimization and Related Topics III, 519 – 538, Peter Lang Verlag, Franfurt am Main 1993.

[15] Weber, G.: Charakterisierung struktureller Stabilität in der nichtlinearen Optimierung. Aachener Beiträge zur Mathematik, Band 5, Aachen 1992.

[16] Weidner, P.: An approach to different scalarizations in vector optimization. Wissenschaftliche Zeitschrift der TH Ilmenau 3(1990), 103 – 110.

[17] Wierzbicki, A. P.: On the Completeness and Constructiveness of Parametric Characterization to Vector Optimization Problem, OR Spektrum 8(1986), 73 – 87.

[18] Wierzbicki, A. P.: Basic Properties of scalarizing functional for multiobjective optimization. Math. Oper.Stat. Optim. **8**, 55-60, 1977.

Optimization and Optimal Control, pp. 185 – 210
P.M. Pardalos, I. Tseveendorj, and R. Enkhbat, Editors

Optimization for Black-box Objective Functions

Hirotaka Nakayama[1] (nakayama@konan-u.ac.jp)
Department of Information Science and Systems Engineering,
Konan University, 8-9-1 Okamoto, Higashinada, Kobe 658-8501, Japan.

Masao Arakawa
Department of Reliability-based Information System Engineering,
Kagawa University, 2217-20 Hayashicho, Takamatsu Kagawa 761-0396, Japan

Koji Washino
Department of Information Science and Systems Engineering,
Konan University, 8-9-1 Okamoto, Higashinada, Kobe 658-8501, Japan.

Abstract

In many practical engineering design problems, the form of objective functions is not given explicitly in terms of design variables. Given the value of design variables, under this circumstance, the value of objective functions is obtained by real/computational experiments such as structural analysis, fluidmechanic analysis, thermodynamic analysis, and so on. Usually, these experiments are considerably expensive. In order to make the number of these experiments as few as possible, optimization is performed in parallel with predicting the form of objective functions. Response Surface Methods (RSM) are well known along this approach. This paper presents a brief review machine learning approaches to RSM such as Radial Basis Function Networks (RBFN) and Support Vector Machines (SVM). One of the most important tasks in this approach is to find effective sample data moderately in order to make the number of experiments as small as possible. Several methods are compared along with numerical examples.

Keywords: Response surface method, RBF networks, Support vector machines, Genetic algorithms, Expected improvement

[1] Author for correspondence

1 Introduction

The aim of this paper is to review optimization of black-box objective functions whose forms are not explicitly known in terms of design variables, but whose values are given by sampled real/computational experiments. In many engineering design problems, several kinds of analyses such as structural analysis, fluidmechanic analysis, thermodynamic analysis, and so on are applied for this purpose. These analyses usually require considerably expensive computational time. Therefore, if these functions are optimized by existing methods, it takes an unrealistic order of time to obtain a solution. For this situation, the number of necessary sampled experiments should be as few as possible.

This kind of problems have been studied in geology as well as in engineering, because our aim is to develop "divining rods" which find effective sample points for optimizing our objective functions. One of most important feature of "divining rods" is that the black-box objective functions should be well approximated by as small number of sampled experiments as possible.

Response Surface Method (RSM) is probably most popular to our aim. Usually, Response Surface Method is a generic name, and it covers a wide range of methods. Above all, methods using experimental design are famous. However, many of them select sample points only on the basis of statisitical analysis of design variable space. They may provide a good approximation of black-box functions with a mild nonlineality. It is clear, however, that in cases in which the black-box function is highly nonlinear, we can obtain better performance by methods taking into account not only the statistical property of design variable space but also that of range space of the black-box function (in other words, the shape of function).

Recently, the authors suggested to apply machine learning techniques sucha as RBF (Radial Basis Function) networks, Support Vector Machines (SVM) for approximating the black-box function. There, the sample points are given by considering both global and local information of the black-box function. On the other hand, Shonlau *et al.* suggested a "figure of merit" for sample points as "Expected Improvement" (EI). This paper reports a comparative study among those methods based on several examples.

2 Incremental Learning in Machine Learning

Since the number of sample points for predicting objective functios should be as few as possible, we adopt some incremental learning techniques which predict blacbox functions by adding learning samples step by step. RBF Networks (RBFN) are effective to this end. The output of RBFN is given by

$$f(\boldsymbol{x}) = \sum_{j=1}^{m} w_j h_j(\boldsymbol{x}),$$

where $h_j,\ j=1,\ldots,m$ are radial basis functions, e.g.,

$$h_j(\boldsymbol{x}) = e^{-\|\boldsymbol{x}-\boldsymbol{c}_j\|^2/r_j}.$$

Given the training data $(\boldsymbol{x}_i, \hat{y}_i),\ i=1,\cdots,p$, the learning of RBFN is usually made by solving

$$\text{Min} \quad E = \sum_{i=1}^{p}(\hat{y}_i - f(\boldsymbol{x}_i))^2 \ + \ \sum_{j=1}^{m}\lambda_j w_j^2$$

where the second term is introduced for the purpose of regularization.

Letting $A = (H_p^T H_p + \Lambda)$, we have

$$A\boldsymbol{w} = H_p^T \hat{\boldsymbol{y}},$$

as a necessary condition for the above minimization. Here

$$H_p^T = [\boldsymbol{h}_1 \ \cdots \ \boldsymbol{h}_p],$$

where $\boldsymbol{h}_j^T = [h_1(\boldsymbol{x}_j), \ldots, h_m(\boldsymbol{x}_j)]$, and Λ is a diagonal matrix whose diagonal components are $\lambda_1 \ \cdots \ \lambda_m$.

Therefore, the learning in RBFN is reduced to finding

$$A^{-1} = (H_p^T H_p + \Lambda)^{-1}.$$

The incremental learninig in RBFN can be made by adding new data and/or a basis function, if necesary. Since the learning in RBFN is equivalent to the matrix inversion A^{-1}, the additional learning here is reduced to the incremental calculation of the matrix inversion. The following algorithm can be seen in [11]:

(i) Adding a New Training Data

Adding a new data $\boldsymbol{x}_{p+1}$, the incremental learning in RBFN can be made by the following simple update formula: Let

$$H_{p+1} = \begin{bmatrix} H_p \\ \boldsymbol{h}_{p+1}^T \end{bmatrix},$$

where $\boldsymbol{h}_{p+1}^T = [h_1(\boldsymbol{x}_{p+1}), \ldots, h_m(\boldsymbol{x}_{p+1})]$.

Then

$$A_{p+1}^{-1} = A_p^{-1} - \frac{A_p^{-1}\boldsymbol{h}_{p+1}\boldsymbol{h}_{p+1}^T A_p^{-1}}{1 + \boldsymbol{h}_{p+1}^T A_p^{-1}\boldsymbol{h}_{p+1}}.$$

(ii) Adding a New Basis Function

In those cases where a new basis function is needed to improve the learning for a new data, we have the following update formula for the matrix inversion: Let

$$H_{m+1} = \begin{bmatrix} H_m & \boldsymbol{h}_{m+1} \end{bmatrix},$$

where $\boldsymbol{h}_{m+1}^T = [h_{m+1}(\boldsymbol{x}_1), \cdots, h_{m+1}(\boldsymbol{x}_p)]$.

Then

$$A_{m+1}^{-1} = \begin{bmatrix} A_m^{-1} & \mathbf{0} \\ \mathbf{0}^T & 0 \end{bmatrix}$$

$$+ \frac{1}{\lambda_{m+1} + \boldsymbol{h}_{m+1}^T (I_p - H_m A_m^{-1} H_m^T) \boldsymbol{h}_{m+1}} \times \begin{bmatrix} A_m^{-1} H_m^T \boldsymbol{h}_{m+1} \\ -1 \end{bmatrix} \begin{bmatrix} A_m^{-1} H_m^T \boldsymbol{h}_{m+1} \\ -1 \end{bmatrix}^T.$$

Remark It is important to decide parameters $\boldsymbol{c}_j$ and r_j of radial basis functions moderately. We assign a basis function at each learning data $(\boldsymbol{x}_i, \hat{y}_i)$ $(i = 1, \cdots, p)$ so that $\boldsymbol{c}_i = \boldsymbol{x}_i$ $(i = 1, \cdots, p)$. Modifying the formula given by [3] slightly, the value of r_j is determined by

$$r = d_{max} / \sqrt[n]{nm},$$

where d_{max} is the maximal distance among the data, n is the dimension of the data, m is the number of basis functions.

3 Selection of Additional Samples

If the current solution is not satisactory, namely if our stopping condition is not satisfied, we need some additional samples in order to improve the approximation of the black-box objective function. Now, how to select such additional samples becomes an important issue.

If the current optimal point is taken as such additional data, the estimated optimal point tends to converge to a local maximum (or minimum) point. This is due to lack of global information in predicting the objective function.

On the other hand, if addtional data are taken far away from the existing data, it is difficult to obtain more detailed information near the optimal point. Therefore, it is hard to obtain a solution with a high precision. This is because of insufficient information near the optimal point.

It is important to get well balanced samples providing both global information and local information on black-box objective functions. The author and his coresearchers suggested a method which gives both global information for predicting the objective function and local information near the optimal point at the same time [10]. Namely, two kinds of additional samples are taken at the same time for relearning the form of the objective function. One of them is selected from a neighborhood of the current optimal point in order to add local information near the (estimated) optimal point. The size of this neighborhood is controlled during the convergence process. The other one is selected far away from the currenct optimal value in order to give a better prediction of the form of the objective function. The former additional data gives more detailed information near the current optimal point. The latter data prevents converging to local maximum (or minimum) point.

The neighborhood of the current optimal point is given by a square S, whose center is the current optimal point, with the length of a side l. Let S_0 be a square, whose center is the current optimal point, with the fixed length of a side l_0. The square S is shrinked according to the number C_x of optimal points appeared continuously in S_0 in the past. Namely,

$$l = l_0 \times \frac{1}{C_x + 1}, \tag{1}$$

The first additional data is selected inside the square S randomly. The second additional data is selected in an area, in which the existing learning data are sparse, outside the square S. An area with sparsely existing data may be found as follows: First, a certain number (N_{rand}) of data are generated randomly outside the square S. Denote d_{ij} the distance from this random data p_i ($i = 1, \ldots, N_{rand}$) to the existing learning data q_j ($j = 1, \ldots, N$). Select the shortest k distances $\tilde{d}_{ij}$ ($j = 1, \ldots, k$) for each p_i, and sum up these k distances, i.e., $D_i = \sum_{j=1}^{k} \tilde{d}_{ij}$. Take p_t which maximizes $\{D_i\}_{(i=1,\ldots,N_{rand})}$ as an addtional data outside S.

The algorithm is sumarized as follows:

Step 1: Predict the form of the objective function by RBFN on the basis of the given training data.

Step 2: Estimate an optimal point for the predicted objective function by GA.

Step 3: Count the number of optimal points appeared continuously in the past in S_0. This number is represented by C_x.

Step 4: Terminate the iteration,

(i) if C_x is larger than or equal to the given C_x^0 a priori,

or

(ii) if the best value of the objective function obtained so far is identical during the last certain number (C_f^0) of iterations.

Otherwise calculate l by (1), and go to the next step.

Step 5: Select an additional data near the current optimal value, i.e., inside S.

Step 6: Select another additional data outside S in a place in which the density of the training data is low as stated above.

Step 7: Go to Step.1.

4 Optimizing Predicted Objective Functions

We can apply any optimization methods for the predicted objective functions, because they are represented explicitly in terms of design variables. In order to obtain an approximate solution to the global optimum, genetic algorithms (GA) can be effectively applied.

In the following exapmles, we applied the BLX-α method which can treat in a simple way continuous design varibles [13]. The method uses continuous values of design valriables as codes of individuals as they are. In crossover, children are generated randomly inside a hyperbox including their parents. It has been observed that the BLX-α method without mutation is effective in problems with continuous design variables. Several sophisticated methods which can treat continuous variables have been developed (e.g., Arakawa *et al.*[1]).

5 Numerical Example 1

Consider an example given by

$$f(x_1, x_2) = 10\exp(-0.01(x_1 - 10)^2 - 0.01(x_2 - 15)^2)\sin x_1 \tag{2}$$

$$(0 \leq x_1 \leq 15, \quad 0 \leq x_2 \leq 20).$$

This function has a maximum value 9.5585 at $x_1 = 7.8960$ and $x_2 = 15.0000$ as shown in Figure 1.

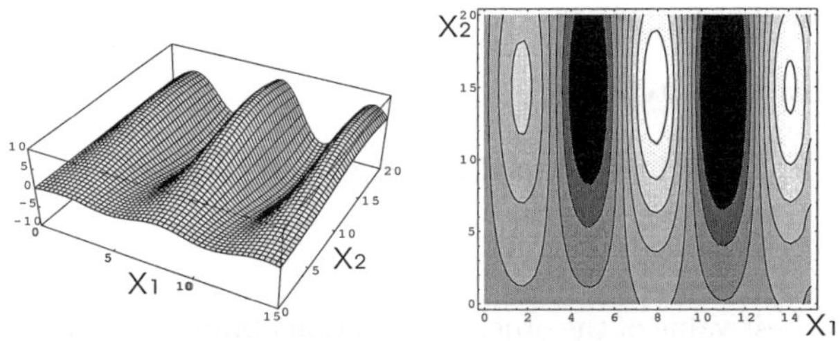

$x_1 = 7.8960$, $x_2 = 15.0000$, $f = 9.5585$

Figure 1: Contour of Example 1

Set λ in RBFN: 0.01, population in GA: 10, generation in GA: 50, l_0: 4.0, C_0: 10, N_{rand}: 50, and k: 2. At the begining, five data ($(x_1, x_2) = (0,0), (15,0), (0,20), (7.5,10)$, $(15,20)$) were taken for the initial learning. The contour of the forecasted objective function by RBFN is shown in Figure 2. In Figures 2-4, here after, "$\odot$" represents

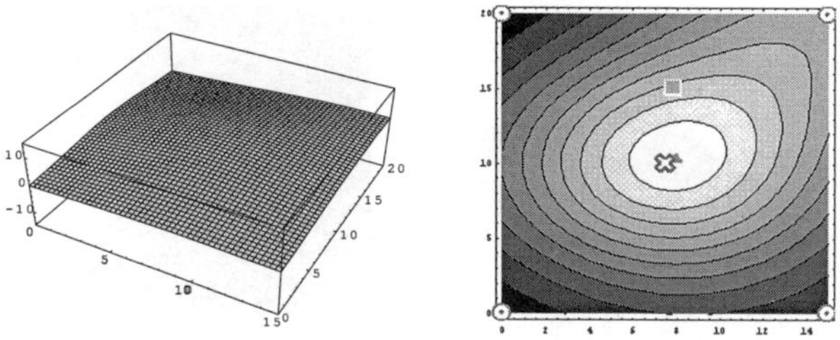

$x_1 = 7.9947$, $x_2 = 10.5039$, $f = 6.8400$

Figure 2: Result after the first iteration with 5 training data in Example 1

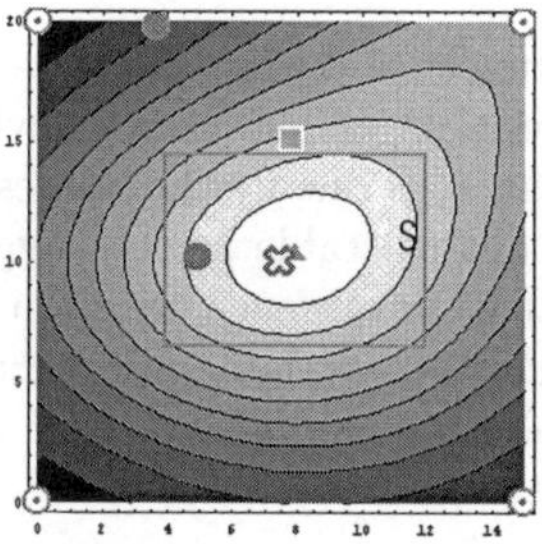

Figure 3: Additional data denoted by • in Example 1

the training data, "×" the current optimal point, and "□" the correct optimal point to (2).

Figure 2 shows the result after the first iteration (5 training data). At this stage, two additional data ("•") are selected according to the method stated in Section 3 (Figure 3). Then, there is no past optimal points in the square S_0. Therefore, the squares S and S_0 are identical at this stage. An additional data is selected randomly inside the square S, and another additional data is outside S in an area where the density of training data is low.

Figure 5 shows the result at an intermediate 31 iterations. After 69 iterations, we have a good estimate as shown in Figure 6.

For the sake of comparison, the results obtained by the suggested method and by only GA are given in Table 1 and Table 9, respectively. We made 10 simulations with differently generated random variables for each method. In optimizing by only GA, when the top 10 individuals with high ranks converge within a given fixed range, the calculation is stopped. "*" shows converging to a local maximum point different from the exact solution. It can be seen in these tables that the average number of

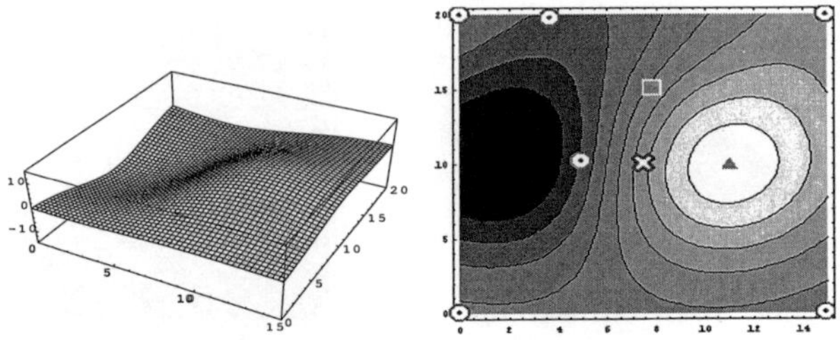

$x_1 = 7.5000$, $x_2 = 10.0000$, $f = 6.8730$

Figure 4: The result of our simulation with 7 training data in Example 1

evaluations of the objective function by the suggested method is less than the half of the one with GA only. Moreover, the standard deviation of errors by the method with GA only is much more than the one by the suggested method. This implies that the method with GA only is not as stable as the suggested method. In the method with GA only, a more restrictive convergence criterion would be necessary for getting a similar stability as the suggested method. In other words, the method with GA only would require more number of analyses in order to attain at a similar degree of stability as the suggested method.

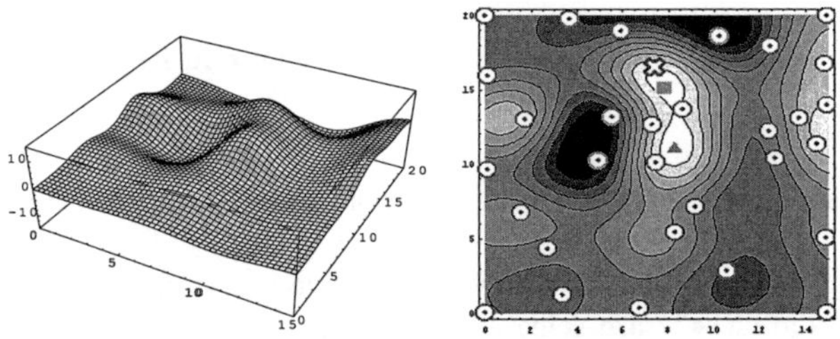

$x_1 = 7.4645$, $x_2 = 16.4721$, $f = 8.4892$

Figure 5: The result of our simulation with 31 training data in Example 1

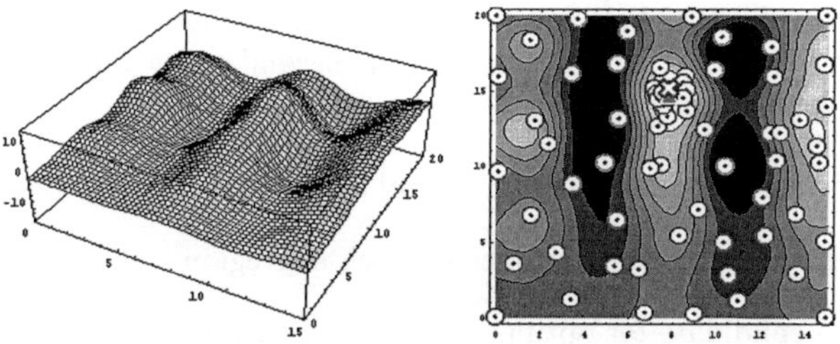

$x_1 = 7.8725$, $x_2 = 15.2023$, $f = 9.5519$

Figure 6: The result of our simulation with 69 training data in Example 1

Table 1: Simulation results of the suggested method in Example 1

trial	# analysis	x_1	x_2	x error	f	f error	$f$error(%)
1	59	8.031	15.188	0.2312	9.4662	0.0923	0.9656
2	81	7.678	12.505	2.5050	8.7646	0.7939	8.3057
3	49	7.962	13.203	1.7985	9.2340	0.3245	3.3949
4	61*	13.803	14.688	5.9149	8.1658	1.3927	14.5703
5	113	7.645	16.139	1.1663	9.1341	0.4244	4.4400
6	89	7.771	12.081	2.9213	8.7084	0.8501	8.8937
7	49	8.228	11.577	3.4394	8.0233	1.5352	16.0611
8	55	7.785	11.180	3.8220	8.2088	1.3497	14.1204
9	47	7.987	14.514	0.4942	9.4959	0.0626	0.6549
10	69	7.873	15.202	0.2037	9.5519	0.0066	0.0690
Average	67.2	—	—	2.2496	—	0.6832	7.1476
Standard Deviation	—	—	—	1.8426	—	0.5870	6.1414

6 Comparison with Polynomial Approximation

Polynomials are often used for approximating the objective funciton in RSM. In particular in many cases, second order polynomials are used in connection with experimental design. Since second order polynomials can approximate in a local sense, for highly nonlinear cases we divide the whole space into several subregions. In this event, how to divide the whole space into subregions causes a problem.

We apply the space partition method using second order polynomial to the above example. The algorithms is as follows:

Step 1. Divide the design space into several subregions.

Step 2. Generate the response surface on each subregion.

Step 3. Find a candidate for optimal solution on each subregion.

Step 4. Select the best one among the candidates.

Step 5. Generate the response surface on some enlarged region around the best candidate, and find the optimal solution for the response surface.

For Example 1 in the previous section, we obtain six subregions by dividing equal three parts along x_1-axis and equal two parts along x_2-axis. Sample points are selected on meshes of 3×3 on each subregion. Response surface is generated by using second order polynomial on each subregion, and the optimal solution for the response surface on each subregion is found (Fig.7). In Fig.7, "⊙" denote sample points, "×" the best candidate among six subregions, and "□" the theoretical optimal solution.

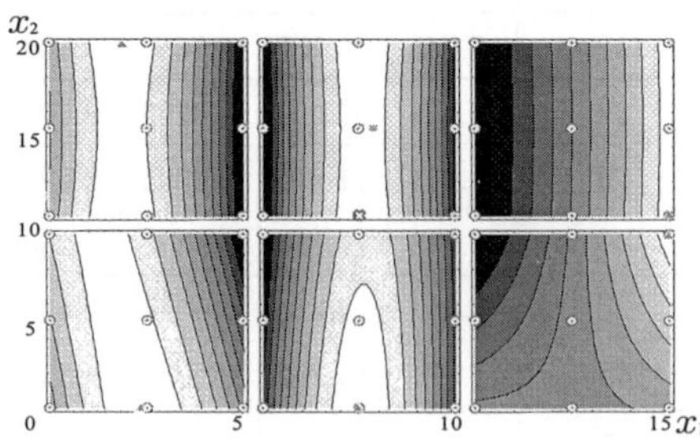

Figure 7: Result by space partition method (i)

The best canditate belongs to the region which contains the real optimal solution. We made an enlarged approximation around the best candidate ("×" in Fig. 7) five times. This brought predicted solutions towards the real optimal solution (Fig. 8). Table 2 and Table 3 show this result. The symbol "*" shows the region containing the predicted optimal solution among 6 subregions.

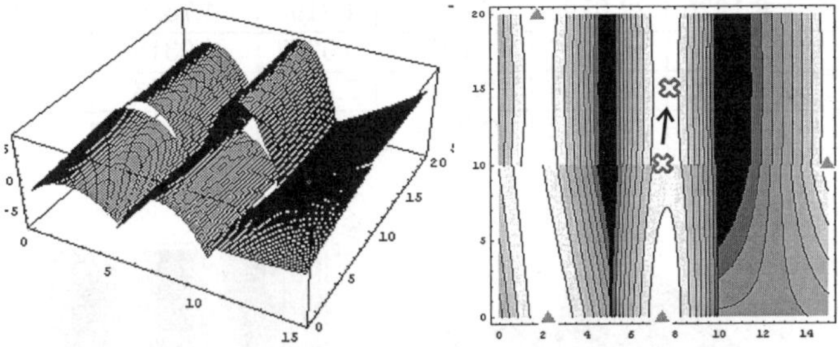

Figure 8: Final result by space partition method (i)

Table 2: Result by space partition method (i)

—	x_1	x_2	error of x	f	error of f	error of f (%)
1	2.3388	0.0062	15.9905	1.8520	7.7065	80.6246
2	1.8693	19.9289	7.7856	3.4139	6.1446	64.2841
3	7.4598	0.0528	14.9536	4.0979	5.4606	57.1282
4*	7.5351	10.0156	4.9974	7.6206	1.9379	20.2741
5	14.9798	9.9869	8.6782	3.7227	5.8358	61.0535
6	14.9998	10.0920	8.6344	4.3390	5.2195	54.6058

We obtained a good approximate optimal solution by the method of space partition in the above example. This seems because the way of space partition is moderate enough to obtain the predicted optimal solution near the real solution. However, the performance may depend on the way of space partition. The next example changes the whole region from $0.0 \leq x_1 \leq 15.0$C$0.0 \leq x_2 \leq 20.0$ into $0.0 \leq x_1 \leq 20.0$C$0.0 \leq x_2 \leq 20.0$ so that the different partition may be brought (Fig. 9). Fig. 10CFig. 11CTable 4 and Table 5 show the result.

Similarly in the previous example, we made an enlarged approximation around the predicted optimal point five times. As a result, the predicted solution moved towards a different point from the real solution along the line "⇒" in Fig. 11. This explains the way of space partition gives a significant effect to the final result.

On the hand, the method using RBFN can provide good approximation even for highly nonlinear cases without any space partition. Fig. 12 shows the result by

Table 3: Result by enlarged approximation (i)

# analysis	x_1	x_2	error of x	f	error of f	error of f (%)
44	6.8783	17.0824	2.3178	7.5067	2.0518	21.4657
53	6.1326	22.0796	7.2959	4.8195	4.7390	49.5789
62	8.7703	16.5017	1.7377	5.5267	4.0318	42.1803
71	7.8803	14.0440	0.9561	7.4452	2.1133	22.1091
80	7.8723	14.9452	0.0597	9.4150	0.1435	1.5013

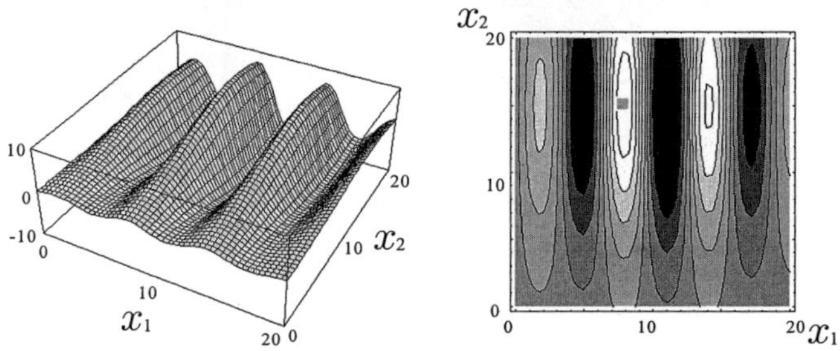

Figure 9: Example 1 with a modified design region

applying RBFN to Example 1 with the design region $0.0 \leq x_1 \leq 20.0$C$0.0 \leq x_2 \leq 20.0$.

7 Using SVM

Recently, Support Vector Machines (SVM's) have gained much popularity in machine learning. The main idea of SVM is the kernel based linear classifier with the maximal margin for pattern classification problems on the basis of statistical learninig. It has been suggested to apply SVM to regression problems. Denote the given sample data by $(\boldsymbol{x_i}, y_i)$ for $i = 1, ..., l$. Suppose that the regression function on the X space is expressed by $f(\boldsymbol{x}) = \sum_{i=1}^{l} w_i \boldsymbol{x_i} + b$. The linear ε loss function is defined by

$$L^{\varepsilon}(\boldsymbol{x}, y, f) = |y - f(\boldsymbol{x})|_{\varepsilon} = \max(0, |y - f(\boldsymbol{x})| - \varepsilon).$$

Similarly the quadratic loss function is defined by

$$L_2^{\varepsilon}(\boldsymbol{x}, y, f) = |y - f(\boldsymbol{x})|_{\varepsilon}^2.$$

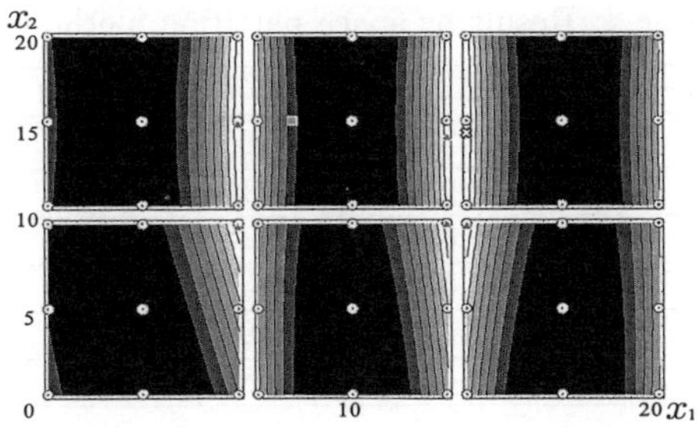

Figure 10: Result by space partition method (ii)

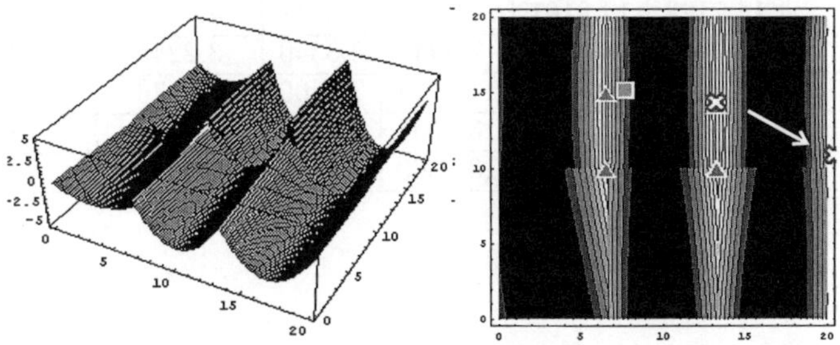

Figure 11: Final result by space partition method (ii)

Both of loss functions ignore errors that are within a certain distance of the true value. In this paper, we employ the quadratic ε loss function.

We can optimize our regressor by minimizing the sum of the quadratic ε loss function.

$$\text{Min} \quad \tfrac{1}{2}\|\boldsymbol{w}\|^2 + C\sum_{i=1}^{l} L_2^{\varepsilon}(\boldsymbol{x_i}, y_i, f)$$

In order to make a regression for general nonlinear cases, the given sample point $(\boldsymbol{x_i}, y_i)$ $i = 1, ..., l$ in the X space are mapped into a high demensional feature space by nonlinear transformation function ϕ. Now our regression problems are reduced to

Table 4: Result by space partition method (ii)

—	x_1	x_2	error of x	f	error of f	error of f (%)
1	6.6537	9.8512	5.2966	2.1760	7.3825	77.2349
2	6.6626	14.8431	1.2433	2.9510	6.6075	69.1270
3	13.3267	9.9878	7.3902	3.5475	6.0110	62.8864
4*	13.3331	14.0858	5.5134	5.4852	4.0733	42.6144
5	13.3347	9.8776	7.4712	3.5596	5.9989	62.7598
6	13.3336	14.2983	5.4827	5.4986	4.0599	42.4742

Table 5: Result by enlarged approximation (ii)

# analysis	x_1	x_2	error	f	error of f	error of f (%)
44	14.7123	14.9161	6.8168	5.5211	4.0374	42.2388
53	15.7895	15.0306	7.8936	6.3340	3.2245	33.7344
62	19.3108	11.5569	11.9228	1.7809	7.7776	81.3684
71	20.0196	9.0763	13.4934	2.0894	7.4691	78.1409
80	20.4505	10.8076	13.2360	2.5623	6.9962	73.1935

optimizing the following problem:

$$\begin{aligned} \text{Min} \quad & \|\boldsymbol{w}\|^2 + C\sum_{i=1}^{l}(\xi^2 + \acute{\xi}_i^2) \\ \text{s.t.} \quad & \sum_{i=1}^{l} w_i\phi(\boldsymbol{x_i}) + b - y_i \leq \varepsilon + \xi_i \\ & y_i - \sum_{i=1}^{l} w_i\phi(\boldsymbol{x_i}) - b \leq \varepsilon + \acute{\xi}_i \\ & \xi_i, \acute{\xi}_i \geq 0 \quad (i = 1, ..., l) \end{aligned} \tag{3}$$

where the slack variable ξ_i is the excess amount from the target value more than ε, and the other $\acute{\xi}_i$ is the unattainment amount below the target more than ε. Minimizing the quadratic programming (3) is equivalent to maximizing the dual problem by intorducing Lagrange multiplier $\alpha, \acute{\alpha}$. In the dual problem to (3) we employ the kernel function $K(\boldsymbol{x_i}, \boldsymbol{x_j})$ which satisfies

$$\phi(\boldsymbol{x_i})^T\phi(\boldsymbol{x_j}) = K(\boldsymbol{x_i}, \boldsymbol{x_j}). \tag{4}$$

Finally we have the following formulation:

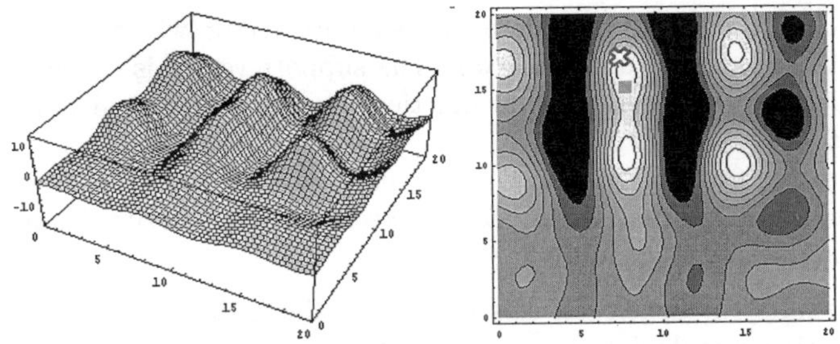

Figure 12: Result by RBFN for Example 1 with a modified design region (69 sampled data)

$$
\begin{aligned}
\text{Max} \quad & -\tfrac{1}{2}\sum_{i,j=1}^{l}(\acute{\alpha}_i-\alpha_i)(\acute{\alpha}_j-\alpha_j)\{K(\boldsymbol{x_i},\boldsymbol{x_j})+\frac{1}{C}\delta_{ij}\} \\
& +\sum_{i=1}^{l}y_i(\acute{\alpha}_i-\alpha_i)-\varepsilon\sum_{i=1}^{l}(\acute{\alpha}_i+\alpha_i) \\
\text{s.t.} \quad & \sum_{i=1}^{l}(\acute{\alpha}_i-\alpha_i)=0 \\
& \acute{\alpha}_i\ ,\ \alpha_i\geq 0 \quad (i=1,...,l)
\end{aligned}
\tag{5}
$$

Here δ_{ij} is Kronecker δ defined to be 1 if $i=j$ and 0 otherwise. The corresponding Kaurush-Kuhn-Tucker complementarity conditions are

$$
\begin{aligned}
\alpha_i(f(\boldsymbol{x_i})-y_i-\varepsilon-\xi_i)=0 \\
\acute{\alpha}_i(y_i-f(\boldsymbol{x_i})-\varepsilon-\acute{\xi}_i)=0 \\
\xi_i\acute{\xi}_i=0,\ \alpha_i\acute{\alpha}_i=0 \quad (i=1,...,l).
\end{aligned}
$$

We solve the dual problem (5) rather than the primal problem (3) with the Guassian kernel

$$K(\boldsymbol{x_i},\boldsymbol{x_j})=\exp(\frac{-\|\boldsymbol{x_i}-\boldsymbol{x_j}\|}{r^2}). \tag{6}$$

As well as in RBFN, the width r can be well estimated by $d_{max}/\sqrt[n]{nm}$, where d_{max} is the maximum distance between sample data, n is the dimension of sample data in the X space, and m is the number of sample data. The regression with Kernel in the high dimensional feature space is given by

$$f(\boldsymbol{x})=\sum_{i=1}^{l}\alpha_i{}^*K(\boldsymbol{x_i},\boldsymbol{x})+b^* \quad (\alpha_i{}^*=\acute{\alpha}_i-\alpha_i). \tag{7}$$

If $\alpha_i{}^* \neq 0$ then the ith sample data is a support vector which is on the band of $\pm\varepsilon$ around the predicted function $f(\boldsymbol{x})$. The support vector is considered to be an important data, since the regression function $f(\boldsymbol{x})$ is represented only in terms of support vectors.

The algorithm using SVM is almost the same as that of RBFN. However, we can utilize the information of α_i for i-th support vector as a sensitivity. Namely, we select additional sample points as follows:

Step 1: Consider the density of existing sample data, that is, find a part in which the existing sample data is sparse. To do this, some candidates of sample data $\boldsymbol{x}_i^c$, for example $i = 1, ..., N$, are generated randomly.

Step 2: Measure the distance between each candidate data and every existing sample data.

Step 3: Take a small number (for example 5) of the candidate sample data having the maximum distance to the closest sample data.

Step 4: Consider the sensitivity α_c^* at each candidate of sample data selected in Step 3.

Step 5: Add a data which has the biggest value of $|\alpha_c^*|$ as a new sample data.

Let x^c be a candidate of new data. Since the true value of $y(x^c)$ is not available, we use the predicted value $f(x^c)$ instead. In order to obtain the sensitivity at x^c, we add the following constraints (8) to the primal problem (3).

$$\begin{aligned} & f(\boldsymbol{x}^c) - \sum_{i=1}^{l+1} w_i \phi(\boldsymbol{x}_i) - b \leq \varepsilon + \acute{\xi}_{l+1} \\ & \sum_{i=1}^{l+1} w_i \phi(\boldsymbol{x}_i) + b - f(\boldsymbol{x}^c) \leq \varepsilon + \xi_{l+1} \\ & \xi_{l+1}, \acute{\xi}_{l+1} \geq 0 \end{aligned} \tag{8}$$

After adding new constraints (8), the dual problem in the high dimensional feature space is given by the formulation (9) below.

$$\begin{aligned} \text{Max} \quad & -\tfrac{1}{2} \sum_{i,j=1}^{l+1} \alpha_i^* \alpha_j^* \{K(\boldsymbol{x_i}, \boldsymbol{x_j}) + \frac{1}{C}\delta_{ij}\} \\ & \sum_{i=1}^{l} y_i \alpha_i^* - \varepsilon \sum_{i=1}^{l+1} \alpha_i^* + f(\boldsymbol{x}^c)\alpha_{l+1}^* \\ \text{s.t} \quad & \sum_{i=1}^{l+1} \alpha_i^* = 0 \\ & \acute{\alpha}_i \ , \ \alpha_i \geq 0 \ \ (i = 1, ..., l+1) \end{aligned} \tag{9}$$

where $\alpha_i^* = \acute{\alpha}_i - \alpha_i$. We select a new data x^c which has the largest value of α_c^* in the solution to (9). The result by using SVM is given in Table 6. In this example, there is no big difference between RBFN and SVM.

Table 6: Result by SVM for Example 1

No	#analysis	$f(x)$	x_1	x_2
1	67	9.5170	7.8911	14.3409
2	71	9.4478	7.7956	14.1936
3	55	9.4751	7.8791	14.0714
4	51	8.4422	14.0750	14.6326
5	59	9.5464	7.8894	14.6467
6	65	9.2321	7.8529	13.1615
7	69	9.5582	7.9013	14.9559
8	47	9.5580	7.8929	14.9265
9	73	9.5320	7.9008	15.5263
10	63	9.4652	7.8989	14.0099
ave	62.0	9.3774	8.4977	14.4465
σ	8.307	0.3248	1.8594	0.6155

8 Using Expected Improvement

Jones *et al.* [5] suggested a method called EGO (Efficient Global Optimization) for black-box objective functions. They applied a stochastic process model for predictor and the expected improvement as a figure of merit for additional sample points. Namely, letting $\boldsymbol{y} = (y_1, \dots, y_p)$ denote the p-vector of observed function values corresponding to the sample points $(\boldsymbol{x}_1, \dots, \boldsymbol{x}_p)$, then the predictor of black-box objective function is given by

$$\hat{y}(\boldsymbol{x}) = \hat{\mu} + \boldsymbol{r}^T R^{-1}(\boldsymbol{y} - \mathbf{1}\hat{\mu})$$

where the i-th element of $\boldsymbol{r}$ is $r_i(\boldsymbol{x}) = \mathrm{Corr}[\varepsilon(\boldsymbol{x}), \varepsilon(\boldsymbol{x}_i)]$ and R is a correlation matrix whose (i, j) entry is

$$\mathrm{Corr}[\varepsilon(\boldsymbol{x}_{(i)}), \varepsilon(\boldsymbol{x}_{(j)})] = -\exp[-\frac{\|\boldsymbol{x}_{(i)} - \boldsymbol{x}_{(i)}\|^2}{r^2}]. \tag{10}$$

Here, $\varepsilon(\boldsymbol{x})$ is the error between the real value of objective function and the predicted value at $\boldsymbol{x}$.

The estimated value of the mean of the stochastic process, $\hat{\mu}$, is given by

$$\hat{\mu} = \frac{\mathbf{1}^T \boldsymbol{R}^{-1} \boldsymbol{y}}{\mathbf{1}^T \boldsymbol{R}^{-1} \mathbf{1}}. \tag{11}$$

In this event, the variation σ^2 is estimated by

$$\hat{\sigma}^2 = \frac{(\boldsymbol{y} - \mathbf{1}\hat{\mu})^T \boldsymbol{R}^{-1} (\boldsymbol{y} - \mathbf{1}\hat{\mu})}{n} \tag{12}$$

The mean squared error of the predictor is estimated by

$$s^2(\boldsymbol{x}) = \sigma^2 [1 - \boldsymbol{r}^T \boldsymbol{R}^{-1} \boldsymbol{r} + \frac{(1 - \mathbf{1}^T \boldsymbol{R}^{-1} \boldsymbol{r})^2}{\mathbf{1}^T \boldsymbol{R}^{-1} \mathbf{1}}] \tag{13}$$

In the following $s = \sqrt{s^2(\boldsymbol{x})}$ is called a standard error.

Using the above predictor on the basis of stochastic process model, Jones *et al.* applied the expected improvemnet for adding new sample point. Let $f_{\min}^p = \mathrm{Min}\{y_1, \dots, y_p\}$ be the current best function value. They model the uncertainty at $y(\boldsymbol{x})$ by treating it as the realization of a normally distributed random variable Y with mean and standard deviation given by the above predictor and its standard error.

For minimization cases, the improvement at $\boldsymbol{x}$ is $I = [\max(f_{\min}^p - Y,\ 0)$. Therefore, the expected improvement is given by

$$\mathrm{E}[I(\boldsymbol{x})] = \mathrm{E}[\max(f_{\min}^p - Y,\ 0)].$$

It has been shown that the above formula can be expanded as follows:

$$E(I) = \begin{cases} (f_{\min}^p - \hat{y})\Phi(\frac{f_{\min}^p - \hat{y}}{s}) + s\phi(\frac{f_{\min}^p - \hat{y}}{s}) & \text{if } s < 0 \\ 0 & \text{if } s = 0 \end{cases} \tag{14}$$

where ϕ is the standard normal density and Φ is the distribution function.

We can add a new sample point which maximizes the expected improvement. Although Jones *et al.* proposed a method for maximizing the expected improvement by using the branch and bound method, we select the best one among several candidates which are generated randomly in the design variable space in the following examples.

Furthermore, Schonlau [15] extended the expected improvement as follows: Letting $I^g = \max((f_{\min}^p - Y)^g,\ 0)$, then

$$E(I^g) = s^g \sum_{i=0}^{g} (-1)^i (\frac{g!}{i!(g-i)!})(f_{min}^{p'})^{g-i} T_i \tag{15}$$

where

$$f_{min}^{p'} = \frac{f_{min}^p - \hat{y}}{s}$$

and

$$T_k = -\phi(f^{p'}_{min})(f^{p'}_{min})^{(k-1)} + (k-1)T_{k-2}.$$

Here

$$T_0 = \Phi(f^{p'}_{min})$$

$$T_1 = -\phi(f^{p'}_{min})$$

It has been observed that larger value of g makes the global search, while smaller value of g the local search. Therefore, we can control the value of g depending our situations. Table 7-9 show the result by SVM using the expected improvement. The result seems good, but it should be noted that it takes an expensive computation time for calculating EI (about 2 hours for this example, while about 5 minutes simply by RBFN and SVM).

Table 7: Result by SVM using EI (g=1)

No	#analysis	f	x_1	x_2
1	54	8.9632	8.0273	17.3548
2	54	9.1618	7.8614	17.0439
3	50	8.9941	8.0466	17.2179
4	54	9.2583	8.0470	16.4205
5	52	9.1278	8.1740	16.9631
6	75	9.0487	7.8480	12.6841
7	74	8.4320	14.0348	15.5009
8	59	9.4245	7.7702	15.7772
9	53	9.2185	7.8960	16.9033
10	45	9.1854	7.9435	16.9167
ave	57.0	9.0814	8.5522	16.2782
σ	9.3702	0.2513	1.8298	1.3300

9 Numerical Example 2 (Design of Pressure Vessels)

As a more practical problem, consider the pressure vessel design problem given by Sandgren (1990), Kannan and Kramer (1994), Qian *et al.* (1993), Hsu, Sun *et al.* (1995), Lin *et al.* (1995), Arakawa *et al.* (1997). A cylindrical pressure vessel is

Table 8: Result by SVM using EI (g=3)

No	#analysis	f	x_1	x_2
1	45	9.5560	7.8860	14.8544
2	60	8.8049	7.6294	17.1364
3	45	9.5124	7.8749	15.6792
4	71	9.3250	7.7630	13.7470
5	45	9.5046	7.9882	15.3608
6	56	9.4879	7.9948	15.4910
7	51	8.7833	7.7936	17.8148
8	48	9.3707	7.8859	16.4068
9	73	8.8962	7.5995	13.3690
10	45	9.2693	7.7899	16.5804
ave	53.9	9.2510	7.8205	15.6444
σ	10.2900	0.2908	0.1266	1.3379

Table 9: Rsult by SVM using EI (g=5)

No	#analysis	f	x_1	x_2
1	47	9.3497	7.7004	15.5026
2	52	9.4851	7.9629	15.7364
3	46	9.4046	7.8226	16.1612
4	47	9.4424	7.8198	15.962
5	48	9.4615	8.0242	14.5809
6	46	9.4247	7.7597	15.6787
7	49	8.6188	7.8467	11.8025
8	61	9.4426	7.8160	14.0547
9	51	9.3666	8.0939	14.9332
10	54	9.3649	8.0724	14.3351
ave	50.1	9.3361	7.8919	14.8747
σ	4.4373	0.2428	0.1296	1.2293

capped at both ends by hemispherical heads as shown in Fig. 13. The shell is to be made in two halves, of rolled plates which are joined by two longitudinal welds to form a cylinder. Each head is forged and then welded to the shell. All welds are single welded butt joints with a backing strip. The material used in the vessel is carbon steel ASME SA203 grade B. The pressure will be oriented such that the axis of the cylindrical shell is vertical. The pressure vessel is to be a compressed air storage tank with a working pressure of 2.068×10^7 Pa and a minimum volume of 1.229×10^{-2} m^3. The objective is to minimize the total cost of manufacturing the pressure vessel, including the cost of material and cost of forming and welding. The design variables are R and L, the inner radius and length of the cylindrical selection, and Ts and Th are integer multiples of 0.0015875 m, the available thickness of rolled steel plates. The constraints g_1, g_2 and g_3 correspond to ASME limits on the geometry while g_4 corresponds to a minimum volume limit.

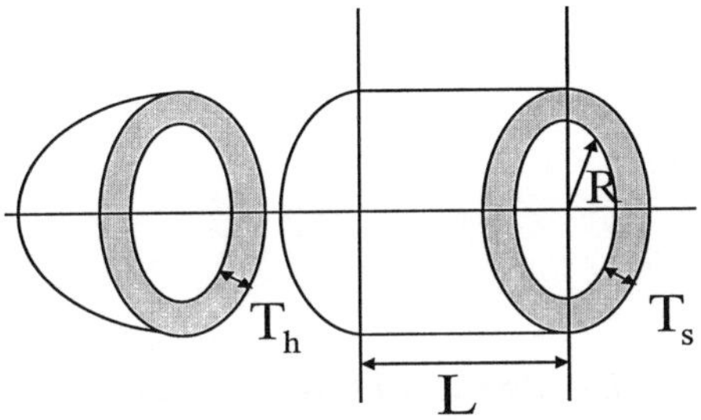

Figure 13: Pressure vessel

We can summarize our problem as follows:

<u>Design variables:</u>

R : inner radius (continuous),

L : length of the cylindrical section (continuous),

Ts : thickness of rolled steel plate (discrete),

Th : thickness of rolled steel plate (discrete).

<u>Constraints:</u>

g_1: minimum shell wall thickness

$$\frac{0.00049022\ R}{Ts} \leq 1.0,$$

g_2: minimum head wall thickness

$$\frac{0.000242316\ R}{Th} \leq 1.0,$$

g_3: maximum length

$$\frac{L}{240} \leq 1.0$$

g_4: minimum volume of tank

$$\frac{-\frac{4}{3}\ \pi\ R^3 + 21.24}{\pi\ R^2\ L} \leq 1.0.$$

Side constraints:

$$R_{LB} \leq R \leq R_{UB},$$
$$L_{LB} \leq L \leq L_{UB},$$
$$Ts_{LB} \leq Ts \leq Ts_{UB},$$
$$Th_{LB} \leq Th \leq Th_{UB}.$$

Objective:

Minimize

$$f = 37982.2\ Ts\ R\ L + 108506.3\ Th\ R^2 + 193207.3\ Ts^2\ L + 1210711 Ts^2\ R.$$

In this example, the objective function is explicitly given in terms of design variables. However, many researchers use this problem as a bench mark, and hence we use the problem in order to see the effectiveness of our method by treating it in such a way that the objective function is not known. The parameters of the RBF network are given by $\lambda = 0.01$, and the width r is decided by the conventional form. In this example, a simple GA is adopted: continuous variables are transformed into binary variables of 11 bit. On the other hand, discrete varibles are transformed into binary ones of 4 bit. The population is 80, the generation 100, and the rate of mutation 10%. The constraint functions are penalized by

$$F = f(x) + \sum_{i=1}^{3} p_j \times [P(g_j(x))]^a,$$

where we set $p_j = 100$(penalty coefficient), $a = 2$(penalty exponent) and $P(y) = max\{y, 0\}$. Our problem is to minimize the augmented objective function F.

The data for initial learning is 17 points at the corners of the region given by the side constraints and its center. Finally, each side of the rectangle controlling the convergence and the neighborhood of the point under consideration is half the upper bound - the lower bound of each side constraint. The results of numerical simulation for various stop conditions are given in the following tables.

Table 9 shows the comparison among all existing methods. In this table, the best solution by each method is listed (Sandgren 1990, Kannan and Kramer 1994, Qian *et al.* 1993, Hsu, Sun *et al.* 1995, Lin Zhang *et al.* 1995, Arakawa *et al.* 1997). Sandgren (1990) and Kannan *et al.* (1994) use gradient type optimization techniques which treat constraints by penalty function or augmented Lagrangean methods. Lin

Table 10: Results by RBFN for Example 2 with $C_x^0 = 60$ and $C_f^0 = 30$

trial	# of data	R (m)	L (m)	Ts (cm)	Th (cm)	f ($)
1	83	1.1515	3.5643	2.2225	1.1112	6092.1030
2	143	0.9855	5.6505	1.9050	0.9525	5862.2705
3	153	1.1499	3.5803	2.2225	1.1113	6099.2701
4	143	0.9700	5.9066	1.9050	0.9525	5958.4034
5	139	1.1499	3.5803	2.2225	1.1113	6099.2701
6	139	1.0677	4.5620	2.0638	1.1113	6118.7685
7	147	1.1499	3.5803	2.2225	1.1113	6099.2701
8	131	1.0677	4.5620	2.0638	1.1113	6118.7685
9	165	1.0693	4.4873	2.0638	1.1113	6060.3558
10	133	1.1515	3.5643	2.2225	1.1113	6092.1030
AVERAGE	137.6					6060.0580
σ	21.6					83.6750

et al. (1995) uses simulated annealing method and genetic algorithm, and Qian *et al.* (1993) and Arakawa *et al.* (1997) use genetic algorithms.

On the other hand, the results by SVM with $C = 1.0$, $\varepsilon = 0.1$ is given in Table 11.

Table 11: Results by SVM for example 2 with $C_x^0 = 60$ and $C_f^0 = 30$

No	#analysis	$	R	L	Th	Ts
1	119	5850.7425	0.9796	5.6238	0.9525	1.893
2	131	5851.0251	0.9869	5.6246	0.9525	1.893
3	125	5851.1846	0.9871	5.6231	0.9525	1.893
4	116	5849.7117	0.9871	5.6211	0.9525	1.893
5	134	5851.4190	0.9869	5.6254	0.9525	1.893
6	113	5852.7372	0.9868	5.6273	0.9525	1.893
7	116	5851.2598	0.9869	5.6253	0.9525	1.893
8	119	5849.6567	0.9871	5.6212	0.9525	1.893
9	116	5850.0106	0.9871	5.6221	0.9525	1.893
10	125	5852.3037	0.9868	5.6272	0.9527	1.893
ave	121.4	5851.0051	0.9870	5.6241	0.9527	1.893
σ	6.6813	0.9749	0.0046	0.0843	0.000	0.000

Table 9 compares the best result for each method so far. Table 9 is the result by GA only. It can be readily seen that GA requires a tremendous number of experiments.

Table 12: Comparison among existing methods

trial	R (m)	L (m)	Ts (cm)	Th (cm)	f ($)
Sandgren	1.212	2.990	2.858	1.588	8129.80
Qian(GAs)	1.481	1.132	2.858	1.588	7238.83
Kannan	1.481	1.198	2.858	1.588	7198.20
Lin	N/A	N/A	N/A	N/A	7197.70
Hsu	1.316	2.587	2.540	1.270	7021.67
Lewis	0.985	5.672	1.905	0.953	5980.95
Arakawa	0.987	5.626	1.905	0.953	5851.78
RBFN	0.986	5.651	1.905	0.953	5862.27
SVM	0.985	5.359	1.905	0.953	5849.67

10 Concluding Remarks

We reviewd several methods for optimizing black-box objective functions. Especially, it has been seen that machine learning approaches to RSM such as RBFN and SVM can be effectively applied to this purpose in particular for highly nonlinear cases. The authors are now applying the method to a practical problem in reinforcement of a cable stayed bridge with several decades of variables and other engineering design problems. These results are promising, and will be reported in the near future.

References

[1] Arakawa, M. and Hagiwara, I. (1997) Nonlinear Integer, Discrete and Continuous Optimization Using Adaptive Range Genetic Algorithms, *Proc. of ASME Design Technical Conferences (in CD-ROM)*

[2] Cristianini, N. and Shawe-Tylor, J. (2000) *An Introduction Support Vector Machines and Other Kernel-based Learning Method*, Cambridge Univ. Press

[3] Haykin, S. (1994), *Neural Networks: A Comprehensive Foundation*, Macmillan College Publishing Company

[4] Hsu, Y.H., Sun, T.L. and Leu, L.H. (1995), A Two-stage Sequential Approximation Method for Nonlinear Discrete Variable Optimization, *ASME Design Engineering Technical Conference*, Boston, MA, 197-202

[5] Jones, D.R., Schonlau, M. and Welch, W.J. (1998) Efficient Global Optimization of Expensive Black-Box Functions, *J. of Global Optimization*, **13**, 455-92

Table 13: Results by GA only for Exapme 2

trial	# of data	R (m)	L (m)	Ts (cm)	Th (cm)	f ($)
1	240	1.1313	4.9142	2.3813	1.1113	7886.5745
2	2880	1.0615	4.5887	2.0638	1.1113	6101.8484
3	4720	1.1469	3.6443	2.2225	1.1113	6147.5434
4	3600	1.1375	3.7083	2.2225	1.1113	6155.3072
5	3200	1.0600	4.6367	2.0638	1.1113	6135.3518
6	2240	1.0507	4.7461	2.0638	1.1113	6172.1275
7	320	0.9964	5.5624	2.3813	0.9525	7332.2005
8	12720	0.9824	5.7732	1.9050	0.9525	5937.7528
9	2960	1.0476	4.7914	2.0638	1.1113	6192.1117
10	2560	1.1500	3.5856	2.2225	1.1113	6104.9587
AVERAGE	3544					6416.5776
σ	3503.41					645.8896

[6] Kannan, B.K. and Kramer, S.N. (1994), An Augmented Lagrange Multiplier Based Method for Mixed Integer Discrete Continuous Optimization and Its Appliations to Mechanical Design, *Transactions of ASME, J. of Mechanical Design*, Vol. 116, 405-411

[7] Lewis, K. and Mistree, F. (1996) Foraging-Directed Adaptive Linear Programming : An Algorithm For Sloving Nonlinear Mixed Discrete/Continuous Design Problems, *Proc. AMSE Design Engeneering Technical Conference*

[8] Myers, R.H. and Montgomery,D.C. (1995), *Response Surface Methodology: Process and Product Optimization using Designed Experiments*, Wiley

[9] Nakayama, H., Yanagiuchi, S., Furukawa, K., Araki, Y., Suzuki, S. and Nakata, M. (1998), Additional Learning and Forgetting by RBF Networks and its Application to Design of Support Structures in Tunnel Construction *Proc. International ICSC/IFAC Symposium on Neural Computation (NC'98)*, 544-550

[10] Nakayama, H., Arakawa, M., Sasaki, K. (2002), Simulation based Optimization for Unknown objective Functions, *Optimization and Engineering*, to appear

[11] Orr, M.J.L. (1996), Introduction to Radial Basis Function Networks, http://www.cns.ed.ac.uk/people/mark.html

[12] Qian, Z., Yu, J. and Zhou, J. (1993), A Genetic Algorithm for Solving Mixed Discrete optimization Problems, *DE-Vol. 65-1, Advances in Design Automation*, Vol. 1, 499-503

[13] Radcliffe, N.J. (1991), Forma Analysis and Random Respectful Recombination, *Proceedings of the Forth International Conference on Genetic Algorithms*, 222-229

[14] Sandgren, E. (1990), Nonlinear Integer and Discrete Programming in Mechanical Engineering Systems, *J. of Mechanical Design*, Vol. 112, 223-229

[15] Schonlau, M. (1997) *Computer Experiments and Global Optimization*, PhD. thesis, Univ.of Waterloo, Ontario, Canada

[16] Zhang, C. and Wang, H.P. (1993), Mixed-Discrete nonlinear Optimization with Simulated Annealing, *Engineering Optimization*, Vol. 21, 277-291

Optimization and Optimal Control, pp. 211 – 231
P.M. Pardalos, I. Tseveendorj, and R. Enkhbat, Editors

D.C. Approach to Bilevel Bilinear Programming Problem: Application in Telecommunication Pricing

T.Q. Nguyen (quang.nguyen@rd.francetelecom.com)
France Telecom R&D, Laboratoire DAC-OAT 38-40, rue du General-Leclerc 92794 Issy les Moulineaux Cedex 9 (France)

M. Bouhtou and J-L. Lutton
(mustapha.bouhtou(jeanluc.lutton)@rd.francetelecom.com)
France Telecom R&D, Laboratoire DAC-OAT 38-40, rue du General-Leclerc 92794 Issy les Moulineaux Cedex 9 (France)

Abstract

In this paper, we consider tariff-traffic equilibria in a two-level hierarchical system where the leader wants to maximize its revenues while the follower strives to minimize its activity costs. The problem is formulated as a bilevel bilinear program (BBP). We provide an algorithm based on a recent non-convex optimization approach. The (BBP) is reformulated as a d.c. program (DCP), i.e. minimizing a difference of convex functions (d.c. functions) on a polyhedral set and a descent method is developed to solve (DCP). Computational results for some numerical examples are reported.

Keywords: d.c. optimization, Bilevel Programming, pricing, telecommunca-tion

1 Introduction

The bilevel bilinear optimization programming problem (BBP) discussed in this paper arises in the study of tariff-traffic equilibria in two-level hierarchical system. This problem has been first proposed and studied by Labbé and co-authors (see [LMS98]).

It has wide application in numerous areas, particularly in network problem. See Vicente and Calamai [VC94] for a recent review of the literature. The mathematical formulation of the (BBP) is as follows:

$$(BBP)\quad \begin{cases} \max\limits_{t} \left\{ t^T x_1 ; t \in T \right\}, \\ \text{where } x_1 \text{ for each value of } t \text{ is the solution} \\ \text{of the lower level problem :} \\ \min\limits_{x_1, x_2} \left\{ t^T x_1 + c_1^T x_1 + c_2^T x_2); x = (x_1, x_2) \in X \right\}, \end{cases}$$

where $t \in \Re^{m_1}, x_1 \in \Re_+^{m_1}, x_2 \in \Re_+^{m_2}, T = [\alpha, \beta], \alpha \in \Re^{m_1}, \beta \in \Re^{m_1}$ and X is a polyhedral set (closed and bounded region of $\Re^{m_1+m_2}$).

Bilevel programming problem can be interpreted in the following way. At the higher level the decision maker has to choose a variable $t \in T$ solution to the leader's problem and at the lower level the users have to select the reaction of the variable $x = (x_1 = x_1(t), x_2 = x_2(t))$, solution of the follower's problem. The decision variables t corresponds to the "tax level" attached to the "activity" vector variables x_1. Each decision t of the leader modifies the objective function of the follower's problem, i.e so as to minimize the lower level objective $t^T x_1 + c_1^T x_1 + c_2^T x_2$ under constraints $(x_1, x_2) \in X$. The (BBP) problem is to find t such that the expected response $x = (x_1(t), x_2(t))$ of the follower maximizes the leader's revenues while the follower strives to minimize its activity costs.

In this paper, we provide an algorithm based on a recent non-convex optimization approach: *d.c. optimization.* The (BBP) is reformulated as a d.c. program (DCP), i.e. minimizing a difference of convex functions (d.c. functions) on a polyhedral set and a descent method is developed to solve (DCP) [Tu95]. Note that Tuy et al have proposed a reformulation of bilevel linear program to an equivalent d.c. program or more precisely into a linear program with an additional reverse concave constraint (in [MPV98], chapter 10). In the next sections, with the aim of having practical interpretation results, we consider a (BBP) formulation of a network tariff-setting problem, however results are always good for the general (BBP) problem. In Section 3, we give an equivalent d.c. reformulation of the problem. D.c. programming preliminaries are defined in Section 4. In Section 5 a d.c. heuristic is described. An illustrative example and preliminary computational experiments are given in Section 6.

2 Formulation of the network tariff-setting problem

Consider a telecommunication traffic network $G = (V, U)$ consisting of a set V of n nodes and a set U of m arcs. Nodes represent origin, destination or transit stations for telecommunication traffics, and arcs represent transmission lines. The arc set of the network G is partitioned into two subsets U_1 and U_2 which are respectively the set of links operated by the leader telecommunication operator and by the competitor operators. The problem is to determine the tariffs of toll arcs $(t_u)_{u \in U_1}$ and the traffic flows $(x_u)_{u \in U}$ such that the amount of traffic that can be transported by the leader is maximal (see [BLMS98] for an extensive description). The network tariff-setting problem is:

$$
(P) \left\{ \begin{array}{lll} \max_t & \sum_{k=1}^{K} t^T x_1^K, & \\ \text{subject to} & t \in T & (1) \\ & & \\ \min_{x_1, x_2} & \sum_{k=1}^{K} \left[\left(t^T + c_1^T \right) x_1^k + c_2^T x_2^k \right] & \\ \text{subject to} & \left(x_1^k, x_2^k \right)_{k=1,\dots,K} \in X & (2) \\ & x^k \in D & (3) \end{array} \right.
$$

where :

- $| V |= n$, the number of nodes, $| U |= m$, the number of arcs, $| U_1 |= m_1$, the number of toll arcs of the leader and $| U_2 |= m_2$ the number of not toll arcs $\quad (m = m_1 + m_2)$,
- $t = (t_1, ..., x_{m_1})^T$ is the decisional variable of the leader (tax on set of toll arcs),
- $x^k = (x_1^k, x_2^k), k = 1, ..., K$, are the decisional variables of the follower,
  - $x_1^k = \left(x_{1,1}^k, ..., x_{1,m_1}^k \right)$: flow vector on set of arcs U_1 for the single commodity k,
  - $x_2^k = \left(x_{2,1}^k, ..., x_{2,m_2}^k \right)$: flow vector on set of arcs U_2 for the single commodity k,
- (1) is the tax constraint: $t \in T = [\alpha, \beta], \alpha \in \Re^{m_1}, \beta \in \Re^{m_1}$,
- (2) is the flow conservation constraint, i.e.

$$
X = \left\{ x^k = (x_1^k, x_2^k) : Ax^k = d^k \delta^k ; x^k \geq 0; k = 1, ..., K \right\},
$$

where $x^k, k = 1, ..., K$, are K single commodities : for $k = 1, .., K, \quad x^k (\in I\!R_+^m)$ is a single commodity with demand $d^k (\in I\!R^+)$ between source s^k and sink r^k, such that the flow conservation is satisfied, i.e $A\varphi^k = d^k\delta^k$, where A is the node-arc incident matrix of the graph G and $\delta^k(\in I\!R^m)$ the vector equal 1 on s^k, -1 on r^k and 0 elsewhere. Like G is partitioned into two subsets U_1 and U_2, the matrix A can be written as $[A_1, A_2]$ where the bloc-matrix A_1 (resp. A_2) is the node-arc incident matrix of (V, U_1), (resp. (V, U_2)).

This constraint has a bloc-diagonal structure:

$$\begin{bmatrix} A_1A_2 & & & & \\ & .. & & & \\ & & A_1A_2 & & \\ & & & .. & \\ & & & & A_1A_2 \end{bmatrix} \begin{pmatrix} x^1 \\ : \\ x^k \\ : \\ x^K \end{pmatrix} = \begin{pmatrix} d^1\delta^1 \\ .. \\ d^k\delta^k \\ .. \\ d^K\delta^K \end{pmatrix}$$

where $x^k = (x_u^k)_{u\in U} = \begin{pmatrix} x_1^k=(x_u^k)_{u\in U_1} \\ x_2^k=(x_u^k)_{u\in U_2} \end{pmatrix}$

- (3) is the capacity constraint on the arc flows : $D = [0, \gamma], \gamma \in \Re_+^m$.

3 The equivalent d.c. reformulation

The above bilevel program (P) can be formulated as a d.c. programming problem. For that, we used the single level equivalent problem of (P) obtained by replacing the lower level problem by its primal-dual optimality constraints ([LMS98]), i.e.

$$(P)\quad \begin{array}{lllr} \max_{t,x,\lambda} & \sum_{k=1}^K t^T x_1^k, & & \\ \text{s.t.} & t \in T & & (1) \\ & A_1x_1^k + A_2x_2^k = d^k\delta^k & k = 1, ..., K & (A) \\ & \left.\begin{array}{l} \left(\lambda^k\right)^T A_1 \le t + c_1 \\ \left(\lambda^k\right)^T A_2 \le c_2 \\ \lambda^k \in I\!R^n \end{array}\right\} & k = 1, ..., K & (B) \\ & \sum_{k=1}^K \left[(t^T + c_1^T)x_1^k + c_2^T x_2^k) - \left(\lambda^k\right)^T d^k\delta^k\right] = 0 & & (C) \\ & x^k \in D & k = 1, ..., K & (3) \end{array}$$

- (A): primal feasibility,
- (B): dual feasibility,
- (C): equality of primal and dual objectives.

Next we penalized the constraint (C). This yields the bilinear program:

$$\begin{array}{llll}
 & \max_{t,x,\lambda} & \sum_{k=1}^{K} \left[t^T x_1^k - M \left[(t^T + c_1^T) x_1^k + c_2^T x_2^k) - \left(\lambda^k\right)^T d^k \delta^k \right] \right] & \\
 & \text{s.t.} & t \in T & (1) \\
(P_M) & & A_1 x_1^k + A_2 x_2^k = d^k \delta^k \qquad k = 1, ..., K & (A) \\
 & & \left.\begin{array}{l} \left(\lambda^k\right)^T A_1 \leq t + c_1 \\ \left(\lambda^k\right)^T A_2 \leq c_2 \\ \lambda^k \in I\!R^n \end{array}\right\} \qquad k = 1, ..., K & (B) \\
 & & x^k \in D \qquad k = 1, ..., K & (3)
\end{array}$$

with $M > 0$ sufficiently large such that any global solution of problem (P_M) is also a global solution of (P). This penalty scheme is said to be exact.

Finally, the d.c. formulation equivalent problem is obtained by expressing the bilinear objective function as the difference of two convex functions, i.e.

$$\begin{aligned}
f(t, x, \lambda) &= \sum_{k=1}^{K} \left[t^T x_1^k - M \left[(t^T + c_1^T) x_1^k + c_2^T x_2^k) - \left(\lambda^k\right)^T d^k \delta^k \right] \right] \\
&= h(x, t) - g(x, t, \lambda)
\end{aligned}$$

with the convex components:

$$\begin{aligned}
g(t, x, \lambda) &= (M - 1) \sum_{k=1}^{K} \left(t^T x_1^k + \rho \left\| t \right\|^2 + \rho \left\| x_1^k \right\|^2 \right) \\
&\quad + M \sum_{k=1}^{K} \left(c_1^T x_1^k + c_2^T x_2^k - \left(\lambda^k\right)^T d^k \delta^k \right)
\end{aligned} \tag{1}$$

and

$$h(t, x) = (M - 1) \sum_{k=1}^{K} \left(\rho \left\| t \right\|^2 + \rho \left\| x_1^k \right\|^2 \right) \tag{2}$$

with $M > 1$ and $\rho \geq 1/2$.

Proof. It is immediate from the relation:

$$t^T x_1^k + \rho \left\| t \right\|^2 + \rho \left\| x_1^k \right\|^2 = \rho \sum_{u=1}^{m_1} (t_u + x_{1,u}^k)^2, \forall \rho \geq 1/2, \tag{3}$$

where $\|.\|$ is the Euclidean norm, i.e. $\left\| x_1^k \right\|^2 = \sum_{u=1}^{m_1} \left(x_{1,u}^k \right)^2$.

4 D.c. programming preliminaries

In this section we define a number of concepts from d.c. optimization, which enter both into theoretical results and into practical conditions imposed on the steps in an optimization algorithm. Consider the following d.c. program (P),

$$\alpha = min\{g(x) - h(x) \quad / \quad x \in X\} \quad (P) \tag{4}$$

where g and h are convex functions on a convex set X in $I\!R^m$, and its d.c. dual formulation (D),

$$\alpha = min\{h^*(y) - g^*(y) \quad / \quad y \in Y\} \quad (D) \tag{5}$$

where g^* denote the conjugate function of g on Y(: dual vectorial space of X) defined by : $g^*(y) = sup\{\langle x, y\rangle - g(x) \quad / \quad x \in X\}$.

This d.c. duality result defined with the conjugate function give an important d.c. optimization relation (Toland, 1978).

Proposition 1 Let $g : I\!R^n \to I\!R \cup \{+\infty\}$ be an arbitrary proper function and $h : I\!R^n \to I\!R \cup \{+\infty\}$ a convex proper l.s.c. (lower semi-continuity). Then

$$\inf_{x \in domg} \{g(x) - h(x)\} = \inf_{y \in domh^*} \{h^*(y) - g^*(y)\} \tag{6}$$

If y^0 is a minorant of $h^*(y) - g^*(y)$ then each $x^\bullet \in \partial g(x^\bullet)$ is a minorant of $g(x) - h(x)$.
Proof.

$$\begin{aligned} \alpha &= \inf\{g(x) - h(x) \quad / \quad x \in X\} \\ &= \inf\{g(x) - sup\{\langle x, y\rangle - h^*(y) \quad / \quad y \in Y\} \quad / \quad x \in X\} \\ &= \inf\{g(x) + inf\{h^*(y) - \langle x, y\rangle \quad / \quad y \in Y\} \quad / \quad x \in X\} \\ &= \inf_x \inf_y \{h^*(y) - \langle x, y\rangle - g(x)\} \\ &= \inf\{h^*(y) - g^*(y) \quad / \quad y \in Y\} \end{aligned}$$

For general d.c functions, a criterion for the global minimum is the following (Hiriart-Urruty, 1989).

Theorem 2 Let g and h as in proposition 4. A point $x^\bullet \in domg \cap domh$ is a global minimum of $g(x) - h(x)$ on $I\!R^n$ if and only if

$$\partial_\epsilon h(x^\bullet) \subset \partial_\epsilon g(x^\bullet) \quad \forall \epsilon > 0 \tag{7}$$

This relation must be verified for all ϵ, which is very difficult to do. It is why we restrict the research to a local optimum.

Definition 3 Let g and h as in proposition 4. A point $x^\bullet \in domg \cap domh$ is a local minimum of $g(x) - h(x)$ on $\mathbb{R}^n$ if and only if

$$g(x) - h(x) \geq g(x^\bullet) - h(x^\bullet) \quad \forall x \in V_{x^\bullet} \tag{8}$$

where V_x denote a neighbourhood of x

The following proposition specifies a necessary condition to look for a local minimum (Hiriart-Urruty, 1989).

Proposition 4 If x is a local minimum g-h then

$$\partial h(x^\bullet) \subset \partial g(x^\bullet) \tag{9}$$

And in the case where h is polyhedral, this necessary condition is also sufficient.

Proposition 5 If $\partial h(x^\bullet) \subset \partial g(x^\bullet)$ then $x^\bullet$ is a critical point.
Indeed, by definition $x^\bullet$ is a critical point $g - h$ if $\partial h(x^\bullet) \cap \partial g(x^\bullet)$ is not empty.

By duality, all the above results are transposed to the dual problem D. For example :

Proposition 6 If y is a local minimum of $h^* - g^*$ then

$$\partial g^*(y^\bullet) \subset \partial h^*(y^\bullet) \tag{10}$$

The research of such critical point in d.c. optimisation framework is the objective of "d.c. subgradiant algorithm" which has first proposed in [PS88]. The fondation of the d.c. subgradiant algorithm is based on Bulatov's work on local search methods in nonconvex optimization, consisting in solving a sequence of linearization problems [Bu77].

In d.c. optimization context, the construction of the local search algorithm exploits the d.c. decomposition. The sequence of the decreasing direction is obtained here by calculating a subdifferential sequence not directly from the function f, but from these convex components h and g. More exactly, the d.c. algorithm consisted in the construction of two sequences $\{x^k\}$ of step and $\{y^k\}$ of decreasing directions. The first sequence dedicated to primal solution and the second to dual solution. These two sequences have been bound by duality and satisfy the following conditions :

- the sequences $\{g(x^k) - h(x^k)\}$ et $\{g^*(y^k) - g^*(y^k)\}$ are decreasing,
- and if $(g - h)(x^{k+1}) = (g - h)(x^k)$ then the algorithm stopped at $(k+1)^{th}$ iteration and the point x^k (resp. y^k) is a critical point of $g - h$ (resp. $h^* - g^*$),
- else all adherence value $x^\bullet$ of $\{x^k\}$ (resp. $y^\bullet$) of $\{y^k\}$ is a critical point of $G - H$ (resp. $h^* - g^*$).

($x^\bullet$ is a critical point of $g - h$ if $\partial g(x^\bullet) \cap \partial h(x^\bullet)$ is not empty)

Scheme of d.c. heuristic: The main idea of the algorithm is to construct $\{x^k\}$ satisfying at each step $\partial h(x^k) \cap \partial g(x^{k+1})$ which converges to a critical point $x^\bullet$ such that subdifferential intersection $\partial h(x^\bullet) \cap \partial g(x^\bullet)$ is not empty and symmetrically in a similar way by duality a sequence $\{y^k\}$ such as $\partial g^*(y^k) \cap \partial h^*(y^{k+1})$.

D.c. heuristic

1. x^0 given.
2. for each k, x^k being known, find $y^k \in \partial h(x^k)$
3. find $x^{k+1} \in \partial g^*(y^k)$

Well definiteness of the sequence: By construction if $x^0 \in dom\partial h$ then $y^0 \in \partial h(x^0)$ is well definiteness. For $k \geq 1$, y^k is defined if and only if x^k is defined and inclose in $dom\partial h$, after then, x^k and y^k are well definiteness if and only if $\partial g^*(y^{k+1}) \cap dom\partial h$ is not empty, this leads to $y^{k+1} \in dom\partial g^*$

5 Solving problem (DCP_M) by d.c. heuristic

The problem (P_M) can be formulated as a d.c. programming problem (DCP_M):

$$(DCP_M) \qquad min\{f(t,x,\lambda) = g(t,x,\lambda) - h(t,x) \ \ / \ \ (t,x,\lambda) \in \mathcal{X}\} \qquad (11)$$

where g and h are respectively defined by (1) and (2), and $\mathcal{X}$ denote the polyhedral set defined by constraints (1), (A), (B), (3).

The convex quadratic functions g and h can be written in the matrix form:

$$
\begin{aligned}
g(t,x,\lambda) &= \frac{1}{2}\begin{pmatrix} x_1 \\ t \end{pmatrix}^T \mathbf{Q} \begin{pmatrix} x_1 \\ t \end{pmatrix} + M\sum_{k=1}^{K}\left(c_1^T x_1^k + c_2^T x_2^k - \left(\lambda^k\right)^T d^k \delta^k\right), \\
h(t,x) &= (M-1)\sum_{k=1}^{K}\left(\rho t^T \mathbf{I} t + \rho\left(x_1^k\right)^T \mathbf{I}\left(x_1^k\right)\right), \qquad \text{with} M > 1,
\end{aligned}
$$

where $x_1 = \left(x_1^k\right)_{k=1\ldots,K}$, $x_1^k = \left(x_{1,u}^k\right)_{u=1,\ldots,m_1}$, $\mathbf{I}$ is the identity matrix $(m_1 \times m_1)$ and the symmetric matrix $\mathbf{Q}$ is defined such that:

$$
\frac{1}{2}\begin{pmatrix} x_1 \\ t \end{pmatrix}^T \mathbf{Q} \begin{pmatrix} x_1 \\ t \end{pmatrix} = (M-1)\sum_{k=1}^{K}\left(t^T x_1^k + \rho\left\|t\right\|^2 + \rho\left\|x_1^k\right\|^2\right). \tag{12}
$$

where ρ real number such that $\mathbf{Q}$ is positive semi-definite.

Using this d.c. decomposition we have:

D.c. Heuristic to solve the penalized problem ($\mathbf{DCP}_M$):

- **Initialization**:
  - let M be given (penalized parameter),
  - let ρ be given (convexification parameter, here we chose $\rho = 1$),
  - let $X^0 = \left(t^0, x_1^0\right)^T$ be given (tax and flow activities on tariff arcs $u \in U_1$).
- **At each iteration, P,**
  - **compute** Y^P :

$$
\begin{aligned}
Y^P &= \nabla_{t,x} h(t,x) \\
&= 2(M-1)\rho\sum_{k=1}^{K}\left(t^P + x_1^{k,P}\right)
\end{aligned}
$$

  - **and compute** $X^{P+1} = \left(t^{P+1}, x_1^{P+1}\right)^T$ by solving the following convex quadratic program with linear constraints:

$$
\begin{aligned}
\min_{t,x,\lambda} \quad & \tfrac{1}{2}\begin{pmatrix} x_1 \\ t \end{pmatrix}^T \mathbf{Q} \begin{pmatrix} x_1 \\ t \end{pmatrix} - 2K(M-1)\rho\left(t^P\right)^T t \\
& + \textstyle\sum_{k=1}^{K}\left[\begin{array}{c} \left(Mc_1 - 2(M-1)\rho x_1^{k,P}\right)^T x_1^k \\ + Mc_2^T x_2^k - M\left(\lambda^k\right)^T d^k \delta^k \end{array}\right] \\
\text{subject to} \quad & (1), (A), (B), (3).
\end{aligned} \tag{13}
$$

- **The stopping criterion is** $\|X^{P+1} - X^P\| \leq \varepsilon$.

As multiple local minima are expected we proposed to used a multistart strategy, where multiple optimization runs are started from different starting points. The starting points can be determined randomly (choose randomly X^0) or by solving the penalized problem (DCP_M) for different values of the parameter M. For example :

Multistart d.c. heuristic:

- **Let a given sequence of penalty parameters:** $1 < M_0 < M_1 < ... < M_Q$
- **For** $q = 1, ..., Q$
  - $\rightarrow$ **Run d.c. algorithm to solve** (P_{Mq})
  - $\rightarrow$ **Check if solution is best, if so store it**
- **End for**

6 Computational studies

The program was coded in C++ language and run on a Sun Sparc Station E450. We used CPLEX Version 6.5 as quadratic solver to compute X^P at iteration P of **D.c. heuristic.** In all this computational studies, we work with the same d.c. decomposition: ρ is always equal to one.

6.1 Illustrative example

This example is taken from [LMS98]. The graph has 11 nodes, 25 arcs with 7 toll arcs (boldfaced arcs on the figure). Table 1 lists origin-destination requirement and tax constraint for the test problems. The first test is the base scenario. In the second test, demand is increased on origin-destination pair 1-3. In the third test, nonnegative constraints on the tolls are considered, i.e. $t_u \in \Re^+ \quad \forall u \in U_1$. In the fourth test, tax constraint are added, i.e. $t_u \in \Re^+ \quad 0 \leq t_u \leq 7$. The optimal solutions are given in Table 2. We can see that it may be interesting to consider the possibility of using negative tolls: when nonnegativity constraints on tolls are imposed (test 1.3) a solution with lower total profit is obtained (cf Table 3). For an extensive discussion about this illustrative example, we refer the reader to [LMS98].

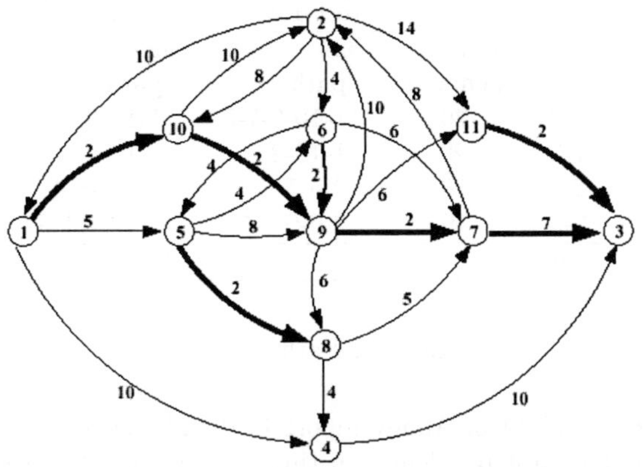

Figure 1: Network used for the numerical tests [LMS98]

	test 1.1	test 1.2	test 1.3	test 1.4
O-D	demand	demand	demand	demand
1-3	12	20	20	20
1-2	8	8	8	8
2-4	16	16	16	16
2-3	10	10	10	10
tax: $\forall u \in U_1$	$t_u \in \Re$	$t_u \in \Re$	$t_u \in \Re^+$	$0 \le t_u \le 7$

Table 1: Test problems

	test 1.1	test 1.2	test 1.3	test 1.4
Total profit $(t^* x_1^*)$	324	354	336	334
Arc $(\in U_1)$	Tax (t^*)	Tax (t^*)	Tax (t^*)	Tax (t^*)
(1,10)	11	-4	0	1
(5,8)	6	6	6	6
(6,9)	4	4	4	4
(7,3)	13	13	7	6
(9,7)	0	-2	0	0
(10,9)	0	0	0	0
(11,3)	14	14	8	7

Table 2: Optimal solutions

		test 1.2	test 1.3
Total profit		354	336
OD	demand	path	path
1-3	20	1-10-9-7-3	1-10-9-11-3
1-2	8	1-10-2	1-10-2
2-4	16	2-6-5-8-4	2-6-5-8-4
2-3	10	2-6-9-7-3	2-6-9-11-3
tolls		$\in \Re$	$\in \Re^+$

Table 3: Path solutions

Tables 11,12,13 and 14 in Annex report in detail the results obtained with d.c. algorithm on the four test problems. In this computational experiment we run d.c. heuristic for different values of the parameter M (> 1) from the same starting point ($X^0 = (t^0, x_1^0)^T$ set to zero). The number of iteration and the CPU time generally grow with the parameter M. Remember that M must be sufficiently large so that any global solution of problem (P_M) is also a global solution of the initial problem (P). So if M is too small the solution becomes suboptimal; the solution optimizes the profit for the leader but not for the follower.

6.2 Numerical test

This example set is taken from [BLMV00]. The graph has 60 nodes, 212 arcs and 10 commodities. Costs were generated randomly.

For each test, table 4 reports computational results for different choices of M values (1.1, 1.8, 2, 3, 4 and 5). The starting point is $X^0 = (t^0, x_1^0)^T$ set to zero. D.c. heuristic find local solutions at 0% to 22.5 % of the known optimal solutions. The faster computational times are obtained for M=1.1 but the solutions are not as good as those obtained for larger values.

For the sake of comparison we have reported the results obtained by [BLMV00] in Table 15 (Annex).

Test	n, m, m_1	M	Total Cost	Profit	CPU	Iter.	Error (%)
2.1	$60, 212, 21$	1.1	6111	2439	37.7	21	3.5
		1.8	6106	2529	113.8	62	0.0
		2.0	6106	2529	145.0	78	0.0
		3.0	6106	2529	302.9	154	0.0
		4.0	6066	2521	670.6	337	3.1
		5.0	6066	2521	781.2	449	3.1
2.2	$60, 212, 17$	1.1	5084	1184	28.0	25	22.5
		1.8	4924	1466	68.8	61	4.1
		2.0	4855	1397	85.6	73	8.6
		3.0	4624	1366	117.0	98	10.6
		4.0	4624	1366	150.1	147	10.6
		5.0	4624	1366	180.5	195	10.6
2.3	$60, 212, 10$	1.1	5986	932	15.0	18	6.3
		1.8	5911	945	38.8	45	5.0
		2.0	5896	930	59.1	70	6.5
		3.0	5865	974	93.6	109	2.1
		4.0	5865	974	141.9	163	2.1
		5.0	5886	995	183.3	234	0.0
2.4	$60, 212, 9$	1.1	4432	599	6.0	9	18.7
		1.8	4432	599	27.6	44	18.7
		2.0	4432	599	48.9	54	18.7
		3.0	4414	737	93.6	106	0.0
		4.0	4414	737	115.0	158	0.0
		5.0	4414	737	131.9	211	0.0

Table 4: Computational results for larger problems

Table 5 presents the numerical results for the test 2.1 with M=1.1, 1.2, ...,2. As expected the number of iteration increases with M. For $M \geq 1.6$ a global solution is found.

Test2.1

M	Total Cost	Profit	CPU	Iter.	Error (%)
1.1	6111	2439	37.7	21	3.5
1.2	6111	2439	39.9	23	3.5
1.3	6111	2439	45.5	25	3.5
1.4	6111	2474	67.2	37	2.1
1.5	6106	2469	76.5	40	2.3
1.6	6106	2529	99.6	55	0.0
1.7	6106	2529	97.9	55	0.0
1.8	6106	2529	113.8	62	0.0
1.9	6106	2529	130.9	70	0.0
2.0	6106	2529	145.3	78	0.0

Table 5: Computational results (test 2.1)

Notation:

n, m, $\mathbf{m}_1$: nodes, arcs, toll arcs,

M: Penalized parameter,

Total Cost: lower level solution,

Profit: higher level solution,

CPU: time in seconds,

Iter.: Number of iterations,

Error (%): 100*|Profit_optimal – Profit| /Profit_optimal.

7 Conclusion

In this study, we have presented a method based on d.c. optimization approach for solving the network-tariff-setting-(BBP) problem, however results are applied also for the general-(BBP) problem. The method finds a local solution. By the use of duality gap, penalty function, and d.c. reformulation, a new bilevel optimization feasible solution is obtained. Some computational results are given. It appears that using the proposed algorithm, for the most part, we are able to find a global solution for problems that have been studied. However, the CPU times are relatively long and the optimal choice of the penalty parameter value remains an open problem. Experiments on different kinds of starting point, penalty parameter and other d.c. decomposition are a topics for further researches.

References

[BLMS98] L. Brotcorne, M. Labbé, P. Marcotte, G. Savard (1998) "A bilevel model and solution algorithm for a freight tariff setting problem", *Technical report.*

[BLMV00] M. Bouthou, J-L. Lutton, V. Martin, A. Van des Kraaij (2000), "Optimization of tariffs on interconnections for telecommunications networks", *Technical report, France Telecom R&D*, NT/FTRD/6883.

[Bu77] V.P. Bulatov (1977) "Methods of Embeddings in Optimization Problems", Nauka (in russian)

[Hi95] J.P. Hiriart-Urruty (1995), "Conditions for global optimality", in Handbook of Global Optimization, R. Horst and Pardalos eds, *Kluwer Academic Publishers*, 1-26.

[LMS98] M. Labbé, P. Marcotte, G. Savard (1998), "A bilevel Model of Taxation and Its Application to Optimal Highway Pricing", *Management Science*, Vol. 44, No. 12, 1609-1622.

[MPV98] A. Migdalas, P. Pardalos, and P. Varbrand, eds., *Multilevel Optimization: Algorithms and Applications*, *Kluwer*, 1998.

[Ng01] Q. Nguyen (2001), "Optimization of tariff-traffic equilibrium in a telecommuncation network", *Technical report, France Telecom R&D*, NT/FTRD/7464.

[PS88] D.T. Pham and E.B. Souad (1988), "Duality in d.c. optimization. Subgradient methods", Trends in Mathematical Optimization, *International Series of Numer. Math.* 84, Birkhauser, 276-294.

[St97] A.S. Strekalovsky (1997) "On Global Optimality Conditions for D.C. Programming Problemsl", Publisher of Irkutsk State University, *seria Optimization and Optimal Control.*

[To78] J.F. Toland (1978), "Duality in nonconvex optimization", *Journal of Mathematical Analysis and Applications* 66, 399-415.

[Tu95] H. Tuy (1995), "D.C. Optimization: Theory, Methods and Applications", in eds. R. Horst and P.M. Pardalos Handbook on Global Optimization, *Kluver Academic Publishers*, 149-216.

[VC94] L. N. Vicente, P. H. Calamai, (1994), "Bilevel and multilevel programming: A bibliography review", *Journal of Global Optimization*, Vol 5, No. 3.

Annex

Test1.1

M	Total Cost	Profit	EPS	CPU	Solution	Iter.
1.1	1044	324	0.001	0.19	*optimal*	5
1.2	1044	324	0.001	0.32	*optimal*	8
1.3	1044	324	0.001	0.48	*optimal*	12
1.4	1044	324	0.003	0.63	*optimal*	17
1.5	1028	318	0.000	0.93	*local*	25
1.6	1031	311	0.759	2.09	*local*	61
1.7	944	234	0.436	4.00	*local*	110
1.8	924	298	0.005	3.88	*local*	102
1.9	924	298	0.004	4.5	*local*	114
2.0	904	280	0.038	4.4	*local*	110

Table 6: Computational results (test 1.1)

Test 1.2

M	Total Cost	Profit	EPS	CPU	Solution	Iter.
1.1	1204	324	0.095	0.2	*local*	6
1.2	1204	324	0.000	0.4	*local*	11
1.3	1204	324	0.000	0.6	*local*	17
1.4	1186	316	0.563	1.3	*local*	32
1.5	1084	354	0.011	2.21	*optimal*	64
1.6	1084	354	0.002	10.1	*optimal*	274
1.7	1084	354	0.003	6.27	*optimal*	165
1.8	1052	332	0.004	3.3	*local*	84
1.9	1048	348	0.032	3.5	*local*	92
2.0	1064	344	0.041	5.1	*local*	110

Table 7: Computational results (test 1.2)

M	Total Cost	Profit	EPS	CPU	Solution	Iter.
			Test 1.3			
1.1	1204	324	0.000	0.2	*local*	6
1.2	1204	324	0.000	0.3	*local*	10
1.3	1204	324	0.002	0.5	*local*	15
1.4	1188	308	0.005	1.1	*local*	31
1.5	1060	180	0.468	0.6	*local*	17
1.6	1056	336	0.003	1.5	*optimal*	38
1.7	1056	336	0.004	1.6	*optimal*	44
1.8	1056	336	0.003	1.8	*optimal*	50
1.9	1056	336	0.000	2.6	*optimal*	70
2.0	1056	336	0.004	3.2	*optimal*	62

Table 8: Computational results (test 1.3)

M	Total Cost	Profit	EPS	CPU	Solution	Iter.
			Test 1.4			
1.1	1102	222	0.000	0.2	*local*	6
1.2	1102	222	0.000	0.4	*local*	10
1.3	1102	222	0.001	0.5	*local*	14
1.4	1102	222	0.000	0.7	*local*	18
1.5	1059	179	0.633	0.5	*local*	16
1.6	1054	334	0.000	1.2	*optimal*	31
1.7	1054	334	0.001	1.3	*optimal*	36
1.8	1054	334	0.001	1.5	*optimal*	41
1.9	1054	334	0.000	2.6	*optimal*	70
2.0	1054	334	0.007	1.8	*optimal*	50

Table 9: Computational results (test 1.4)

The test problems

Test	n, m, m_1	Method	Best* Time	Best* Error (%)	Profit_opt
2.1	$60, 212, 21$	$MCH1$	14.4	77	2529
		$MCH3$	9.7	53	
		MXP	440.7	8	
		$SimAn$	8.7	26	
2.2	$60, 212, 17$	$MCH1$	15.2	100	1529
		$MCH3$	9.7	75	
		MXP	257.4	9	
		$SimAn$	5.6	10	
2.3	$60, 212, 10$	$MCH1$	12.4	100	995
		$MCH3$	6.4	53	
		MXP	132.7	0	
		$SimAn$	5.3	9	
2.4	$60, 212, 9$	$MCH1$	15.3	100	737
		$MCH3$	2.7	76	
		MXP	156.3	0	
		$SimAn$	3.6	8	

Table 10: Computational results [BLMV00] (test 2)

Notation:

M: Penalized parameter,

EPS: $EPS = \left|f(X^{P+1}) - f(X^P)\right|$,

CPU: time in seconds,

Iter.: Number of iterations,

n, m, m$_1$: nodes, arcs, toll arcs,

Best*: best solution from several runs (set of penalty start and penalty step), see [BLMV00],

Error (%): 100*|Profit_optimal - Profit| /Profit_optimal.

Test1.1

M	Total Cost	Profit	EPS	CPU	Solution	Iter.
1.1	1044	324	0.001	0.19	*optimal*	5
1.2	1044	324	0.001	0.32	*optimal*	8
1.3	1044	324	0.001	0.48	*optimal*	12
1.4	1044	324	0.003	0.63	*optimal*	17
1.5	1028	318	0.000	0.93	*local*	25
1.6	1031	311	0.759	2.09	*local*	61
1.7	944	234	0.436	4.00	*local*	110
1.8	924	298	0.005	3.88	*local*	102
1.9	924	298	0.004	4.5	*local*	114
2.0	904	280	0.038	4.4	*local*	110

Table 11: Computational results (test 1.1)

Test 1.2

M	Total Cost	Profit	EPS	CPU	Solution	Iter.
1.1	1204	324	0.095	0.2	*local*	6
1.2	1204	324	0.000	0.4	*local*	11
1.3	1204	324	0.000	0.6	*local*	17
1.4	1186	316	0.563	1.3	*local*	32
1.5	1084	354	0.011	2.21	*optimal*	64
1.6	1084	354	0.002	10.1	*optimal*	274
1.7	1084	354	0.003	6.27	*optimal*	165
1.8	1052	332	0.004	3.3	*local*	84
1.9	1048	348	0.032	3.5	*local*	92
2.0	1064	344	0.041	5.1	*local*	110

Table 12: Computational results (test 1.2)

Test 1.3

M	Total Cost	Profit	EPS	CPU	Solution	Iter.
1.1	1204	324	0.000	0.2	*local*	6
1.2	1204	324	0.000	0.3	*local*	10
1.3	1204	324	0.002	0.5	*local*	15
1.4	1188	308	0.005	1.1	*local*	31
1.5	1060	180	0.468	0.6	*local*	17
1.6	1056	336	0.003	1.5	*optimal*	38
1.7	1056	336	0.004	1.6	*optimal*	44
1.8	1056	336	0.003	1.8	*optimal*	50
1.9	1056	336	0.000	2.6	*optimal*	70
2.0	1056	336	0.004	3.2	*optimal*	62

Table 13: Computational results (test 1.3)

Test 1.4

M	Total Cost	Profit	EPS	CPU	Solution	Iter.
1.1	1102	222	0.000	0.2	*local*	6
1.2	1102	222	0.000	0.4	*local*	10
1.3	1102	222	0.001	0.5	*local*	14
1.4	1102	222	0.000	0.7	*local*	18
1.5	1059	179	0.633	0.5	*local*	16
1.6	1054	334	0.000	1.2	*optimal*	31
1.7	1054	334	0.001	1.3	*optimal*	36
1.8	1054	334	0.001	1.5	*optimal*	41
1.9	1054	334	0.000	2.6	*optimal*	70
2.0	1054	334	0.007	1.8	*optimal*	50

Table 14: Computational results (test 1.4)

The test problems

Test	n, m, m_1	Method	Best* Time	Best* Error (%)	Profit_opt
2.1	$60, 212, 21$	$MCH1$	14.4	77	2529
		$MCH3$	9.7	53	
		MXP	440.7	8	
		$SimAn$	8.7	26	
2.2	$60, 212, 17$	$MCH1$	15.2	100	1529
		$MCH3$	9.7	75	
		MXP	257.4	9	
		$SimAn$	5.6	10	
2.3	$60, 212, 10$	$MCH1$	12.4	100	995
		$MCH3$	6.4	53	
		MXP	132.7	0	
		$SimAn$	5.3	9	
2.4	$60, 212, 9$	$MCH1$	15.3	100	737
		$MCH3$	2.7	76	
		MXP	156.3	0	
		$SimAn$	3.6	8	

Table 15: Computational results [BLMV00] (test 2)

Notation:

M: Penalized parameter,

EPS: $EPS = \left| f(X^{P+1}) - f(X^P) \right|,$

CPU: time in seconds,

Iter.: Number of iterations,

n, m, m$_1$: nodes, arcs, toll arcs,

Best*: best solution from several runs (set of penalty start and penalty step), see [BLMV00],

Error (%): 100*|Profit_optimal - Profit| /Profit_optimal.

Optimization and Optimal Control, pp. 233 – 249
P.M. Pardalos, I. Tseveendorj, and R. Enkhbat, Editors

Fuzzy Optimization Problems Concerning Oil Equations of L–fuzzy Numbers

Seiji SAITO (saito@ist.osaka-u.ac.jp)
Graduate School of Information Science and Technology
Osaka University
Suita, Osaka, 565-0871, Japan

Hiroaki ISHII (ishii@ist.osaka-u.ac.jp)
Graduate School of Information Science and Technology
Osaka University
Suita, Osaka, 565-0871, Japan

Abstract

In this study we give a new representation of fuzzy numbers with bounded supports and also we show that a fuzzy number means a bounded continuous curve in the two-dimensional metric space. Our aims of this research are to discuss optimization problems with objective functions and constraints both of which are L–fuzzy functions and to consider oil well equations which are represented by fuzzy differential equations $C_L'(t) + D_L C_L(t) = 0$, where t is the time, $0 \in \mathbf{R}$, $C_L(t)$ an L–fuzzy function and D_L a constant L–fuzzy number by applying the above criteria of L–optimization problems. Furthermore we discuss boundary value problems of fuzzy differential equations on finite intervals and infinite intervals.

Keywords: L-fuzzy number, fuzzy differential equation, lambda-max order, fuzzy optimization problem, oil well equation.

In this study we introduce the set of fuzzy numbers with compact supports, denoting $\mathcal{F}_{\mathbf{b}}^{st}$ which has membership functions identified with fuzzy numbers. Moreover we consider a set of L–fuzzy numbers, denoting $\mathcal{F}_L$, with a total order relation over $\mathcal{F}_L$. In Chapter 3 we discuss a criterion for optimization problems of L–fuzzy numbers. In Chapter 4 we show a result ([8]) of oil well equations of $\mathbf{R}$–valued functions.

Chapter 5 deals with fuzzy functions and we give definitions of the differential and integral of fuzzy functions. In Chapter 6 we treat oil well equations of $L-$fuzzy functions and we get a result of L-fuzzy type of optimization problems concerning oil well equations. We give concluding remarks in which we mention that investigation of optimization problems with an objective function and the infinite number of constraints of $L-$fuzzy numbers is one of important objectives in the theory of optimization and control. In Chapter 7 is devoted to the study of boundary value problems, *i.e.* BVP, of fuzzy differential equations on finite intervals and infinite intervals. It's expected that the theory of BVP of fuzzy differential equations plays an important role in analyzing control theory.

1 Set of fuzzy numbers

Let $I = [0,1]$. We define the following set of fuzzy numbers, where a fuzzy number x is characterized by a membership function μ_x as follows (cf. [2, 3]):

Definition 1 *Denote*

$$\mathcal{F}_{\mathbf{b}}^{st} = \{\mu_x : \mathbf{R} \to I \textit{ satisfying } \text{(i)} - \text{(iv)} \textit{ below}\}.$$

(i) *There exists a unique* $m \in \mathbf{R}$ *such that* $\mu_x(m) = 1$;

(ii) *The support set* $supp(\mu_x) = cl(\{\xi \in \mathbf{R} : \mu(\xi) > 0\})$ *is bounded in* $\mathbf{R}$;

(iii) *Let* $J = \{\xi \in \mathbf{R} : 0 < \mu_x(\xi)\}$. μ_x *is strictly quasi-convex on* J, *i.e.,* $\mu_x(\lambda\xi_1 + (1-\lambda)\xi_2) > \min[\mu_x(\xi_1), \mu_x(\xi_2)]$ *for* $0 < \lambda < 1$ *and* $\xi_1, \xi_2 \in J$ *such that* $\xi_1 \neq \xi_2$;

(iv) μ_x *is upper semi-continuous on* $\mathbf{R}$.

In usual case a fuzzy number x satisfies quasi-convex on $\mathbf{R}$, *i.e.*, $\mu_x(\lambda\xi_1 + (1-\lambda)\xi_2) \geq \min[\mu_x(\xi_1), \mu_x(\xi_2)]$ for $0 \leq \lambda \leq 1$ and $\xi_1, \xi_2 \in \mathbf{R}$. Condition (iii) plays an important role in proving properties of membership function μ_x in Theorem 1, where we show significant properties concerning the end-points of the α-cut set $L_\alpha(\mu_x) = \{\xi \in \mathbf{R} : \mu_x(\xi) \geq \alpha\}$ by considering the following two remarks.

Remark 1 *In the case that* $supp(\mu_x) = \{m\}$, *a fuzzy number* $x \in \mathcal{F}_{\mathbf{b}}^{st}$ *means a real number. Thus it follows that* $\mathcal{F}_{\mathbf{b}}^{st} \supset \mathbf{R}$. *When* $supp(\mu_x) \neq \{m\}$ *and the above Condition* (ii) *holds the following statements* (a)-(d) *are equivalent each other:*

(a) $\mu_x(\cdot)$ *is fuzzy convex on* $\mathbf{R}$;

(b) $L_\alpha(\mu_x)$ *is convex for each* $\alpha \in I$;

(c) *There exist real numbers* $m_1 \leq m_2$ *such that* $\mu_x(\cdot)$ *is non-decreasing on* $(-\infty, m_1]$ *and that* $\mu_x(\cdot)$ *is non-increasing on* $[m_2, \infty)$;

(d) $L_\alpha(\mu_x) \subset L_\beta(\mu_x)$ *for* $\alpha > \beta$.

It is clear that (a) means (b). If we suppose that (a) doesn't hold but (b) hold, this leads to a contradiction. It can be seen that (c) leads to (d) and the converse holds. Suppose that for any $m_1 m_2 \in \mathbf{R}$ such that $m_1 < m_2$ there exist $y_i, i = 1, \cdots, 4$, such that $y_1 < y_2 \le m_1$ and $m_2 \le y_3 < y_4$ such that $\mu_x(y_1) > \mu_x(y_2)$ and $\mu_x(y_3) < \mu_x(y_4)$ under Condition (ii) and (a). Then it leads to a contradiction. From (c), it follows that (a) holds.

Remark 2 *In the similar way as the above discussion it follows that the following statements are equivalent under Condition* (i) *of Definition 1.*

- μ *is strictly fuzzy convex on* $\mathbf{R}$;
- μ *is strictly increasing on* $(-\infty, m]$ *and strictly decreasing on* $[m, \infty)$.

In the similar way as [2, 3] we consider the following parametric representation of $\mu_x \in \mathcal{F}^{st}_{\mathbf{b}}$ such that

$$x_1(\alpha) = \min L_\alpha(\mu_x), \quad x_2(\alpha) = \max L_\alpha(\mu_x)$$

for $0 < \alpha \le 1$ and that $x_1(0) = \min cl(supp(\mu_x)), x_2(0) = \max cl(supp(\mu_x))$. In what follows we denote a fuzzy numbers x by (x_1, x_2), *i.e.*, $x = (x_1, x_2)$.

Extension principle

In order to give parametric representation of binary operations of fuzzy numbers we discuss an extension principle of fuzzy operations of $F : \mathcal{F}^{st}_{\mathbf{b}} \times \mathcal{F}^{st}_{\mathbf{b}} \to \mathcal{F}^{st}_{\mathbf{b}}$. For example, letting $z = (x, y) \in \mathcal{F}^{st}_{\mathbf{b}} \times \mathcal{F}^{st}_{\mathbf{b}}$, the addition $F(z) = x + y$, the product $F(z) = xy$ and the square $F(z) = x^2$.

Let f be an $\mathbf{R}$–valued function defined $\mathbf{R}^2$ and $\Xi = (\xi_1, \xi_2) \in \mathbf{R}^2$. Denote conditions of membership functions μ_x, μ_y with respect to f by C_f. We define the membership function

$$\mu_{F(z)}(\xi) = \sup_{\xi = f(\Xi)} \Big(\min_{j=1,2 \text{ and } C_f} [\mu_x(\xi_1), \mu_y(\xi_2)] \Big).$$

In the case that $F(z) = x + y$, we consider $f(\Xi) = \xi_1 + \xi_2, C_f = \emptyset$ and also

$$\mu_{x+y}(\xi) = \sup_{\xi = \xi_1 + \xi_2} (\min_{j=1,2} [\mu_x(\xi_1), \mu_y(\xi_2)]).$$

When $F(z) = xy$, then we have $f(\Xi) = \xi_1 \xi_2, C_f = \emptyset$ and also

$$\mu_{xy}(\xi) = \sup_{\xi = \xi_1 \xi_2} \Big(\min_{j=1,2} [\mu_x(\xi_1), \mu_y(\xi_2)] \Big).$$

When $F(z) = x^2$, then $f(\Xi) = \xi_1 \xi_2, C_f = \{\mu_x = \mu_y\}$ and

$$\mu_{x^2}(\xi) = \sup_{\xi = \xi_1 \xi_2} \Big(\min_{j=1,2 \text{ and } C_f} [\mu_x(\xi_1), \mu_y(\xi_2)] \Big) = \sup_{\xi = \xi_1 \xi_2} \Big(\min_{j=1,2} [\mu_x(\xi_1), \mu_y(\xi_2)] \Big).$$

By applying the above extension principle and the representation of fuzzy numbers we get the following results.

1) Addition. Let $x = (x_1, x_2),\ y = (y_1, y_2) \in \mathcal{F}_{\mathbf{b}}^{st}$. We get the addition

$$\begin{aligned}
\mu_{x+y}(\xi) &= \sup_{\xi=\xi_1+\xi_2} \min[\mu_x(\xi_1), \mu_y(\xi_2)] \\
&= \sup\{\alpha \in I : \xi = \xi_1 + \xi_2, \quad \xi_1 \in x_\alpha, \xi_2 \in y_\alpha\} \\
&= \sup_{\xi \in [x_1(\alpha)+y_1(\alpha), x_2(\alpha)+y_2(\alpha)]} \alpha,
\end{aligned}$$

which means that $x + y = (x_1 + y_1, x_2 + y_2)$. Here $x_\alpha = L_\alpha(\mu_x)$ etc.

2) Subtraction. It follows that

$$\mu_{x-y}(\xi) = \sup\{\alpha \in I : \xi = \xi_1 - \xi_2, \quad \xi_1 \in x_\alpha, \xi_2 \in y_\alpha\}$$

means that $x - y = (x_1 - y_2, x_2 - y_1)$.

3) Product. It follows that

$$\mu_{xy}(\xi) = \sup\{\alpha \in I : \xi = \xi_1\xi_2, \quad \xi_1 \in x_\alpha, \xi_2 \in y_\alpha\}$$

means that the following relation.

$$xy = \begin{cases}
(x_1y_1, x_2y_2) & (0 \le x_1, 0 \le y_1) \\
(x_2y_1, x_2y_2) & (0 \le x_1, y_1 \le 0 \le y_2) \\
(x_2y_1, x_1y_2) & (0 \le x_1, y_2 \le 0) \\
(x_1y_2, x_2y_2) & (x_1 \le 0 \le x_2, 0 \le y_1) \\
(\min\{x_2y_1, x_1y_2\}, \max\{x_1y_1, x_2y_2\}) & (x_1 \le 0 \le x_2, y_1 \le 0 \le y_2) \\
(x_2y_1, x_1y_1) & (x_1 \le 0 \le x_2, y_2 \le 0) \\
(x_1y_2, x_2y_1) & (x_2 \le 0, 0 \le y_1) \\
(x_1y_2, x_1y_1) & (x_2 \le 0, y_1 \le 0 \le y_2) \\
(x_2y_2, x_1y_1) & (x_2 \le 0, y_2 \le 0)
\end{cases}$$

By the above parametric representation of fuzzy numbers we get the following theorem concerning properties of end-points.

Theorem 1 *Denote $x = (x_1, x_2) \in \mathcal{F}_{\mathbf{b}}^{st}$, where $x_1, x_2 : I \to \mathbf{R}$. Then the following properties* (i)-(iii) *hold:*

(i) $x_i \in C(I), i = 1, 2$. *Here $C(I)$ is the set of all the continuous functions on I;*

(ii) *There exists a unique $m \in \mathbf{R}$ such that $x_1(1) = x_2(1) = m$ and $x_1(\alpha) \le m \le x_2(\alpha)$ for $\alpha \in I$;*

(iii) *One of the following statements* (a) *and* (b) *holds;*

  (a) *Functions x_1, x_2 are non-decreasing, non-increasing on I, respectively, with $x_1(\alpha) < x_2(\alpha)$ for $0 \le \alpha < 1$;*

(b) $x_1(\alpha) = x_2(\alpha) = m$ *for* $0 < \alpha \leq 1$.

Conversely, under the above conditions (i) -(iii), *if we denote*

$$\mu_x(\xi) = \sup\{\alpha \in I : x_1(\alpha) \leq \xi \leq x_2(\alpha)\} \tag{1.1}$$

then μ_x *is the membership function of* x, *i.e.*, $\mu_x \in \mathcal{F}_{\mathbf{b}}^{st}$.

Proof. (i) Let $x = (x_1, x_2) \notin \mathbf{R}$. From the definition of $\mu_x \in \mathcal{F}_{\mathbf{b}}^{st}$ it follows that $supp(\mu_x) = [a_1, b_1]$, where $a_1 \leq m \leq b_1$. Let $\lim_{n\to\infty} \alpha_n = \alpha_0$. Denote $A_0 = \liminf_{n\to\infty} x_1(\alpha_n)$. If $A_0 < x_1(\alpha_0)$, then a contradiction arises. Thus $x_1(\alpha)$ is lower semi-continuous on I. Denote $A_1 = \limsup_{n\to\infty} x_1(\alpha_n)$. Then $a_1 \leq A_1 \leq m$. Suppose that $A_1 > x_1(\alpha_0)$. For each $\varepsilon > 0$ there exists an integer $\ell_\varepsilon > 0$ such that $A_1 + \varepsilon > \sup_{n\geq \ell_\varepsilon} x_1(\alpha_n) \geq A_1$, *i.e.*, there exists a subsequence $\{\alpha_m\}$ of the sequence $\{\alpha_n\}$ such that $x_1(\alpha_m) \geq A_1 > x_1(\alpha_0)$. From Remark 2 we get $\mu(x_1(\alpha_m)) > \mu(A_1) > \mu(x_1(\alpha_0)) = \alpha_0$, so that we have $\alpha_m \geq \mu(A_1) > \alpha_0$ for any m. This contradicts $\lim_{n\to\infty} \alpha_n = \alpha_0$. Therefore we have $A_1 \leq x_1(\alpha_0)$. Then $x_1(\alpha)$ is lower semi-continuous and upper semi-continuous, *i.e.*, continuous on I. In the same way it can be proved that $x_2(\alpha)$ is continuous on I. It is easily seen that (ii) and (iii) hold by Remark 2.

Conversely (1.1) means that $L_\beta(\mu_x) = [x_1(\beta), x_2(\beta)] \subset \mathbf{R}$ for $\beta \in I$. From (1.1) it follows that if $\xi \in [x_1(\alpha), x_2(\alpha)]$ then $\mu_x(\xi) \geq \alpha$ and that $\xi \notin [x_1(\mu_x(\xi) + \varepsilon), x_2(\mu_x(\xi) + \varepsilon)]$ for each $\varepsilon > 0$. Then it can be seen that $[x_1(\beta), x_2(\beta)] \subset L_\beta(\mu_x)$. When $\mu_x(\xi) = \beta$, from (1.1), it follows that a $\xi \in [x_1(\beta), x_2(\beta)]$, because of Remark 2. When $\mu_x(\xi) > \beta$, then there exists an $\alpha \in I$ such that $\xi \in [x_1(\alpha), x_2(\alpha)]$ and $\alpha \geq \beta$, which means that $\xi \in [x_1(\alpha), x_2(\alpha)] \subset [x_1(\beta), x_2(\beta)]$ by Remark 2. Therefore we have $L_\beta(\mu_x) = [x_1(\beta), x_2(\beta)]$.

From (1.1) it is immediately seen that (i) and (ii) of Definition 1 hold. The $\alpha-$cut set $L_\alpha(\mu_x)$ is closed for $\alpha \in I$, *i.e.*, the function μ_x is upper semi-continuous on $\mathbf{R}$. For $\alpha \in I$ we have $L_\alpha(\mu_x)$ is convex, *i.e.*, the function μ_x is fuzzy convex on $\mathbf{R}$. See, *e.g.*, [9]. We shall prove that μ_x is strictly fuzzy convex. From (1.1), $\mu_x(\xi) = \overline{\alpha}$ means that $\xi = x_1(\overline{\alpha})$ or $\xi = x_2(\overline{\alpha})$. If suppose that $x_1(\overline{\alpha}) < \xi < x_2(\overline{\alpha})$, which means that $\mu_x(\xi) > \overline{\alpha}$. Suppose that there exist $\xi_1, \xi_2 \in J$ and λ such that $\xi_1 \neq \xi_2, 0 < \lambda < 1$ and $\mu_x(\xi_3) = \mu_x(\overline{\xi})$, where $\xi_3 = \lambda\xi_1 + (1-\lambda)\xi_2$ and $\mu_x(\overline{\xi}) = \min[\mu_x(\xi_1), \mu_x(\xi_2)]$. Then we have $\xi_3 \neq \overline{\xi}$ and $\xi_3 = x_1(\mu_x(\overline{\xi}))$ or $\xi_3 = x_2(\mu_x(\overline{\xi}))$. Thus there exists a constant number c such that $\mu_x(\xi) = c$ on $[\xi_1, \xi_3]$ or $[\xi_3, \xi_2]$, which contradicts the continuity of $x_i, i = 1, 2$, and (1.1). Therefore μ_x is strictly fuzzy convex.

Q.E.D.

Let a metric between $x = (x_1(\cdot), x_2(\cdot)), y = (y_1(\cdot), x_2(\cdot))$ be defined as follows.

$$d(x, y) = \sup_{\alpha \in I} \sqrt{|x_1(\alpha) - y_1(\alpha)|^2 + |x_2(\alpha) - y_2(\alpha)|^2}$$

Then we get the following result immediately.

Theorem 2 *$(\mathcal{F}_{\mathbf{b}}^{st}, d)$ is complete metric space.*

We shall prove that the metric space $(\mathcal{F}_{\mathbf{b}}^{st}, d)$ is complete. Suppose that a sequence $\{y_k = (a_k, b_k) : k = 1, 2, \cdots\}$ is a Cauchy sequence in $\mathcal{F}_{\mathbf{b}}^{st}$. Then there exists a $y_0 = (a_0, b_0) \in C(I) \times C(I)$ such that $\lim_{k\to\infty} d(y_k, y_0) = 0$ and that (i) and (ii) of Theorem 1 hold. And also a_0, b_0 are non-decreasing, non-increasing on I, respectively. Denote an interval A by $A = \{\alpha \in I : a_0(\alpha) \equiv constant\}$. If $A = (0, 1]$, then we have $a_0(\alpha) = b_0(\alpha) = y_0$ and $y_0 \in \mathbf{R}$. If $A \neq (0, 1]$, then $A = \emptyset$ because of (iii) of Definition 1. Thus (iii) of Theorem 1 holds and $(\mathcal{F}_{\mathbf{b}}^{st}, d)$ is a complete metric space.

2 Set of $L-$fuzzy numbers

Denote a shape function by $L : \mathbf{R} \to I$, where L is upper semi-continuous and satisfies the following properties (i) - (iv):

(i) $L(0) = \max_{\mathbf{R}} L(\xi) = 1$; (ii) $L(\xi)$ is strictly decreasing in $\xi \geq 0$;

(iii) $L(-\xi) = L(\xi)$ for $\xi \geq 0$; (iv) $\sup\{\xi \in \mathbf{R} : L(\xi) > 0\} = 1$.

In what follows we consider a set of $L-$fuzzy numbers $\mathcal{F}_L = \{\mu \in \mathcal{F}_{\mathbf{b}}^{st} : (a) \text{ or } (b) \text{ hold.}\}$ Let $m \in \mathbf{R}$, $\ell \geq 0$. There exist two typical types (a) and (b) of $\mathcal{F}_L$.

(a) $\ell > 0$ and $\mu(\xi) = \begin{cases} L(\frac{m-\xi}{\ell}) & \text{for } \xi \leq m \\ L(\frac{\xi-m}{\ell}) & \text{for } \xi > m \end{cases}$

(b) $\ell = 0$ and $\mu(\xi) = \begin{cases} 1 & \text{for } \xi = m \\ 0 & \text{for } \xi \neq m \end{cases}$

In this section we introduce a total order relation λ-fuzzy max order $\preceq_\lambda$ over $\mathcal{F}_L$. Here $0 \leq \lambda \leq 1$ is given by decision makers. Let $x = (x_1, x_2), y = (y_1, y_2) \in \mathcal{F}_L$ with the center $x_1(1)$, the spread $\ell_x = x_1(1) - x_1(0) \geq 0$ and center $y_1(1)$, spread $\ell_y = y_1(1) - y_1(0) \geq 0$. We define that $x \preceq_\lambda y$ if and only if the following statements (i)-(iii) ([1]):

(i) $|\ell_y - \ell_x| \leq y_1(1) - x_1(1)$ for $y_1(1) \geq x_1(1)$;

(ii) $\lambda|\ell_y - \ell_x| \leq y_1(1) - x_1(1) < |\ell_y - \ell_x|$ for $y_1(1) > x_1(1)$ and $\ell_y \neq \ell_x$;

(iii) $|y_1(1) - x_1(1)| < \lambda(\ell_y - \ell_x)$ for $\ell_y - \ell_x > 0$.

Furukawa [1] gives the following theorem so that two any $L-$fuzzy numbers can be compared to each other.

Theorem F1 *For any $x, y \in \mathcal{F}_L$ it follows that one of the relations $x \preceq_\lambda y$, and $y \preceq_\lambda x$ hold. Thus $\preceq_\lambda$ is a total order relation over $\mathcal{F}_L$.*

The following theorem plays an important role in comparing two $L-$fuzzy numbers.

Theorem F2 *For* $x = (x_1, x_2), y = (y_1, y_2) \in \mathcal{F}_L$ *satisfying* $\ell_x = x_1(1) - x_1(0) \geq 0,\ \ \ell_y = y_1(1) - y_1(0) \geq 0$, *it follows that* $x \preceq_\lambda y$ *means that* (i) or (ii) *hold.*

(i) $\lambda\ell_x + x_1(1) < \lambda\ell_y + y_1(1)$ *for* $\ell_y > \ell_x$;

(ii) $\lambda\ell_x + x_1(1) \leq \lambda\ell_y + y_1(1)$ *for* $\ell_y \leq \ell_x$.

3 Fuzzy optimization problems

Let $0 \leq \lambda \leq 1$. In this section we show criteria concerning the following $L-$fuzzy optimization problem

$$\text{minimize } f(z) \ \text{ subject to } z \in \mathcal{C}_\lambda^\delta. \qquad (P_\lambda^\delta)$$

where $\mathcal{C}_\lambda^\delta$ is a feasible set in $\mathcal{F}_L^n$ or $\mathcal{F}_L^\infty$ and $f : \mathcal{C}_\lambda^\delta \to \mathcal{F}_L$ is an objective function. We denote $z \in \mathcal{F}_L^\infty$ by

$$z = (z_1, z_2, z_3, \cdots) \text{ where } z_i \in \mathcal{F}_L \text{ for } i = 1, 2, \cdots.$$

In what follows we consider

$$\mathcal{C}_\lambda^\delta = \{z \in \mathcal{F}_L^\infty : g_j(z) \preceq_\lambda (0, \delta_j)_L, j = 1, 2, \cdots\}.$$

where $g_j : \mathcal{C}_\lambda^\delta \to \mathcal{F}_L$, $(0, \delta_j)_L \in \mathcal{F}_L$, and $\delta_j > 0$ are constants for $j = 1, 2, \cdots$. Let $\delta = (\delta_1, \delta_2, \cdots) \in \mathbf{R}^\infty$. If $z^* \in \mathcal{C}_\lambda^\delta$ satisfies $f(z^*) = \min\{f(z) : z \in \mathcal{C}_\lambda^\delta\}$ then z^* is called an optimal solution of (P_λ^δ).

In order to analyze (P_λ^δ) which is general case, we may consider an $\mathbf{R}-$valued optimization problem

$$\text{minimize } f(z) \text{ subject to } z \in \mathcal{C}_\lambda^\delta \cap \mathbf{R}. \qquad (P_0^0)$$

Then, letting $f_0^* = \min\limits_{z \in \mathcal{C}_\lambda^\delta \cap \mathbf{R}} f(z)$, we get $f_0^* \in \mathbf{R}$ which gives the optimal value in $\mathbf{R}$ as follows.

Corollary 1 *Let* $f^* \in \mathbf{R}$. *Then there exists no* $L-$*fuzzy number* $f \in \mathcal{F}_L \backslash \mathbf{R}$ *such that* $f = f^*, i.e., f \preceq_\lambda f^*$ *and* $f^* \preceq_\lambda f$.

It can be seen that the above corollary holds. Suppose that $f = (f^c, \ell_f)_L \in \mathcal{F}_L$ with $f = f^*$, where $f^c \in \mathbf{R}$ is the center and $\ell_f > 0$ is the spread. Then we have $f^c + \lambda\ell_f = f^*$. From $f^* \preceq_\lambda f$ and Theorem F2(i), it follows that $f^* < f^c + \lambda\ell_f$, *i.e.*, $f^* < f^*$, which means a contradiction. This completes the proof.

From the above corollary we get the following lemma immediately.

Lemma 1 *Denote*

$$f_1^\delta = \min\{f(z) : z \in \mathcal{C}_\lambda^\delta\},\ f_2^\delta = \min\{f(z^c) : z \in \mathcal{C}_\lambda^\delta\} \in \mathbf{R},$$

$$f_3^\delta = \min\{f(z)^c : z \in \mathcal{C}_\lambda^\delta\} \in \mathbf{R},\ f_0^* = \min\{f(z) : z \in \mathcal{C}_0^0\} \in \mathbf{R},$$

where $z^c \in \mathbf{R}, f(z)^c \in \mathbf{R}$ *are centers of* $z, f(z)$*, respectively, and* $\mathcal{C}_0^0 = \mathcal{C}_\lambda^\delta \cap \mathbf{R}$.
If there exist $f_i^\delta, i = 1,2,3$ *and* f_0^**, then it follows that* $f_1^\delta \in \mathbf{R}$ *and that*

$$f_1^\delta = f_2^\delta = f_3^\delta \leq f_0^*.$$

If $\delta = 0$*, then* $f_1^0 = f_2^0 = f_3^0 = f_0^*$.

Remark 3 *It follows that*

$$\mathcal{C}_0^0 = \{z \in \mathbf{R}^\infty : g_j(z) \leq 0, j = 1,2,\cdots\} = \mathcal{C}_\lambda^\delta \cap \mathbf{R}.$$

When $\delta_j > 0$ *for some integer* j*, there exists an example such that* $f_1^\delta = f_2^\delta = f_3^\delta < f_4^\delta$. *See Exmaple 1 below.*

Proof of Lemma 1. We shall prove $f_2^\delta = f_3^\delta$. Denote the membership function of $f(z)$ by $\mu_{f(z)}(\xi) = L(\frac{\xi - f(z)^c}{\ell_{f(z)}})$. Then we have $\mu_{f(z^c)}(f(z)^c) = L(\frac{0}{\ell_{f(z^c)}}) = 1$, so that $f(z)^c$ is the center of $L-$fuzzy number $f(z^c)$. Thus $f(z^c)^c = f(z)^c$. By $f(z^c) \in \mathbf{R}$, we get $f(z^c) = f(z)^c$. Therefore $f_2^\delta = f_3^\delta$.

We shall prove $f_1^\delta = f_3^\delta$. From the above definitions it follows that $(f_1^\delta)^c \preceq_\lambda f_1^\delta$ and $(f_1^\delta)^c = f_3^\delta \in \mathbf{R}$. Then we have $f_3^\delta \leq (f_1^\delta)^c + \lambda\ell_{f_1^\delta}$. Suppose that $\ell_{f_1^\delta} > 0$. Then f_3^δ is the real optimal value of (P_λ^δ). From the above corollary there exists no $f \in \mathcal{F}_L \backslash \mathbf{R}$ such that $f = f_3^\delta$, which leads to $\ell_{f_1^\delta} = 0$. It follows that $f_1^\delta = f_3^\delta$.

By Remark 3 we get $f_1^\delta \preceq_\lambda f_0^*$. From $f_1^\delta = f_2^\delta \in \mathbf{R}$ and $f_0^* \in \mathbf{R}$ it follows that $f_1^\delta \leq f_0^*$. It's clear that $f_1^0 = f_2^0 = f_3^0 = f_0^*$.

Q.E.D.

In case that there exists an optimal solution of $L-$fuzzy optimization problems, by the above lemma, the solution means a real number.

Theorem 3 *Denote* $f^* = \min\{f(z) : z \in \mathcal{C}_\lambda^\delta\}$ *and* $f_0^* = \min\{f(z) : z \in \mathcal{C}_\lambda^\delta \cap \mathbf{R}\}$. *Suppose that* (P_λ^δ) *has at least one optimal solution in* $\mathcal{C}_\lambda^\delta$. *Then there exist* f^*, f_0^* *in* $\mathbf{R}$, *which satisfy* $f^* = f_0^*$.

$L-$fuzzied numbers

In what follows we introduce an idea of $L-$fuzzied numbers generalized by $\mathcal{F}_{\mathbf{b}}^{st}$. Let $x \in \mathcal{F}_L$. The quadratic x^2 of an $L-$fuzzy number x isn't necessarily $L-$fuzzy number but fuzzy number in $\mathcal{F}_{\mathbf{b}}^{st}$ (see [5]). For $x = (x_1, x_2) \in \mathcal{F}_L$ and $\alpha \in I$, we have the following three cases:

- $x^2 = (x_1^2, x_2^2)$ if $x_1(\alpha) \geq 0$;
- $x^2 = (x_1 x_2, \max[x_1^2, x_2^2])$ if $x_1(\alpha) \leq 0 \leq x_2(\alpha)$;
- $x^2 = (x_2^2, x_1^2)$ if $x_2(\alpha) \leq 0$.

In this study we consider the left portion of the membership function μ_{x^2} is more significant than the right portion of μ_{x^2}. Denote an operator $(\cdot)_L : \mathcal{F}_{\mathbf{b}}^{st} \to \mathcal{F}_L$ such that $(x)_L = (x_1(1), x_1(1) - x_1(0))_L$ for $x = (x_1, x_2) \in \mathcal{F}_{\mathbf{b}}^{st}$. We call that $(x)_L$ is an *L–fuzzized number*. Here the membership function of x is $\mu_x(\xi) = L(\frac{x_1(1)-\xi}{x_1(1)-x_1(0)})_+$ for $\xi \in \mathbf{R}$, $L : \mathbf{R} \to \mathbf{R}_+$ is a shape function and $\xi_+ = \max(\xi, 0)$ if $\xi \in \mathbf{R}$. For $x \in \mathcal{F}_L$ we get the L–fuzzied number

$$(x^2)_L = (x_1(1)^2, x_1(1)^2 - x_i(0)x_j(0)))_L,$$

where $i = 1, j = 2$ if $x_1(0)x_2(0) \leq 0$, $i = j = 1$ if $x_1(0)x_2(0) > 0$ and $|x_1(0)| < |x_2(0)|$, $i = j = 2$ if $x_1(0)x_2(0) > 0$ and $|x_1(0)| \geq |x_2(0)|$.

Let a shape function be $L(\xi) = (1 - |\xi|)_+$. For an L–fuzzy number $x = (\xi_0, \ell)_L$ with $|\xi_0| \leq \ell$, which has the membership function $\mu_x(\xi) = L(\frac{\xi_0 - \xi}{\ell})_+$ for $\xi \in \mathbf{R}$. Then we get the membership function

$$\mu_{x^2}(\xi) = \begin{cases} \left(1 - \frac{\sqrt{\xi_0^2 - \xi}}{\ell}\right)_+ & \text{for } \xi < \xi_0^2; \\ \left(1 - \frac{\xi_0 - \sqrt{\xi}}{\ell}\right)_+ & \text{for } \xi \geq \xi_0^2. \end{cases}$$

In this case we construct an L–fuzzy numbers $(x^2)_L$ with the same portion as the left one of μ_{x^2}. It follows that $(x^2)_L = (\xi_0^2, \ell^2)_L$. For $x \in \mathcal{F}_L$ and $k \in \mathbf{R}$ we have $(kx)_L = kx$.

In the following example we consider L–fuzzy optimization problem with a fuzzy objective function and fuzzy constraints.

Example 1 *Let $z = (u, v) \in \mathcal{F}_L^2$ and $\lambda \in I$. Fuzzy functions $F, g_j, j = 1, 2, 3$, are as follows (P_λ^δ):*

$$\begin{aligned} F(z) &= -u - v; \\ g_1(z) &= -u \preceq_\lambda (0, \delta_1)_L; \\ g_2(z) &= -v \preceq_\lambda (0, \delta_2)_L; \\ g_3(z) &= (u^2)_L + (v^2)_L \preceq_\lambda (1, \delta_3)_L. \end{aligned}$$

Here $(0, \delta_1)_L, (0, \delta_2)_L, (1, \delta_3)_L$ are L–fuzzy numbers and $(u^2)_L = (u_1(1)^2, \ell_{u^2})_L$, $(v^2)_L = (v_1(1)^2, \ell_{v^2})_L$ are L–fuzzized numbers.

In order to find an optimal solution $z^* = (u^*, v^*) \in \mathcal{F}_L^2$ we consider the Lagrangian function $\mathcal{L}(w) = -u - v + k(w)$, where $w = (z, \eta), \eta = (\eta_1, \eta_2, \eta_3)^T \in \mathbf{R}_+^3$, and

$k(w) = \eta_1[-u_1(1)+\lambda(\ell_u-\delta_1)]+\eta_2[-v_1(1)+\lambda(\ell_v-\delta_2)]+\eta_3[u_1(1)^2+v_1(1)^2-1+\lambda(\ell_{u^2}+\ell_{v^2}-\delta_3)]$. Denote $w^* = (u^*, v^*, 0, 0, 0)$. We find conditions of $u^* = (u_1(1), \ell_{u^*})_L, v^* = (v_1(1), \ell_{v^*})_L$ satisfying the inequality $\mathcal{L}(z^*, \eta) \preceq_\lambda \mathcal{L}(w^*) \preceq_\lambda \mathcal{L}(z, 0, 0, 0)$ for $\eta \in \mathbf{R}^3_+$ and $(u, v) \in \mathcal{C}$. Then it follows that

$$-u^* - v^* \preceq_\lambda -u - v.$$

Since saddle points of $\mathcal{L}$ are optimal solutions of (P^δ_λ), we get

(i) $-u_1^*(1) - v_1^*(1) + \lambda(\ell_{u^*} + \ell_{v^*}) \le -u_1(1) - v_1(1) + \lambda(\ell_u + \ell_v)$ for $\ell_{u^*} + \ell_{v^*} \ge \ell_u + \ell_v$,

(ii) $-u_1^*(1) - v_1^*(1) + \lambda(\ell_{u^*} + \ell_{v^*}) < -u_1(1) - v_1(1) + \lambda(\ell_u + \ell_v)$ for $\ell_{u^*} + \ell_{v^*} < \ell_u + \ell_v$.

From conditions of constraints we get the feasible set $\mathcal{C}^\delta_\lambda = \{(u, v) \in \mathcal{F}^2_L : u = (u_1(1), \ell_u)_L, v = (v_1(1), \ell_v)_L$ satisfy the following conditions (c-i) - (c-iii) below $\}$.

(c-i) $u_1(1) \ge \lambda(\ell_u - \delta_1)$,
(c-ii) $v_1(1) \ge \lambda(\ell_v - \delta_2)$,
(c-iii) $u_1(1)^2 + v_1(1)^2 \le 1 + \lambda[\delta_3 - \ell_{u^2} - \ell_{v^2}]$.

Here

$$\ell_{u^2} = \begin{cases} 2u_1(1)\ell_u - (\ell_u)^2 & (u_1(1) \ge \ell_u) \quad (U_1) \\ (\ell_u)^2 & (|u_1(1)| \le \ell_u) \quad (U_2) \\ -2u_1(1)\ell_u - (\ell_u)^2 & (u_1(1) \le -\ell_u) \quad (U_3) \end{cases}$$

$$\ell_{v^2} = \begin{cases} 2v_1(1)\ell_v - (\ell_v)^2 & (v_1(1) \ge \ell_v) \quad (V_1) \\ (\ell_v)^2 & (|v_1(1)| \le \ell_v) \quad (V_2) \\ -2v_1(1)\ell_v - (\ell_v)^2 & (v_1(1) \le -\ell_v) \quad (V_3) \end{cases}$$

The set $\mathcal{C}^\delta_\lambda$ is non-empty since the point $(u_1(1), v_1(1))^T = (\lambda(\ell_u - \delta_1), \lambda(\ell_v - \delta_2))^T$ satisfies (c-iii) in case that $u_1(1) \ge \ell_u, v_1(1) \ge \ell_v$.

Conditions (c-i) - (c-iii) leads to

$(\mathrm{c-i})'$ $u_1(1) \ge -\lambda\delta_1$,
$(\mathrm{c-ii})'$ $v_1(1) \ge -\lambda\delta_2$,
$(\mathrm{c-iii})'$ $u_1(1)^2 + v_1(1)^2 \le 1 + \lambda\delta_3$.

So the set $C_e = \{(u_1(1), v_1(1))^T \in \mathbf{R}^2 : (\mathrm{c-i})' - (\mathrm{c-iii})'$ hold $\}$ is compact. It can be easily seen that the set $S_{pq} = \{(\ell_u, \ell_v) \in \mathbf{R}^2_+ : (U_p)$ and (V_q) hold $\}, p = 1, 3; q = 1, 3,$ are compact. In case that $p = 1$ and $q = 1, 2, 3$, it follows that

$$\lambda(\ell_u)^2 \le 1 + \lambda\delta_3 \text{ and } \lambda\ell_{v^2} \le 1 + \lambda\delta_3.$$

The latter inequality means that $\ell_v \le |v_1(1)|$ or $\lambda(\ell_v)^2 \le 1 + \lambda\delta_3$, which show that $S_{pq}, p = 2; q = 1, 2, 3$, are compact. In the similar way it follows that $S_{pq}, p = 1, 3; q = 2$, are compact. Thus, from the compactness of $C_e \subset \mathbf{R}^2$ and $S_{pq} \subset \mathbf{R}^2_+, p, q = 1, 2, 3$, the feasible set $\mathcal{C}^\delta_\lambda$ is compact in $\mathcal{F}_L$.

From $(\mathrm{c-i})$ and $(\mathrm{c-ii})$ we have

$$-u_1(1) - v_1(1) + \lambda(\ell_u + \ell_v) \le \lambda(\delta_1 + \delta_2),$$

so $f(z) \preceq_\lambda (0, \delta_1 + \delta_2)_L$. From (c – iii) it follows that $-u_1(1) - v_1(1) \geq -\sqrt{2(1+\lambda\delta_3)}$ and the minimum is attained at $u_1(1) = v_1(1) = (-\sqrt{\frac{1+\lambda\delta_3}{2}}, 0)_L$, which means that $\min_z f(z) = (-\sqrt{\frac{1+\lambda\delta_3}{2}}, 0)_L$ and $u^* = v^* = (-\sqrt{\frac{1+\lambda\delta_3}{2}}, 0)_L$. When $\lambda = 0$ and $\delta_j = 0, j = 1, 2, 3$, then the real type of optimization problem (P_0^0) gives $-\sqrt{2} \leq f(z) \leq 0$ in $\mathbf{R}$ and $u^* = v^* = 1/\sqrt{2} \in \mathbf{R}$.

This example shows that there exists a unique optimal solution of $L-$fuzzy number of fuzzy optimization problem (P_λ^δ) with a fuzzy coefficient, where (P_λ^δ) is an optimization problem with $\mathbf{R}-$valued coefficients if $\ell_z = 0$ and (P_λ^δ) is fuzzy type if $\ell_z \neq 0$, where ℓ_z is the spread of $z \in \mathcal{C}_\lambda^\delta$. Therefore the optimal solution to the real type (P_0^0) is the same as solution to the fuzzy type (P_λ^δ) concerning $\lambda = 0$ and $\ell_z = 0$.

In what follows we show an existence criterion to (P_λ^δ) with compact feasible sets. By theorems in Section 3 we get the following existence criterion for real optimal solutions of $L-$fuzzy optimization problems.

In [3] the authors discuss the following relation in the sense of means defined by membership functions.

Note. In considering a relation $\leq_m$, where $x \leq_m y$ means that

$$\int_0^1 \alpha(x_1(\alpha) + x_2(\alpha))d\alpha \leq \int_0^1 \alpha(y_1(\alpha) + y_2(\alpha))d\alpha$$

for$x = (x_1, x_2), y = (y_1, y_2) \in \mathcal{F}_\mathbf{b}^{st}$, we have the following statements (i) - (iii). Let $x, y, z \in \mathcal{F}_\mathbf{b}^{st}$.

(i) $x \leq_m x$.

(ii) If $x \leq_m y$, and $y \leq_m z$, then we have $x \leq_m z$.

(iii) If $x \leq_m y$, and $y \leq_m x$, then it follows that x is equal to y in the sense of mean. However they aren't necessarily equal each other in the sense of membership functions. Thus the relation $\leq_m$ isn't an order relation over $\mathcal{F}_\mathbf{b}^{st}$.

4 Oil well equations with R-valued functions

In [8] they discuss exponential decay problems, *e.g.*, machine replacement and oil well extraction, etc. They analyze optimization problems for each oil well to determine its optimal replacement schedule. In order to give a mathematical model we introduce the following notations.

- $C(t)$: the quality remaining in the well at time t
- $D > 0$: rate of oil extraction
- P : unit profit of oil

As the oil reserves get depleted, the rate of extraction eventually decreases to uneconomic levels, making it worthwhile to abandon the well and drill a new one at a cost $f(\nu)$. Here ν is capacity of the well, f is continuously differentiable function. Assume that $V \in \mathbf{R}$ and that $0 \leq \nu \leq V,\ f(0)=0,\ f'(\nu) \geq 0$.

Then they get the following rate of oil extraction $C'(t) = -DC(t)$ with $C(0)=\nu$. Then $C(t) = \nu e^{-Dt}$.

Moreover they discuss deterministic discounted models in case of horizon models with a continuous discount rate $r > 0$. Let $\mathcal{T} = \{t_i : i = 1, 2, \cdots\}$ be a sequence of drilling times such that $0 \leq t_i < t_{i+1}$and $\mathcal{V} = \{\nu_i : i = 1, 2, \cdots\}$ a sequence of corresponding oil well capacity such that $\leq \nu_i \leq V$. They get a value of the net profit function

$$J(\mathcal{T}, \mathcal{V}) = \sum_{i=1}^{\infty}[\int_{t_i}^{t_{i+1}} PD\nu_i e^{-D(t-t_i)} e^{-r(t-t_i)} dt - f(\nu_i)] e^{-rt_i}.$$

They consider maximizing problems of $J(\mathcal{T}, \mathcal{V})$ and show the following results.

Theorem STU The following statements 1) and 2) hold:

1) There exist optimal sequences $\mathcal{T}^* = \{t_i^* : i = 1, 2, \cdots\}$ and $\mathcal{V}^* = \{\nu_i^* : i = 1, 2, \cdots\}$ such that

$$\max_{\mathcal{T}, \mathcal{V}} J(\mathcal{T}, \mathcal{V}) = J(\mathcal{T}^*, \mathcal{V}^*).$$

2) It follows that $t_1^* = 0,\ t_2^* = t_{i+1}^* - t_i^*,\ \nu_1^* = \nu_i^*$ for $i = 1, 2, \cdots$.

Let $T = t_2^*,\ \nu = \nu_1^*$. Then

$$J = J(\mathcal{T}^*, \mathcal{V}^*) = \int_0^T PD\nu e^{-(D+r)t} dt - f(\nu) + Je^{-rT},$$

$$\text{or } J = \frac{\nu}{1 - e^{-rT}}\Big[\frac{PD(1 - e^{-(D+r)T})}{D + r} - \frac{f(\nu)}{\nu}\Big].$$

5 Fuzzy functions

Consider a function $x(t) : \mathbf{R} \to \mathcal{F}_{\mathbf{b}}^{st}$. Then $x(t)$ is said to be a fuzzy function. In [4] we find the following definition of fuzzy functions.

$$\begin{aligned} x(t, \cdot) &= \{(x_1(t, \alpha), x_2(t, \alpha))^T \in \mathbf{R}^2 : \alpha \in I\} \\ &= (x_1(t, \cdot), x_2(t, \cdot)) \end{aligned}$$

for $t \in [t_1, t_2]$. Denote $x(t) = (x_1(t), x_2(t))$.

An L-fuzzy function $x(t) = (x_1(t), x_2(t)) : \mathbf{R} \to \mathcal{F}_{\mathbf{b}}^{st}$ is H-differentiable at t in the sense of Hukuhara if there exists an $\eta \in \mathcal{F}_{\mathbf{b}}^{st}$ such that (i) and (ii) hold as $h \to +0$.

(i) $x(t+h) = x(t) + h\eta + o(h)$;   (ii) $x(t) = x(t-h) + h\eta + o(h)$.
Here $o(h) = (o_1(h), o_2(h)) \in C[0,\varepsilon] \times C[0,\varepsilon]$ with $\varepsilon > 0$, which means that

$$\lim_{|h|\to 0} \frac{d(o(h), 0)}{|h|} = 0.$$

Then $x(t) = (x_1(t), x_2(t))$ is H-differentiable at t if and only if $x_1(t,\alpha), x_2(t,\alpha)$ are differentiable in t for each $\alpha \in I$ such that $\eta = (\frac{\partial x_1}{\partial t}, \frac{\partial x_2}{\partial t}) \in \mathcal{F}^{st}_{\mathbf{b}}$.

In [4] the author discuss the integration of fuzzy function $x(t)$.

Definition 2 *An L-fuzzy function $x(t,\cdot) = (x_1(t,\cdot), x_2(t,\cdot))$ is called integrable over $[t_1, t_2]$ if $x_1(t,\alpha)$ and $x_2(t,\alpha)$ are integrable over $[t_1,t_2]$ for $\alpha \in I$. Define*

$$\int_{t_1}^{t_2} x(s,\cdot)ds = \{(\int_{t_1}^{t_2} x_1(s,\alpha)ds, \int_{t_1}^{t_2} x_2(s,\alpha)ds)^T \in \mathbf{R}^2 : \alpha \in I\}.$$

6 Oil well equations with Ł-fuzzy functions

In the same way as analyzing oil well equations of $\mathbf{R}$−valued functions we consider the following problem with $0 \le \lambda \le 1$. We consider the rate of oil extraction D_L as a constant L−fuzzy number $D_L = (D_1, D_2) \in \mathcal{F}_L$ such that $0 \preceq_\lambda D_L$, where $D_1(\alpha)$ is the left end-point of the α−cut set and $D_1(\alpha) > 0$ for $\alpha \in I$. Then we assume that the oil quality and unit profit of oil are L-fuzzy function and L−fuzzy number, respectively.

- $C_L(t) = (C_1(t), C_2(t)) \in \mathcal{F}_L$: L−fuzzy function which means the quality remaining in the well at time t
- $P_L = (P_1, P_2) \in \mathcal{F}_L$: unit profit of oil with $0 \preceq_\lambda P_L$

In this section we consider the following notations. $\nu \in \mathbf{R}$ is a capacity of the well, $f(\nu)$ is renewal cost which is continuously differentiable with $f(0) = 0$ and $f'(\nu) \ge 0$ and V is the upper bound such that $0 \le \nu \le V$.

Then we get an initial value problem of L−fuzzy differential equation $\frac{dC_L}{dt}(t) + D_L C_L(t) = 0$ with $C_L(0) = \nu$, where $0 \in \mathbf{R}$. It follows that as long as $C_1(t) \ge 0$

$$\begin{aligned} C_L'(t) + D_L C_L(t) &= (C_1'(t), C_2'(t)) + (D_1, D_2)(C_1(t), C_2(t)) \\ &= (C_1'(t) + D_1 C_1(t), C_2'(t) + D_2 C_2(t)). \end{aligned}$$

Thus we have the initial value problem of ordinary differential equation

$$C_1'(t) + D_1 C_1(t) = 0$$
$$C_2'(t) + D_2 C_2(t) = 0$$

with $C_1(0) = C_2(0) = \nu$. Therefore

$$C_1(t, \alpha) = \nu e^{-D_1(\alpha)t}, \ C_2(t, \alpha) = \nu e^{-D_2(\alpha)t}$$

for $t \geq 0, \alpha \in I$. We can solve some cases of the above fuzzy differential equations (see [5]). Without using the above information about the $L-$fuzzy function $C_L(t)$ we can find optimal solutions for maximizing problems of values of the net profit function with $L-$fuzzy value concerning $\mathcal{T} = \{t_i \in \mathbf{R} : i = 1, 2, \cdots\}$ such that $0 \leq t_i < t_{i+1}$ and $\mathcal{V} = \{\nu_i \in \mathbf{R} : i = 1, 2, \cdots\}$ such that $0 \leq \nu \leq V$ as follows.

$$\tilde{J}(\mathcal{T}, \mathcal{V}) = \sum_{i=1}^{\infty} [\int_{t_i}^{t_{i+1}} P_L D_L \nu_i C_L(t - t_i) e^{-r(t-t_i)} dt - f(\nu_i)] e^{-rt_i}.$$

It can be seen that $\tilde{J}(\mathcal{T}, \mathcal{V}) = (J_1, J_2) \in \mathcal{F}_L$, where

$$\begin{aligned} J_1 &= \sum_{i=1}^{\infty} [\int_{t_i}^{t_{i+1}} P_1 D_1 \nu_i e^{-D_1 t} e^{-r(t-t_i)} dt - f(\nu_i)] e^{-rt_i}, \\ J_2 &= \sum_{i=1}^{\infty} [\int_{t_i}^{t_{i+1}} P_2 D_2 \nu_i e^{-D_2 t} e^{-r(t-t_i)} dt - f(\nu_i)] e^{-rt_i}. \end{aligned}$$

Here $J_1(\mathcal{T}, \mathcal{V}, \alpha), J_2(\mathcal{T}, \mathcal{V}, \alpha)$ are $\mathbf{R}-$valued functions defined on $\mathcal{Z} \times \mathcal{K} \times I$. Denote

$$\begin{aligned} \mathcal{Z} &= \{\mathcal{T} = (t_1, t_2, \cdots) : t_i \in \mathbf{R}, \ 0 \leq t_i < t_{i+1}, \ i = 1, 2, \cdots\}, \\ \mathcal{K} &= \{\mathcal{V} = (\nu_1, \nu_2, \cdots) : \nu_i \in \mathbf{R}, \ 0 \leq \nu < V, \ i = 1, 2, \cdots\}. \end{aligned}$$

Consider the maximizing problem of $\tilde{J}(\mathcal{T}, \mathcal{V})$. We get the following theorem by applying Theorem 3 and STU.

Theorem 4 *The following statements 1) and 2) hold.*

1) There exists at least one pair of optimal sequences $\mathcal{T}^* = \{t_i^* : i = 1, 2, \cdots\}$ and $\mathcal{V}^* = \{\nu_i : i = 1, 2, \cdots\}$ such that

$$\max_{\mathcal{T}, \mathcal{V} \in \mathcal{Z} \times \mathcal{K}} \tilde{J}(\mathcal{T}, \mathcal{V}) = \tilde{J}(\mathcal{T}^*, \mathcal{V}^*) \in \mathbf{R}.$$

2) It follows that $t_1^* = 0, \ t_2^* = t_{i+1}^* - t_i^*, \nu_1^* = \nu_i^* \in \mathbf{R}$ for $i = 1, 2, \cdots$. Let $T = t_2^*, \ \nu = \nu_1^*$. Then

$$\tilde{J}(\mathcal{T}^*, \mathcal{V}^*) = \frac{\nu}{1 - e^{-rT}} \Big[\frac{PD(1 - e^{-(D+r)T})}{D + r} - \frac{f(\nu)}{\nu} \Big].$$

Here $P, D \in \mathbf{R}$ are respective centers of $L-$fuzzy numbers P_L, D_L.

The aim of this study is to discuss optimization problems arising from oil well equations concerning $L-$fuzzy functions by applying a fuzzy optimization problem (P_λ^δ) which is an extension of real type of optimization problem.

We are interested in an criteria what are sufficient conditions for the existence of fuzzy optimization problems that consists of the objective function $\tilde{J}(\mathcal{T}, \mathcal{V})$ and the infinite numbers of constraints that

$$t_1 \geq 0,\ t_{i+1} - t_i > 0,\ V \geq \nu_i \geq 0 \tag{C}$$

for $i = 1, 2, \cdots$. The maximizing problem of $\tilde{J}(\mathcal{T}, \mathcal{V})$ in the previous section can be reduced to

$$\text{maximize } f(z) \text{ such that } z \in \mathcal{Z} \times \mathcal{K}.$$

Here $z = (\mathcal{T}, \mathcal{V}) \in \mathcal{Z} \times \mathcal{K}$ and $f(z) = \tilde{J}(z)$.

When we treat the existence of optimal sequences $z^* = (\mathcal{T}^*, \mathcal{V}^*)$, some type of algorithms of nonlinear mathematical programming seems to be valid for solving the problems in the similar way as nonlinear analysis for finite dimensional spaces (cf. [5]). From now on it's expected that optimization problems with an objective and the infinite number of constraints, *e.g.,*(C), of $L-$fuzzy functions are investigated by applying methods of nonlinear functional analysis and convex analysis.

7 Fuzzy boundary value problems

Boundary value problems (BVP) of fuzzy differential equations appears in studies of civil engineering and analysis of architecture structures.

In [6] we treat a BVP of a fuzzy differential equation on a finite interval $J = [a, b] \subset \mathbf{R}$ as follows:

$$\begin{aligned} &(F) && \frac{d^2x}{dt^2}(t) = f(t, x, x'), \\ &(1) && x(a) = A, \\ &(2) && x(b) = B, \end{aligned}$$

where $t \in J,\ x = (x_1, x_2), A = (A_1, A_2), B = (B_1, B_2) \in \mathcal{F}_{\mathbf{b}}^{st}$ and f is a continuous fuzzy function. Then we get the following BVP of ordinary differential equations

$$\begin{aligned} &\frac{d^2x_1}{dt^2}(t) = f_1(t, x_1, x_2, x_1', x_2') \\ &\frac{d^2x_2}{dt^2}(t) = f_2(t, x_1, x_2, x_1', x_2') \\ &x_1(a) = A_1, \quad x_2(a) = A_2 \\ &x_1(b) = B_1, \quad x_2(b) = B_2 \end{aligned}$$

with conditions that $x_j^{(i)}(t, \cdot), i = 0, 1, 2; j = 1, 2,$ satisfy conditions of Theorem 1.

In analyzing the above BVPs of fuzzy differential equations we have BVPs of ordinary differential equations. See [6] in detail.

In the case that $J = \mathbf{R}_+ = [0, \infty)$ we get some results on BVPs of fuzzy differential equations. Consider

$$\begin{aligned} &\frac{dx}{dt}(t) = p(t)x + F(t,x), \quad t \in \mathbf{R}_+ \\ &x(\infty) = c, \end{aligned}$$

where $t \in \mathbf{R}_+$, $p = (p_1, p_2)$ an $\mathcal{F}_{\mathbf{b}}^{st}$−valued continuous function defined on $\mathbf{R}_+$, $F = (F_1, F_2)$ an $\mathcal{F}_{\mathbf{b}}^{st}$−valued continuous function defined on $\mathbf{R}_+ \times \mathcal{F}_{\mathbf{b}}^{st}$ and $c = (c_1, c_2) \in \mathcal{F}_{\mathbf{b}}^{st}$. In the similar way as $J = [a, b]$, letting $x(t) = (x_1(t), x_2(t)) \in \mathcal{F}_{\mathbf{b}}^{st}$, we have the following BVP of ODE as follows:

$$\begin{aligned} &\frac{dx_1}{dt}(t) = p_i(t)x_j + F_1(t, x_1, x_2) \\ &\frac{dx_2}{dt}(t) = p_k(t)x_\ell + F_2(t, x_1, x_2), \\ &i, j, k, \ell = 1, 2, \\ &x_1(\infty) = c_1, \quad x_2(\infty) = c_2. \end{aligned}$$

See [7] in detail.

References

[1] N. FURUKAWA (1999), *Mathematical Methods of Fuzzy Optimization*(in Japanese), Morikita Pub., Tokyo.

[2] Jr. R. GOETSCHEL, W. VOXMAN (1983), "Topological Properties of Fuzzy Numbers," *Fuzzy Sets and Systems* , vol 9, 87–99.

[3] Jr. R. GOETSCHEL, W. VOXMAN (1986), "Elementary Fuzzy Calculus," *Fuzzy Sets and Systems* , vol 18, 31–43.

[4] O. KALEVA (1990), "The Cauchy Problem for Fuzzy Differential Equation," *Fuzzy Sets and Systems* , vol 35, 389-396.

[5] S. SAITO (in press), "On Some Topics of Fuzzy Differential Equations and Fuzzy Optimization Problems via a Parametric Representation of Fuzzy Numbers," *"Differential Equations and Applications, Vol. 3"* , Nova Science Publishers, Inc., New York.

[6] S. SAITO, " On Boundary Value Problems of Fuzzy Differential Equations," *Proceedings of the Second Vietnam-Japan Symposium on Fzuuy Systems and Applications*, ed. N. H. PHOUNG and K. YAMADA.

[7] S. SAITO (Preprint), " Boundary Value Problems of Fuzzy Differential Equations on Infinite Intervals."

[8] S. P. SETHI, G. L. THOMPSON and V. UDAYABHANU (1985), "Profit Maximization Models for Exponential Decay Processes," *European J. OR* , vol 22, 101–115.

[9] H.TUY (1998), *Convex Analysis and Global Optimization*, Kluwer Academic Publ.

Optimization and Optimal Control, pp. 251 – 268
P.M. Pardalos, I. Tseveendorj, and R. Enkhbat, Editors

A Genetic Algorithm by Use of Virus Evolutionary Theory for Combinatorial Problems

S. Saito (ssaito@ms.kuki.tus.ac.jp)
School of Management
Tokyo University of Science
Kuki city, Saitama prefecture, JAPAN

Abstract

A genetic algorithm that simulates the virus evolutionary theory(GAV) has been developed and applied to combinatorial optimization problems. The algorithm in this study uses only one individual and a population of viruses. The individual is attacked, infected and improved by the viruses. The viruses are composed of two genes (a top gene and a tail gene). If the individual is improved by the attack, infection occurs. To escape from local minima, an infection allowance is set. After the infection, the tail genes are mutated. If the same virus attacks several times and fails to infect, the top genes of the virus are mutated. The individual can be improved effectively using this mutation. To obtain the optimal solution, "sub-attack" are also useful. GAV with immunologic mechanism and simulated annealing are investigated as well.

Keywords: virus evolutionary theory, genetic algorithm, combinatorial problem, vehicle routing problem, job-shop problem.

1 Introduction

Recently, several evolutionary theories such as the virus evolutionary theory [1] have been proposed in biology. The virus evolutionary theory states that the evolution of life is triggered by an epidemic caused by viruses. It also states that evolution occurs by virus infection and by the reconstruction of the genes of the individual.

In this study a genetic algorithm (GA) that simulates the evolutionary theory by virus infection (GAV) has been developed and reported [2][3][4].

A standard GA is a simulation based on the mechanism of natural selection also known as the evolutionary theory of Darwin. The standard GA is carried out by generating a number of individuals (chromosomes) and performing selection, crossover and mutation on them.

On the other hand, GAV developed in this study uses only one individual and a population of viruses. The individual (chromosome) is attacked and infected by the viruses and the individual is improved without using the selection and the crossover and so on.

The comparison between standard GA and GAV was reported in the former report [2] concerning TSP.

In this paper, the GAV is carried out for solving vehicle routing problem and job-shop problem, and an algorithm for obtaining the optimal solution is investigated.

The GAV is also applied to facility layout problem as described in the former report [4].

2 The Virus Evolutionary Theory

Imanishi's evolutionary theory [5] posits that the evolution of species occurs drastically in a short period (horizontal evolution) after keeping its characteristics stable for a relatively long period. It is also presumed that the mechanism of Imanishi's evolutionary theory can be explained by the virus evolutionary theory.

The virus evolutionary theory in biology assumes that the genes of a virus or transposons of another individual are buried on the chromosome of an individual by the virus and bacillus. Horizontal evolution occurs in a short period.

The method developed here uses the conception of horizontal evolution in Imanishi's theory and it also assumes that virus infection causes the evolution.

3 Vehicle Routing Problem

3.1 The chromosome

The chromosome (individual) in this algorithm is expressed by the string of the genes as shown in Table 1. Standard GAs use many individuals. On the other hand, the GAV developed in this study uses a single individual.

The chromosome of Table 1 shows the route that is described by the sequence of the numerals. The phenotype of the gene can be used in this algorithm. Therefore the encoding is rather simple. The numerals are put randomly at an initial state and the route is improved gradually by the virus attacks.

Table 1: A chromosome for route

Route	3	8	5	4	21	EEE

3.2 The genes of the virus

The virus is composed of two genes, the top gene and the tail gene, as shown in Figure 1. Although there is only a single chromosome (individual), a population of viruses is generated in this algorithm. These viruses are mutated after the attack on the chromosome.

3.3 The virus attack

If one of the viruses attacks the individual, the genes of the chromosome are attacked as shown in Figure 1. The virus attacks the gene of the individual that is same to the top gene.

If the individual is improved by the attack, the infection occurs by reversing the segment of the chromosome between the next gene after the top gene and the tail gene. After the infection, the tail genes are mutated.

If the same virus attacks several times and fails to infect, the top genes of the virus are mutated. The mutation of viruses is very important for the improvement of the individual. By these performing these procedures, virus attack effectively improved the individual.

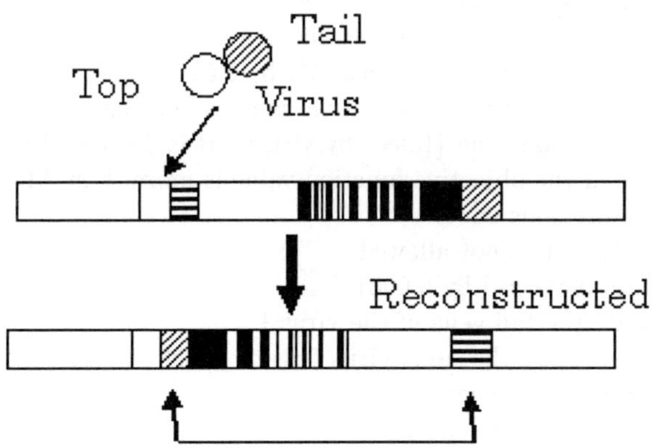

Figure 1: The virus attack

3.4 Escape from the local minima

Solutions determined by the procedure stated above fall easily into local minima as in other search methods. To escape from the local minimum, two methods are proposed. One method introduces an allowance into the infection process. This allowance allows an infection to occur even if it worsens the evaluation value at a small rate. Allowances for infection were also used in the former reports [2][3] in this study. In this report, the value of the allowance is modified according to whether it is infected by the attack. The allowance is decreased by 0.01 if it is infected, and if it is not infected, it is increased by 0.0001. Therefore, the proper value of the allowance is set automatically.

Another method is to allow multiple partial attacks as described later in section 3.6.1. This method is very useful for escaping from local minima. This attack is referred to as a "sub-attack" in this paper.

3.5 The flow of the basic GAV algorithm

The following is the basic GAV algorithm:

```
Start;
    Generate the individual;
     Generate a population of the viruses;
      Repeat until a satisfactory solution is reached;
       Attack the virus;
         Repeat sub-attack of the virus;
         Evaluate the objective function;
          if objective function increased
            {Infect by virus; Decrease AR by 0.01 ;}
            else, if the objective function value is less than AR times
            the previous one {Infect by virus; AR is decreased by 0.01; }
            else, if the objective function value is more than AR times
            the previous one@
            {Infection is not allowed.;
          AR is increased by 0.0001;
         Mutate the tail gene of the virus;}
       if the attack by the same virus fails to infect several times,
      {Mutate the top gene; }
     End of repeat of sub-attack;
     End of repeat;
End.
```

AR is the allowance rate. If the evaluation value multiplied by AR is lower than the previous evaluation value, the infection is admitted.

The phenotype of the chromosome can be used instead of genotype. and therefore the programming of this algorithm is simple compared with standard GAs.

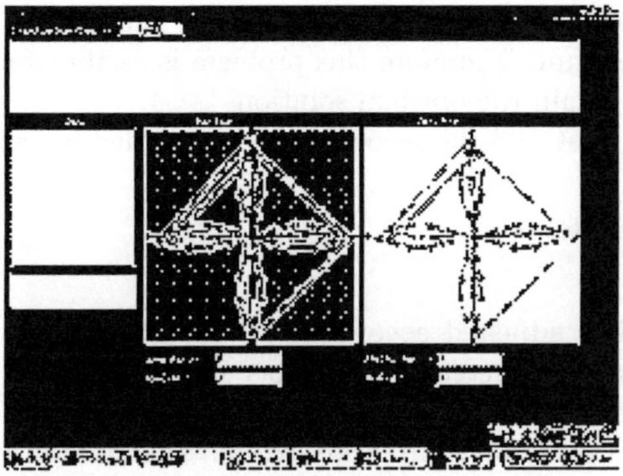

Figure 2: The initial condition of the program

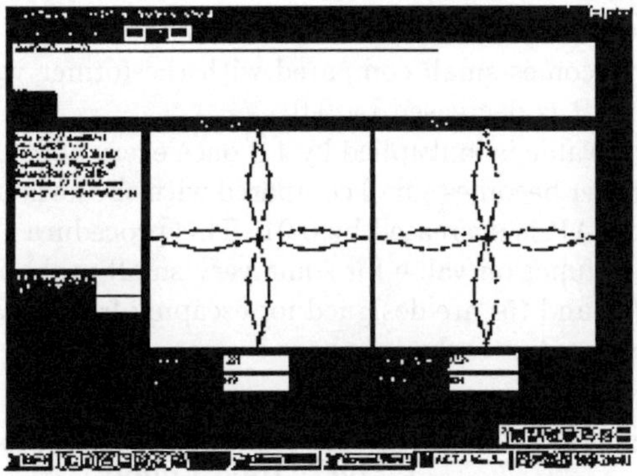

Figure 3: The optimal solution

3.6 Results of GAV

Figure 2 shows the initial condition of the execution for the vehicle routing program. There are forty delivery points and the center point is the depot. The optimal value of the solution is 800.

Figure 3 shows the optimum solution is obtained.

It looks easy to obtain the optimal condition. Actually it is so easy that we are able to obtain the optimal solution at a glance. As for a computer, if the nearest neighbor method is used, it is not difficult to solve it. But to make GAV a general method that is applicable to other combinatorial problems, the nearest neighbor method is not used in this algorithm. Therefore this problem is easily trapped in local minima and it is difficult to obtain the optimal solution.

Two models (the first and the second model) are tried in this algorithm.

3.6.1 The first model

The value of AR is adjusted according to the infection or non-infection of the viruses in the first model. Therefore the proper value of the AR is set automatically. A fixed value of AR and a value that was set according to the objective function value were used in previous reports in this study [2] [3]. Therefore it was rather difficult to obtain a suitable value of AR in the previous reports.

The conditions in the first model are as follows;

(1) The initial value of AR is 1.05.

(2) If the infection is allowed, AR is decreased by 0.01.

(3) If the infection is not admitted, AR is increased by 0.0001, 0.0002 or 0.0005.

(4) The evaluation value is multiplied by AR once every 100 attacks.

If the multiplier becomes small compared with the former value, the infection is also allowed and the AR is decreased by 0.01.

(5) The evaluation value is multiplied by 1.3 once every 500, 1000 or 2000 attacks, and then if the multiplier becomes small compared with the former value, the infection is also allowed and the AR is decreased by 0.01. This procedure allows infections that decrease the objective function value for some very small probability.

The procedures (4) and (5) are designed for escaping from local minima. The first model does not include sub-attacks.

Table 2: Average values of the evaluation by the first model

	500	**1000**	**2000**
0.0001	897.3	912.2	874.4
0.0003	825.5	826.9	829.1
0.0005	865.5	887.1	882.1

Table 2 shows the results of the first model. The numerals in the upper row of the table indicate the frequencies of procedure (5). The figure in the left column is the decrease of AR for procedure (3).

As shown in Table 2, the first model does not reach the optimal condition, though it can obtain satisfactory solutions.

This means that the problem is likely to fall into local minima and it is very difficult to escape from local minima. Therefore some improvements to the first model are necessary.

Among the value of 0.0001, 0.0003 and 0.0005 in the left column, 0.0003 shows the best result for the average evaluation value. The best value is 802, which is obtained at the value of 0.001, but it is likely to fall into the local minimum and it will be very difficult to escape from it with only a value of 0.001. Based on these results, an improvement was attempted.

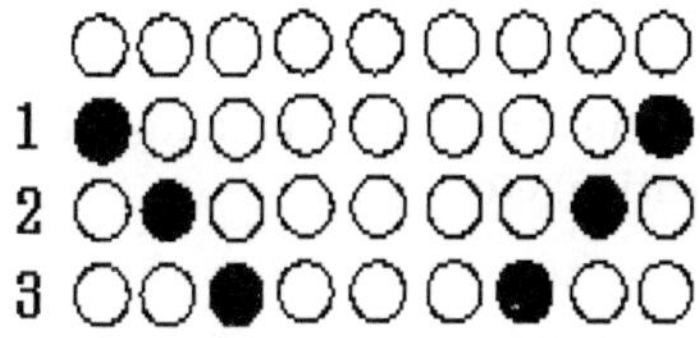

Figure 4: The sub-attack

Table 3: Average attack frequency

A1	5	10	20	30	40
AAF	7062	6621	7197	7825	9369

3.6.2 The second model

The second model is an improvement of the first model. The main points of improvement are as follows;

(1) To prevent the local minima trap, the second model allows multiple partial virus attacks. This attack is called a "sub-attack" in this report. The number of multiple sub-attacks is denoted by A1 in Table 3. Figure 4 shows the improvement. No1 shows the first attack and No2 and No3 do the second and the third attacks respectively. This partial attack or sub-attack is repeated at a certain frequency A1 (5,10,20,,,,) as described in Table 3.

(2) It decreases the AR by 0.0001 when the infection fails at a frequency greater than B1 (Table4) and by 0.00005 at attack number of over a certain number B2 (Table 4). This is in order to decrease the rate of infection that makes the objective function value worse as the attack frequency increases. The purpose of this procedure is to minimize departures from as an excellent solution approaches the optimal solution.

By implementing these improvements, an optimal solution is obtained at almost every trial.

Table 3 shows the results of the second model where AAF in Table 3 is the Average Attack Frequency for each A1.

In this case, A1=10 shows the best result. The best value of A1 is modified according to the problem size.

Table 4 shows a comparison between the values of B1 and B2. By implementing this improvement, useless infections can be decreased. The combination of 1000 for B1 and 3000 for B2 shows the best result. But it begins to fall easily into the local minima as the value of B1 and B2 becomes smaller than 1000 and 3000 respectively. This is because the chance for escaping from the local minima decreases for a small B1 and B2.

This algorithm is also useful for other combinatorial problems such as TSP, job-shop scheduling, facility location, etc

Table 4: The influence of B1 and B2

B1	5000	3000	2000	1000
B2	10000	8000	5000	3000
AAF	6621	5875	6071	3091

3.7 The comparison between the first and the second model

Figure 5 and Figure 6 show the time series of the objective function value of the first model and the second model. Figure 6 shows the condition up to approximately 4000 attacks of viruses and Figure 5 depicts more than 4000. The dotted lines in Figure 5 and 6 indicate the optimal value of evaluation. The first model shows that there are a lot of useless attacks and the appearance of convergence is not observed.

On the other hand, for the second model, the evaluation value changes significantly at attack frequencies less than 3000 (see Figure 6) and it is recognized that the useless attacks are reduced at attack frequencies higher than 3000. It is also recognized that the sub-attack in the second model is effective for converging to an optimal solution.

Figure 5: The time series of the evaluation value of the first model

Figure 6: The time series of the second attack, A1=10, B1=1000, B2=3000

4 Job-shop Scheduling Problem

4.1 Chromosome for job-shop scheduling problem

In this study, the famous job-shop problem (Muth and Thomson's problem [6], Table 5) is used as the simulation of GAV.

The chromosome (individual) is basically expressed by the row of a couple of gene that are job number and the order of processing as shown in Table 6. Table 6 shows the case of six jobs and six machines. The upper row is composed of the number of 1 6. Each number indicates the job number. Therefore there are 36 numerals from 1 to 6 for six jobs and six machines. The scheduling (Gantt chart) is determined in order of these numbers. The numbers of the lower raw show the occurrences count of the same job that is the order of processing of the same job. These pairs of the numbers are the genes of the chromosome.

Phenotype of chromosome can be used in this scheme as well. Therefore, the encoding of the chromosome for the GAV is easy.

Table 5: Job-shop problem(Muth and Thompson, 1963)

Job	Machine	Optimum value
6	6	55
10	10	930
20	5	1165

Table 6: A chromosome for job-shop problem

Job number	3	1	4	2	6	5	5	3	3	4	6	2	1	EEE
Order of processing	1	1	1	1	1	1	2	2	3	2	2	2	3	EEE

Table 7: A example of genes of virus

	JOB	NUM
Top	3	1
Tail	4	2

Table 8: Attack of the virus to the chromosome

Job number	1	2	5	4	**3**	5	6	6	**4**	1	3
Order of processing	1	1	1	1	**1**	2	1	2	**2**	2	2

4.2 The genes of virus

The viruses are composed of two parts, top genes and tail genes as shown in Table 3. Top genes are expressed by the gene (JOB) which is the job number and gene (NUM) which is the order of processing. Tail genes are similarly expressed. If the virus in Table 7 attacks the individual in Table 8, the screened genes showed in Table 8 are attacked.

Two ways of the infection are tried. One is the insert method and another is the exchange method.

The insert method is to insert the gene of the chromosome corresponding to the tail genes after the gene of the chromosome corresponding to the top genes. The virus attacks to the gene of the individual that is matched to the top gene of the virus. If the individual is improved by the attacking, the infection occurs by embedding the tail genes after the attacked gene and removing the corresponding genes from the string of the chromosome.

The exchange method is a method of exchanging the genes specified with top genes and tail genes. If the individual is improved by the attacking, the infection occurs. After the infection, the tail genes are mutated. If the same virus attacks several times and fails to infect, the top genes of the virus are mutated. The mutation of the viruses is very important for the improvement of individual. By these procedures, the viruses attack effectively i.e. the individual is improved effectively.

The solutions by the procedure stated above fall easily into local minima as other search methods. To avoid the local minima, the allowance of the infection is set i.e. the infection that makes the individual worse in a small amount is admitted in a small provability as well. By this procedure, it reaches to the acceptable or satisfactory solution easily though it may not be the optimum solution. One of the examples of the simulation procedure is described below.

```
Start;
    Generate the individual;
     Generate the population of the viruses;
      Repeat until the satisfactory solution is reached;
       Attack the virus;
        Evaluation of the objective function;
         if evaluation becomes higher
          {Infect by virus;}
          else, if the evaluation value is less than (1+AR) times the previous one
```

in a small probability below PR,
{Infect by the virus;
Mutate the tail gene of the virus;}
else, if the attack of the same virus fails to infect over several times
{Mutate the top gene;}
End of repeat;
End.

Where AR is the allowance rate and the infection that lower the evaluation value by (1+AR) times is admitted. PR is the probability rate and AR is applied in a probability below PR.

The phenotype can be used and the programming of this algorithm is simple as compared with the standard GA.

Figure 7: The Gantt chart of 10 jobs and 10 machines

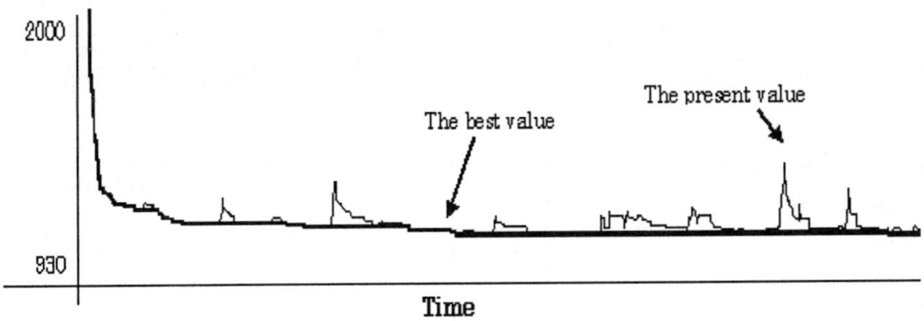

Figure 8: Time series of the evaluation value(AR=3%,PR=1%

4.3 Result of GAV

Figure 7 shows the time series data of the evaluation value. The present value becomes worse sometimes than the best value that is the best one obtained at that time. This is because the allowance is set.

Figure 8 shows the Gantt chart of 10 jobs and 10 machines derived by GAV As for the six jobs and six machines, the optimum solution can be obtained easily. Table 9 shows the results of 10 jobs and 10 machines obtained by GAV of exchange method. Although it failed to obtain the optimum value (930) by this method, it is considered that the better value can be obtained as compared with result obtained by standard GAs.

81% for AR in table 9 means that the value of AR changes from 8% to 1% as the evaluation value becomes better. 18% for PR means that the value of PR increases as the evaluation value increases.

The setting that AR decreases and the PR increase according to the evaluation value showed the best result in the simulation (screened cells in table 9). AR 18% and PR 81% that is an opposite pattern shows good result in the same way as average 998.7 and standard deviation 19.3. That is to say, the chance of infection (PR) should be increased instead of reducing AR as approaching optimum value or it is necessary to decrease the chance of infection (PR) in case of decreasing AR instead.

Table 9: Results by GAV (10 Jobs, 10 Machines)

	1	2	3	4	5	6	7	8	9	10
AR	81%	81%	18%	18%	18%	81%	1%	8%	1%	8%
PR	18%	81%	18%	81%	4%	4%	11%	8%	8%	1%
The best value	937	977	971	971	968	975	1001	1021	999	981
Average value	971.6	1013	1009.7	998.7	1005.5	1005.2	1026	1047	1031	1020
Standard deviation	18.79	18.51	21.48	19.31	28.51	19.92	16.58	21.07	18.87	35.71

Table 10: The genes of Virus and individual for the immunologic mechanism (The case that the infection succeeded)

(a)Virus before attack

	JOB	**NUM**	**Insert Position**
Top	3	2	Search
Tail	1	1	Search

(b)Chromosom(individual) before it is attacked

JOB	2	3	**1**	4	5	**3**	2	4	4	3	EEE
Occurence number	1	1	**1**	1	1	**2**	2	2	3	4	EEE
Position	1	2	3	4	5	6	7	8	9	10	EEE
Antibody	3	3	3	3	3	2	0	0	0	0	EEE

(c)Mutated virus after attack

	JOB	**NUM**	**Insert Position**
Top	**3**	**2**	Search
Tail	x	x	Search

(d)Chromosome (Infected)

JOB	2	3	**3**	4	5	**1**	2	4	4	3	EEE
Occurence number	1	1	**2**	1	1	**1**	2	2	3	4	EEE
Position	1	2	3	4	5	6	7	8	9	10	EEE
Antibody	3	3	**0**	3	3	**0**	0	0	0	0	EEE

5 GAV with the Immunologic Mechanism (Job-Shop Scheduling)

5.1 Algorithm

The influence of immunity (antibody) on evolution of individual is investigated. Table 10, 11 shows the explanation of GAV with immunologic mechanism. The antibody value is given to the gene of chromosome. As this antibody is one related to the connection of the genomic location and Job number, the antibody value returns to 0 when the position of Job changes as shown in Table 10. The antibody value is enlarged when the infection of the virus does not succeed (Table 11).

The virus searches the gene that can be attacked while avoiding this antibody and attacks the gene that does not have the antibody (Table 11). This process is for the virus to attack the chromosome effectively.

A strong virus that disregards the antibody is made to appear sometimes also.

Table 11: The genes of Virus and individual for the immunologic mechanism (The case that the infection failed)

(a)Virus before attack

	JOB	**NUM**	**Insert Position**
Top	3	2	Search
Tail	1	1	Search

(b)Chromosom(individual) before it is attacked

JOB	2	3	**1**	4	5	**3**	2	4	4	3	EEE
Occurence number	1	1	**1**	1	1	**2**	2	2	3	4	EEE
Position	1	2	3	4	5	6	7	8	9	10	EEE
Antibody	3	3	3	3	3	**2**	0	0	0	0	EEE

(c)Mutated virus after attack

	JOB	**NUM**	**Insert Position**
Top	**3**	**2**	Search
Tail	x	x	Search

(d)Chromosome (not infected)

JOB	2	3	1	4	5	3	**2**	4	4	3	EEE
Occurence number	1	1	1	1	1	2	**1**	2	3	4	EEE
Position	1	2	3	4	5	6	7	8	9	10	EEE
Antibody	3	3	3	3	3	**3**	0	0	0	0	EEE

This is to keep the variety of the gene change and to avoid the local minimum.

The cording of the virus is similar to usual GAV. The top gene and the tail gene have the attack destination and the replacement gene information respectively.

If the insert method is used, all following antibodies from the burial position are deleted. Therefore the exchange method is used in this scheme.

5.2 Results of GAV with the immunologic mechanism

Table 12 shows the results of GAV with the immunologic mechanism (10 jobs, 10 machines). Though a good result was expected, it was not better result than GAV without the immunologic mechanism. This process is rather complicated as compared with the usual GAV. Therefore it is not efficient in computer simulation, which is processed sequentially. It is considered that the better result can be obtained by improving the attack of the virus in GAV with the immunologic mechanism. In the natural world, the attack of viruses and generation of antibody are processed in parallel. Therefore it cannot be asserted that the immunity is not related in evolution.

Table 12: Results by GAV with the immunologic mechanism (10 Jobs, 10 Machines)

Random rate of selection of virus	0%	25%	50%	75%
MIN	1028	1020	1006	1024
Average	1043.2	1026.2	1023.4	1061.6
Standard deviation	10.11	4.26	21.66	35.25

6 GAV with Simulated Annealing

6.1 1 Algorithm

The solutions by the procedure stated above fall easily into local minima as other search methods. To avoid the local minima, the allowance of the infection is set i.e. the infection that makes the individual worse in a small amount is admitted in a small provability as well in the former section.

In this section, simulated annealing is used instead of the allowance in a small provability.

The following is the algorithm of GAV with simulated annealing.

Start;
    The generation of the individual;
     The generation of the population of the viruses;
      Repeat until the satisfactory solution is reached;
        Attacking of the virus;
        Evaluation of the objective function;
         if evaluation becomes higher
           { Infection of virus; }
         else, if $e^{-\frac{\Delta f}{T}} > R$ //**simulated annealing**
           {Infection of the virus;
             Mutation of the tail gene of the virus; }
         else, if the attack of the same virus fails
              to infect over several times
              { Mutation of the top gene; }
              if En is reached {
               if $\frac{\bar{f}-\bar{f}'}{\bar{f}'} > \epsilon$ {
               $T = \alpha T$ } }
      End of repeat;

End.

Where En; epoch number, Δf ; increment of the objective function by the virus attack,$\bar{f}$;the average of f at the present epoch, $\bar{f}'$; the avaverage of f, T ; the temperature, R ; the random number, *alpha*; the rate which lower temperature,

6.2 Results of GAV with simulated annealing

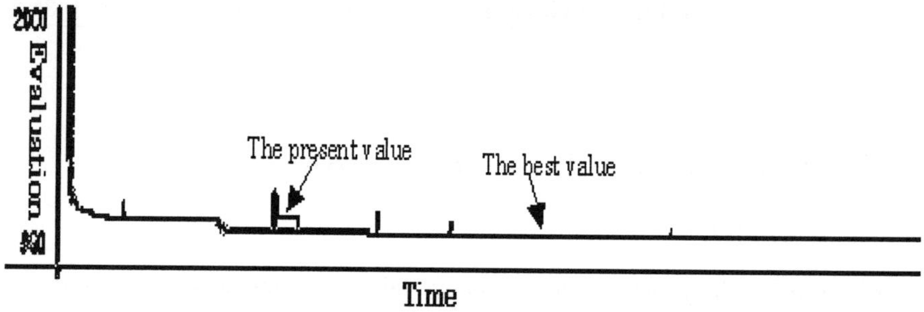

Figure 9: Time series of the evaluation value(allowance rate is set.)

Figure 9 shows the time series of the evaluation value by the former algorithm in this study. The present value becomes worse sometimes than the best value that is the best one obtained at that time. This is because the allowance is set.

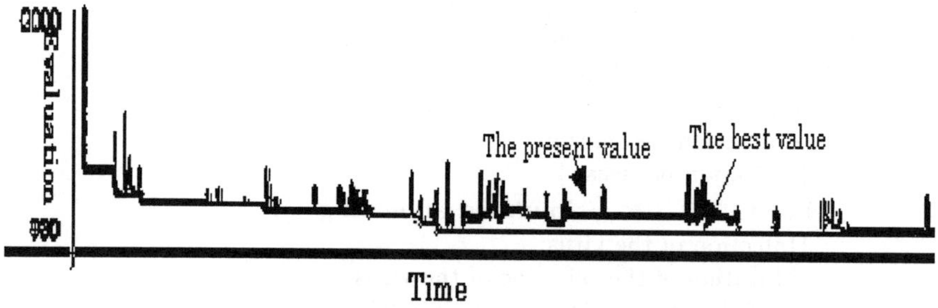

Figure 10: Time series of the evaluation value (simulated annealing)

Figure 10 shows the same one with the Figure 9. But the simulated annealing is used in this case. As compared with Figure 9, the escape from the local minimum is done well. Table 13 shows the result of GAV with simulated annealing.

This simulation obtained comparatively good results for the combinatorial problem.

Table 13: Result

	6m6j	10m10j	5m20j
Initial Temperater	30	1100	100
En	5	5	5
α	0.9	0.99	0.99
The best solution	55	935	1208
average	55	1001.1	1220.8
Standard deviation	0	40.35	8.82

7 Conclusions

An algorithm that expresses the combinatorial problems by chromosomes and solves them by improving the chromosome by attacking it with many viruses has been developed. This algorithm is called "GAV" in this paper. The validity of the algorithm has been recognized through application to the vehicle routing problem and job-shop problems.

Usually, genotype of chromosomes is used for the algorithm in standard GAs and therefore the encoding is rather difficult requiring the special technique. On the other hand phenotype of chromosome can be used in this scheme of GAV. Therefore, the encoding of the chromosome for the GAV is easy.

The important process of GAV is the mutation of viruses and effective improvements of the individual.

This method (GAV) has one defect; it falls into local minima like in other search methods. Some means are pursued to escape from these local minima. The improvements are carried out by allowing infections that worsen the objective function's value by setting an allowance rate (AR) at a small probability.

In the vehicle rooting problem , the value of AR is set automatically according to the virus infection or the failure of the infection. The variation of AR is decreased at a certain attack frequency. This is useful for preventing degradation away from an excellent potential solution. The sub-attack is also useful for escaping from the local minima and for obtaining an optimal solution.

In the job-shop problems, the effects of the immunologic mechanism and the simulated annealing are investigated in addition to the allowance rate. It is recognized that some improvements are needed for the immunologic mechanism and comparatively good result is obtained by simulated annealing.

In this report, the vehicle routing problem and job-shop scheduling are described. This algorithm is also applicable to JSP, facility location problem and so on.

References

[1] Anderson,N., (1970), "Evolutionary Significance of Virus Infection," *Nature*, vol.l227, 1347.

[2] Saito,S., (1998), "A Genetic Algorithm by Use of Virus Evolutionary Theory," *Proc. of 2nd Asia-Pacific Conf. on Industrial Engineering and Management System*, 335-338.

[3] Saito,S., (2000), "A Genetic Algorithm by Use of Virus Evolutionary Theory for Combinatorial Problem" *Proc. of APORS (CD-ROM)*, .

[4] Saito.S., (2002), "A Genetic Algorithm by Use of Virus Evolutionary Theory for Combinatorial Problem," *Proc of 2002 JUSFA*, vol.l1, 87-92.

[5] Harstead,B,. (1985), "Anti-Darwinian Theory in Japan," *Nature*, vol.317, 587.

[6] Muth,J.F. and Thompson,G.L.,(1963), *Industrial Scheduling*, Printice-Hall, Englewood Cliffs, N.J Each.

Optimization and Optimal Control, pp. 269 – 286
P.M. Pardalos, I.Tseveendorj, and R.Enkhbat, Editors

New Variants of Gradient Type Methods in Optimal Control Problems

V.A. Srochko (srochko@math.isu.runnet.ru)
Institute of Mathematics and Economics,
Irkutsk State University,
1, Karl Marx street,
664003, Irkutsk, Russia

Abstract

A standard optimal control problem - one with general functional, nonlinear phase system and convex compact bounding set on control function, is considered. On the basis of optimization of a variation procedure and constructive use of nonclassical approximation of a cost functional, new variants of gradient methods are constructed. The quality of the proposed modifications is determined by nonlocal descent property for bilinear problems and by possibility of improvement for stationary controls in nonconvex problems.

Keywords: Optimal control, iterative methods.

In the paper a gradient approach to construction and justification of iterative methods is further developed according to the basic optimal control problem. In connection with gradient procedures for optimal control, it is necessary to reference the related scientific monographs and textbooks [1]-[7].

The optimization problem of a nonlinear dynamic system with a convex compact constraint set on a control is considered. In this case, the basic optimality condition is differential maximum principle (DMP), standard improvement technique is conditional gradient method, universal variation procedure is generalized convex combination of a pair of admissible controls, where variation function is used as a parameter.

In section 1, a comparable analysis of variation procedures associated with different parametrizations of the variation function is given. Normalizations in L_∞, L_1 spaces lead to the standard weak and needle-shaped variations. By applying L_2-norm, a nonclassical procedure of combined variation is constructed. This procedure combines elements of standard procedures (segments of weak and needle-shaped variation) and enhances constructive abilities of control improvement.

In section 2 quasigradient methods are developed and studied on the basis of quasivariation of the functional and procedures of weak and combined variation. These methods use only first derivatives of the functional elements of the problem with respect to control and phase variables.

The main qualitative characteristics of the proposed procedures are the following:
- local improvement of nonstationary controls in general problems;
- nonlocal descent in bilinear problems and convergence property with respect to modified residual of DMP;
-possibility of improvement for stationary controls because of discontinuity of the variation procedure.

These properties greatly exceed standard potential of the conditional gradient method.

1 Problem formulation and optimization of variation procedure

Let us define a basic optimal control problem:

$$\begin{gathered}\Phi(u) = \varphi(x(t_1)) + \int_T F(x,u,t)dt \to \min, \\ \dot{x} = f(x,u,t), \quad x(t_0) = x^0, \\ u(t) \in U, \quad t \in T.\end{gathered} \tag{1}$$

We introduce a set of admissible controls

$$V = \{u \in L_\infty^r(T) \ : \ u(t) \in U, \ \ t \in T\},$$

which includes measurable and bounded on T vector-functions $u(t)$ with the inclusion type constraints with respect to a convex compact set $U \subset R^r$. We assume that for each admissible control $u \in V$ there exists only one phase trajectory $x(t,u)$ (absolutely continuous vector-function), and the family of these trajectories is bounded on V

$$x(t,u) \in X, \quad t \in T, \quad u \in V,$$

where $X \subset R^n$ is a convex compact set.

Let us assume that in problem (1) the following conditions hold:

1) terminal function $\varphi(x)$ is continuously differentiable on X;

2) integral function $F(x,u,t)$ and vector-function $f(x,u,t)$ are continuous with respect to their arguments on $X \times U \times T$ together with derivatives with respect to (x,u) up to second order.

Let us isolate in (1) a bilinear problem with respect to (x, u) that is characterized by the bilinear functional

$$\Phi(u) = \langle c, x(t_1)\rangle + \int_T (\langle a(u,t), x\rangle + \langle b(t), u\rangle)\, dt,$$

$$a(u,t) = a^0(t) + \sum_{j=1}^{r} a^j(t) u_j,$$

which is associated with the bilinear phase system

$$\dot{x} = A(u,t)x + B(t)u + c(t), \quad x(t_0) = x^0,$$

$$A(u,t) = A_0(t) + \sum_{j=1}^{r} A_j(t) u_j.$$

Then we introduce a conjugate variable $\psi \in R^n$ and form the Pontryagin function as

$$H(\psi, x, u, t) = \langle \psi, f(x,u,t)\rangle - F(x,u,t).$$

The conjugate problem is defined by

$$\dot{\psi} = -H_x(\psi, x, u, t), \quad \psi(t_1) = -\varphi_x(x(t_1)).$$

Let $u(t),\ t \in T$ be an admissible control, and $x(t,u), \psi(t,u)$ are the corresponding trajectories of the phase and conjugate systems, respectively. Using the antigradient

$$H_u[t,u] = H_u(\psi(t,u), x(t,u), u(t), t), \quad t \in T$$

of the functional Φ with respect to a control $u \in V$, we formulate the corresponding optimality condition.

The Differential Maximum Principle: If a control $u \in V$ is optimal in problem (1), then following condition holds:

$$u(t) = \arg\max_{v \in U} \langle H_u[t,u],\ v\rangle, \quad t \in T.$$

Let us define a variation procedure for the control u. Let $v \in V$ be an admissible control, $\chi \in L_\infty(T)$ be a variation function with condition $\chi(t) \in [0,1],\ t \in T$. We form a set of varied controls as a generalized convex combination of the pair u, v

$$u_{v,\chi}(t) = u(t) + \chi(t)(v(t) - u(t)), \quad t \in T. \tag{2}$$

The weak variation of Φ on $u, u_{v,\chi}$ has the form [6]:

$$\delta_0 \Phi(u, u_{v,\chi}) = -\int_T \chi(t)\langle H_u[t,u],\ v(t) - u(t)\rangle\, dt.$$

We fix $\chi(t)$ and formulate the subproblem for the control u:

$$\delta_0\Phi(u, u_{v,\chi}) \to \min, \quad v \in V.$$

Its solution for any χ is defined by the maximality condition

$$\bar{u}(t) = \arg\max_{v\in U} \langle H_u[t,u],\ v\rangle, \quad t \in T.$$

Introduce notations

$$u_\chi(t) = u(t) + \chi(t)(\bar{u}(t) - u(t)), \qquad g(t) = \langle H_u[t,u],\ \bar{u}(t) - u(t)\rangle, \quad t \in T.$$

The weak variation on the pair u, u_χ is the linear functional:

$$\delta_0\Phi(u, u_\chi) = -\int_T \chi(t)\ g(t)\ dt, \qquad \chi(t) \in [0,1],\ g(t) \geq 0.$$

Now we shall consider the question of optimal choice of the variation function. Using L_p - normalization, $p = 1, 2, \infty$ with parameter $\alpha \in (0,1]$, we introduce a set of variation functions

$$\mathcal{X}_\alpha^{(p)} = \{\chi \in L_\infty(T) : \chi(t) \in [0,1],\ (\int_T \chi^p(t)\ dt)^{1/p} = \alpha(t_1 - t_0)^{1/p}\}. \qquad (3)$$

Note that $\alpha \in \mathcal{X}_\alpha^{(p)}$, and $\mathcal{X}_1^{(p)}$ consists of the single function $\chi(t) = 1,\ t \in T$.

For a fixed $\alpha \in (0,1)$, we define the subproblem for searching a variation function using the steepest descent principle

$$\delta_0\Phi(u, u_\chi) \to \min, \quad (\int_T \chi(t)g(t)\ dt \to \max), \quad \chi \in \mathcal{X}_\alpha^{(p)}. \qquad (4)$$

We state the solution of this problem for different normalizations.

1. Let $p = \infty$, i.e. the norm in $L_\infty(T)$ is used: ess $\sup_{t\in T} |\chi(t)| = \alpha$. Then $\mathcal{X}_\alpha^{(\infty)}$ is defined by the restrictions $0 \leq \chi(t) \leq \alpha,\ t \in T$.

The solution to problem (4) has a simple form $\chi_\alpha(t) = \alpha,\ \ t \in T$ and is unique if problem (4) is nondegenerate: $g(t) > 0,\ t \in T$. For $g(t) = 0,\ t \in T_0 \subset T$, the solution is not unique and it is defined by the condition $\chi_\alpha(t) \in [0,\alpha],\ t \in T_0$.

As a result, we get the standard procedure of weak variation by the conditional gradient method

$$u_\alpha(t) = u(t) + \alpha(\bar{u}(t) - u(t)), \quad \alpha \in (0,1],\ t \in T. \qquad (5)$$

The corresponding value of the functional variation is

$$\delta_0\Phi(u, u_\alpha) = -\alpha \int_T g(t)\ dt. \qquad (6)$$

2. Let in problem (4) $p = 1$. Applying the maximum principle to normal form ($\lambda_0 = 1$), we get λ - parameter family of variation functions

$$\chi(t,\lambda) = \begin{cases} 0, & g(t) < \lambda, \\ 1, & g(t) > \lambda. \end{cases}$$

The function $\chi(t,\lambda)$ takes value on $[0,1]$ at singular points $g(t) = \lambda$.

A decision parameter λ_α is defined by the integral condition

$$\int_T \chi(t,\lambda)\, dt = \alpha(t_1 - t_0)$$

and it isolates the following solution of problem (4): $\chi_\alpha(t) = \chi(t,\lambda_\alpha), \quad t \in T$.

As a result α -parametric control family is formed using the needle-shaped variation technique

$$u_\alpha(t) = \begin{cases} u(t), & g(t) < \lambda_\alpha, \\ \bar{u}(t), & g(t) > \lambda_\alpha. \end{cases} \tag{7}$$

For $g(t) = \lambda_\alpha$, we have $u_\alpha(t) \in conv\{u(t), \bar{u}(t)\}$.

Since $\chi(t) \equiv \alpha, \quad t \in T$ is an admissible function for problem (4), the weak variation decrease is evaluated as

$$\delta_0 \Phi(u, u_\alpha) \leq -\alpha \int_T g(t)\, dt. \tag{8}$$

In comparison with (6), we conclude that in a theoretical aspect the needle-shaped variation (7) is preferable (with respect to the speed of decrease of Φ) to weak variation (5).

3. Let us consider problem (4) for $p = 2$ in nondegenerated case: $g(t) > 0$, $t \in T$. According to the maximum principle optimal variation function is defined by a solution of the following problem with multiplier λ

$$\chi g(t) - \frac{1}{2}\lambda\chi^2 \to \max, \quad \chi \in [0,1].$$

If then $\lambda < 0$, the solution $\chi(t,\lambda) = 1$ cannot be admissible because of the integral restriction for $\alpha \in (0,1)$. The problem makes sense for $\lambda > 0$ and here the solution is

$$\chi(t,\lambda) = \begin{cases} \frac{g(t)}{\lambda}, & g(t) < \lambda, \\ 1, & g(t) \geq \lambda. \end{cases} \tag{9}$$

Since $g(t) \geq 0$, for $\lambda = 0$ a solution to (9) is $\chi(t,0) = 1$.

For singular points $g(t) = 0$, we have $\chi(t,\lambda) = 0$ which corresponds to $\lambda > 0$.

Then the decision parameter λ_α is defined by

$$\int_T \chi^2(t,\lambda)\, dt = \alpha^2(t_1 - t_0)$$

and the solution of problem (4) is $\chi_\alpha(t) = \chi(t,\lambda_\alpha)$, $t \in T$. Note that $\lambda_1 = 0$.

Hence, in this case the variation procedure is as (a generalized convex combination for $g(t) < \lambda_\alpha$, maximizing control for $g(t) \geq \lambda_\alpha$)

$$u_\alpha(t) = \begin{cases} u(t) + \frac{g(t)}{\lambda_\alpha}(\bar{u}(t) - u(t)), & g(t) < \lambda_\alpha, \\ \bar{u}(t), & g(t) \geq \lambda_\alpha. \end{cases} \tag{10}$$

The corresponding evaluation of the functional variation decrease is the same as for the previous case ($\chi(t) \equiv \alpha, \quad t \in T$ is an admissible function) given by (8).

Now we consider the improvement properties with respect to functional Φ on family u_α under the assumption that initial control $u(t)$ does not satisfy DMP in problem (1), i.e.

$$\delta_0(u) = \int_T g(t)dt > 0.$$

The increment and weak variation of functional Φ on the pair u, u_α has the relationship [6]

$$\Delta_{u_\alpha}\Phi(u) = \delta_0\Phi(u, u_\alpha) + \eta_0,$$

where

$$\eta_0 = \Delta_1 + \Delta_2 + \Delta_3,$$

$$\Delta_1 = -\int_T o_H(\|\Delta u(t)\|)\, dt, \quad \Delta_2 = -\int_T \langle \Delta_{u_\alpha} H_x[t, u],\ \Delta x(t) \rangle\, dt,$$

$$\Delta_3 = -\int_T o_H(\|\Delta x(t)\|)\, dt + o_\varphi(\|\Delta x(t_1)\|).$$

According to assumption 2) in problem (1), function $H(\psi, x, u, t)$ is twice continuously differentiable with respect to u, and vector-function $H_x(\psi, x, u, t)$ has continuous derivative $H_{xu}(\psi, x, u, t)$. Hence, using standard procedures and facts (representation of the error term in Taylor expansion, generalized mean-value theorem) we get the estimates

$$|\Delta_1| \leq C_1 \int_T \|\Delta u(t)\|^2\, dt, \qquad |\Delta_2| \leq C_2 \int_T \|\Delta u(t)\|\ \|\Delta x(t)\|\, dt. \tag{11}$$

Also, because the set of the admissible processes $\{u(t), x(t, u)\}$ of problem (1) is bounded, for the increments $\Delta u(t),\ \Delta x(t)$ the following inequality holds:

$$\|\Delta x(t)\| \leq C \int_T \|\Delta u(t)\|\, dt, \quad t \in T. \tag{12}$$

For $p = \infty$ (weak variation) the improvement is obvious

$$\|\Delta u(t)\| \sim \alpha, \quad \|\Delta x(t)\| \sim \alpha, \quad t \in T, \quad \eta_0 = o(\alpha),$$

i.e. control u_α guarantees a local descent with respect to Φ :

$$\Phi(u_\alpha) - \Phi(u) = -\alpha\delta_0(u) + o(\alpha).$$

Let $p = 1$ (needle-shaped variation). Let us estimate the term η_0 with respect to α. Since $\Delta u(t) = \chi_\alpha(t)(\bar{u}(t) - u(t)), \quad t \in T$,

$$\int_T \|\Delta u(t)\|\, dt \le D \int_T \chi_\alpha(t)\, dt = \alpha D(t_1 - t_0),$$

where $D = diamU$ - diameter of U.

Hence, by virtue of (12), we have $\|\Delta x(t)\| \le C_3\alpha, \quad t \in T$, therefore, in the expression for η_0

$$|\Delta_2| \le C_4\alpha^2, \quad \Delta_3 = o(\alpha). \tag{13}$$

As for Δ_1, in frames of estimate (11) (which is unimprovable) it is not possible to make a conclusion about its order of vanishing with respect to α.

This situation might be clarified for problems (1) that are linear with respect to a control function. In this case $\Delta_1 = 0$, i.e. $\eta_0 = o(\alpha)$, and control u_α guarantees a local improvement. However, in problems that are linear with respect to the control with a convex compact set U, the differential maximum principle coincides with the maximum principle, i.e. we have standard procedure of needle-shaped improvement.

Therefore, the needle-shaped variation based on the differential maximum principle in problems that are nonlinear with respect to control does not have, generally speaking, the improvement property.

Let us consider the case of $p = 2$ (combined variation). Here, as before, we get

$$\int_T \|\Delta u(t)\|\, dt \le D \int_T \chi_\alpha(t)\, dt.$$

Then, using the inequality

$$\int_T \chi_\alpha(t)\, dt \le \sqrt{t_1 - t_0}\Big(\int_T \chi_\alpha^2(t)\, dt\Big)^{1/2},$$

and taking into account L_2 - normalization for $\chi_\alpha(t)$, we obtain the same result as before

$$\int_T \|\Delta u(t)\|\, dt \le \alpha D(t_1 - t_0). \tag{14}$$

Clearly, estimations (13) follow from (14). Now we will determine order of Δ_1. If we note that

$$\|\Delta u(t)\|^2 = \chi_\alpha^2(t)\|\bar{u}(t) - u(t)\|^2 \le \chi_\alpha^2(t) D^2,$$

then

$$\int_T \|\Delta u(t)\|^2\, dt \le \alpha^2 D^2(t_1 - t_0).$$

This guarantees the second order of vanishing with respect to α for Δ_1. Thus, $\eta_0 = o(\alpha)$ and we obtain a local improvement as follows:

$$\Phi(u_\alpha) - \Phi(u) \le -\alpha\delta_0(u) + o(\alpha).$$

So, the combined variation procedure (10) is an alternative to the standard conditional gradient scheme for the considered problem.

In conclusion, we note that practical realization of procedure (10) implies variation with respect to $\lambda > 0$ in order to decrease functional Φ on the set of controls

$$u(t,\lambda) = \begin{cases} u(t) + \frac{g(t)}{\lambda}(\bar{u}(t) - u(t)), & g(t) < \lambda, \\ \bar{u}(t), & g(t) \geq \lambda. \end{cases}$$

Such variation is quite appropriate for problems (1) when optimal control has interior and boundary values relatively to the set U.

To simplify the previous formula one may introduce the cut off function

$$sat_1 x = \begin{cases} x, & x < 1, \\ 1, & x \geq 1. \end{cases}$$

Then the combined variation procedure has a form

$$u(t,\lambda) = u(t) + (sat_1 \frac{g(t)}{\lambda})(\bar{u}(t) - u(t)).$$

Let us illustrate the above relation in the following simple problem.

Example 1. (comparison of the variation procedures)

$$\Phi(u) = x_1^2(3) + x_2(3) \to \min,$$

$$\dot{x}_1 = u, \ \ \dot{x}_2 = x_1, \ x_1(0) = 0, \ x_2(0) = -1,$$
$$|u(t)| \leq 1, \ t \in [0,3].$$

In this problem

$$H = \psi_1 u + \psi_2 x_1, \quad H_u = \psi_1.$$
$$\dot{\psi}_1 = -\psi_2, \quad \dot{\psi}_2 = 0, \quad \psi_1(3) = -2x_1(3), \quad \psi_2(3) = -1.$$

Consider control $u(t) = 0$ with trajectories $x_1(t,u) = 0$, $x_2(t,u) = -1$, $\psi_1(t,u) = t-3$, $\psi_2(t,u) = -1$, where $\Phi(0) = -1$. This control does not satisfy DMP. We will improve it by the conditional gradient procedure (5) and modification (10).

For (5) we have

$$\bar{u}(t) = sign\ (t-3) = -1, \quad u_\alpha(t) = -\alpha,$$

$$x_1^\alpha(t) = -\alpha t, \ x_2^\alpha(t) = -\alpha\frac{t^2}{2} - 1.$$

It is easy to get

$$\Phi(u_\alpha) = 9\alpha^2 - \frac{9}{2}\alpha - 1, \ \alpha \in [0,1].$$

The minimizing value of parameter $\alpha_1 = 1/4$, which leads to $u_1 = -1/4$ and $\Phi(u_1) = -1\frac{9}{16} < \Phi(0)$.

Now we apply the combined variation procedure (10). Take $\lambda \in [0,3]$. In this case, because of monotonicity of function $g(t) = -H_u[t,u] = 3-t$, it is useful to take a parameter $\tau \in [0,3]$ according to the rule $g(\tau) = \lambda$. Then

$$u_\tau(t) = \begin{cases} -1, & 0 \le t < \tau, \\ \frac{t-3}{3-\tau}, & \tau < t \le 3. \end{cases}$$

By solving the phase system, we get

$$x_1^\tau(t) = \begin{cases} -t, & 0 \le t < \tau, \\ \frac{1}{2(3-\tau)}(t^2 - 6t + \tau^2), & \tau < t \le 3; \end{cases}$$

$$x_2^\tau(t) = \begin{cases} -\frac{t^2}{2} - 1, & 0 \le t < \tau, \\ \frac{1}{2(3-\tau)}(\frac{t^3}{3} - 3t^2 + \tau^2 t) - \frac{\tau^3}{6(3-\tau)} - 1, & \tau < t \le 3. \end{cases}$$

The corresponding functional value

$$\Phi(u_\tau) = 1/12(5\tau^2 + 6\tau - 21), \quad \tau \in [0,3].$$

Minimal value on $[0,3]$ reaches at $\tau = 0$ and leads to the final control

$$u_2(t) = \frac{1}{3}(t-3), \quad t \in [0,3], \quad \Phi(u_2) = -1\frac{3}{4}.$$

Now consider the case when $\lambda > 3$. Here $u_\lambda(t) = \frac{1}{\lambda}(t-3)$. The corresponding solution of the phase system is

$$x_1^\lambda(t) = \frac{1}{\lambda}(\frac{t^2}{2} - 3t), \quad x_2^\lambda(t) = \frac{1}{\lambda}(\frac{t^3}{6} - \frac{3t^2}{2}) - 1.$$

Then the functional value

$$\Phi(u_\lambda) = \frac{81}{4\lambda^2} - \frac{9}{\lambda} - 1, \quad \lambda > 3.$$

It is not difficult to see that the minimal value of $\Phi(u_\lambda)$ reaches at $\lambda = \frac{9}{2}$, that give the final control

$$u_3(t) = \frac{2}{9}(t-3), \quad t \in [0,3], \quad \Phi(u_3) = -2.$$

Therefore, the combined variation procedure is preferable in this case with respect to the improvement quality: $\Phi(u_3) < \Phi(u_1)$.

2 Quasigradient methods

Let us modify the improvement procedures obtained in section 1. As a basic approximation we take a quasivariation of the functional defined by the following relations [6]

$$\delta_1\Phi(u,w) = -\int_T \langle H_u(\psi(t,u), x(t,w), u(t), t), w(t) - u(t)\rangle dt.$$

$$\Delta_w\Phi(u) = \delta_1\Phi(u,w) + \eta_1, \tag{15}$$

$$\eta_1 = o_\varphi(\|\Delta x(t_1)\|) - \int_T o_H(\|\Delta x(t)\|)dt - \int_T o_H^{(1)}(\|\Delta u(t)\|)dt.$$

Unlike classical variation $\delta_0\Phi(u,w)$ here the derivative H_u is calculated along a perturbed trajectory $x(t,w)$. As a result, approximation (15) is complete for linearization of the functional Φ with respect to Δx : the error term η_1 has the order of $o(\|\Delta x\|)$. It means that in the bilinear problem approximation (15) is exact: $\eta_1 = 0$.

The corresponding methods are connected with controls that are defined by the extremal condition

$$u^*(\psi, x, t) = \arg\max_{v\in U}\langle H_u(\psi, x, u(t), t), v\rangle, \quad \psi, x \in R^n, \ t \in T,$$

where $u(t)$ is a nominal control.

Suppose that the structure of problem (1) allows analytical realization of the extremal control $u^*(\psi, x, t)$ (explicit formula).

For $u(t)$, $t \in T$, organize a variation according to the convex combination rule (2) with functional parameters $\chi(t) \in [0,1]$ (variation function) and $v(t) \in U$ (auxiliary control). By definition, quasivariation of the functional Φ on controls $u, u_{v,\chi}$ has the form

$$\delta_1\Phi(u, u_{v,\chi}) = -\int_T \chi(t)\langle H_u(\psi(t,u), x(t,u_{v,\chi}), u(t), t), v(t) - u(t)\rangle \, dt.$$

Fix $\chi(t)$, and substitute an unknown trajectory $x(t, u_{v,\chi})$ by functional parameter $x(t)$ (absolutely continuous vector-function), and solve the subproblem for the control $v(t)$ as

$$\int_T \chi(t)\langle H_u(\psi(t,u), x(t), u(t), t), v(t)\rangle \, dt \to \max, \quad v \in V.$$

Its solution for any χ has a form

$$v^*(t, x(t)) = u^*(\psi(t,u), x(t), t), \ t \in T.$$

Introduce the nonnegative function

$$g(t, x(t)) = \langle H_u(\psi(t,u), x(t), u(t), t), v^*(t, x(t)) - u(t)\rangle$$

and state the subproblem for the variation function $\chi(t)$

$$\int_T \chi(t) g(t, x(t))\, dt \to \max, \quad \chi \in \mathcal{X}_\alpha^{(p)} \quad (p = 2, \infty). \tag{16}$$

Here $\mathcal{X}_\alpha^{(p)}$ is a set of variation functions with normalizing parameter $\alpha \in (0, 1]$, defined by (3).

For $p = \infty$ the solution of this auxiliary problem takes the simple form $\chi(t, \alpha) = \alpha$, $t \in T$ which defined α - parameter family of the variation controls:

$$u(t, x, \alpha) = u(t) + \alpha(v^*(t, x) - u(t)). \tag{17}$$

This solution is unique if problem (16) is nondegenerated, i.e. $g(t, x(t)) > 0, \ t \in T$.

For $p = 2$ the solution of the auxiliary problem in the nondegenerated case is (see formula (9))

$$\chi(t, x(t), \alpha) = \begin{cases} \frac{g(t,x(t))}{\lambda(x,\alpha)}, & g(t, x(t)) < \lambda(x, \alpha), \\ 1, & g(t, x(t)) \geq \lambda(x, \alpha). \end{cases} \tag{18}$$

Here the Lagrange multiplier $\lambda(x, \alpha) \geq 0$ guarantees normalization condition

$$\int_T \chi^2(t, x(t), \alpha)\, dt = \alpha^2(t_1 - t_0).$$

The corresponding family of varied controls is formed as

$$u(t, x, \alpha) = u(t) + \chi(t, x, \alpha)(v^*(t, x) - u(t)). \tag{19}$$

Then we find the solution $x_\alpha(t)$ of the phase system

$$\dot{x} = f(x, u(t, x, \alpha), t), \quad x(t_0) = x^0 \tag{20}$$

together with the generating control $u_\alpha(t) = u(t, x_\alpha(t), \alpha), \quad t \in T$. Let

$$v_\alpha(t) = v^*(t, x_\alpha(t)), \quad g_\alpha(t) = g(t, x_\alpha(t)), \quad \lambda_\alpha = \lambda(x_\alpha, \alpha), \quad \chi_\alpha(t) = \chi(t, x_\alpha(t), \alpha).$$

Then $u_\alpha(t), \ \alpha \in (0, 1]$ is determined by one of the formulas

$$u_\alpha(t) = u(t) + \alpha(v_\alpha(t) - u(t)), \quad t \in T \quad (p = \infty); \tag{21}$$

$$u_\alpha(t) = \begin{cases} u(t) + \frac{g_\alpha(t)}{\lambda_\alpha}(v_\alpha(t) - u(t)), & g_\alpha(t) < \lambda_\alpha, \\ v_\alpha(t), & g_\alpha(t) \geq \lambda_\alpha. \end{cases} \quad (p = 2). \tag{22}$$

The quasivariation of the functional is

$$\delta_1 \Phi(u, u_\alpha) = -\int_T \chi_\alpha(t) g_\alpha(t)\, dt,$$

where $\chi_\alpha(t),\ t\in T$ is the solution of the problem

$$\int_T \chi(t)g_\alpha(t)\ dt \to \max, \quad \chi \in \mathcal{X}_\alpha^{(p)}. \tag{23}$$

Therefore, for $p=\infty$ we have

$$\delta_1\Phi(u,u_\alpha) = -\alpha\int_T g_\alpha(t)\ dt,$$

for $p=2$ the following estimate holds ($\chi(t)=\alpha$ - an admissible function for problem(23)):

$$\delta_1\Phi(u,u_\alpha) \le -\alpha\int_T g_\alpha(t)\ dt. \tag{24}$$

Let us introduce notation

$$\delta_1(u,\alpha) = \int_T g_\alpha(t)\ dt.$$

Note that $\delta_1(u,\alpha)\ge 0$. Now we shall find how this value is connected with the residual of DMP $\delta_0(u)$.

Lemma. *The following estimate is true*

$$|\delta_1(u,\alpha)-\delta_0(u)| \le C\alpha, \quad C=const. \tag{25}$$

Proof. By definition

$$\delta_1(u,\alpha)-\delta_0(u) = \int_T \langle H_u(\psi(t,u),x_\alpha(t),u(t),t), v_\alpha(t)-u(t)\rangle\ dt-$$

$$-\int_T \langle H_u(\psi(t,u),x(t,u),u(t),t), \bar{u}(t)-u(t)\rangle\ dt =$$

$$= \int_T \langle H_u[t,u,\alpha], v_\alpha(t)-u(t)\rangle\ dt - \int_T \langle H_u[t,u], \bar{u}(t)-u(t)\rangle\ dt.$$

Taking into account the maximality condition for $\bar{u}(t)$, we have

$$\langle H_u[t,u],\bar{u}(t)\rangle \ge \langle H_u[t,u], v_\alpha(t)\rangle, \quad t\in T.$$

Hence, the following inequality holds

$$\delta_1(u,\alpha)-\delta_0(u) \le \int_T \langle H_u[t,u,\alpha]-H_u[t,u], v_\alpha(t)-u(t)\rangle\ dt. \tag{26}$$

Since U is a restricted set

$$\|\Delta_\alpha u(t)\| = \|v_\alpha(t)-u(t)\| \le C_1, \quad t\in T.$$

Moreover, by virtue of estimates (12),(14) we conclude that the phase increment has order α

$$\|\Delta_\alpha x(t)\| = \|x_\alpha(t) - x(t,u)\| \leq C_2\alpha, \quad t \in T.$$

Then, under the assumption 2) for problem (1), the vector-function $H_u(\psi(t,u), x, u(t), t)$ has a continuous derivative with respect to x on $T \times X$. Hence, the Lipschitz condition holds

$$\|H_u[t,u,\alpha] - H_u[t,u]\| \leq C_3\|x_\alpha(t) - x(t,u)\|.$$

As a result, from (26) we get the first estimate

$$\delta_1(u,\alpha) - \delta_0(u) \leq C_1C_2C_3(t_1 - t_0)\alpha = C\alpha. \tag{27}$$

Now consider the difference $\delta_0(u) - \delta_1(u,\alpha)$ using maximality condition for $v_\alpha(t)$

$$\langle H_u[t,u,\alpha], v_\alpha(t)\rangle \geq \langle H_u[t,u,\alpha], \bar{u}(t)\rangle, \quad t \in T.$$

Hence

$$\delta_0(u) - \delta_1(u,\alpha) \leq \int_T \langle H_u[t,u] - H_u[t,u,\alpha], \bar{u}(t) - u(t)\rangle \, dt.$$

The proof is the same as previous. Finally, we get the second estimate $\delta_0(u) - \delta_1(u,\alpha) \leq C\alpha$, which together with (27), proves the lemma.

Let us formulate the main result on improvement of $u(t)$ in frames of the variation procedures (21), (22).

Theorem. *If control $u \in V$ does not satisfy DMP for problem (1), then $\Phi(u_\alpha) < \Phi(u)$ for any small $\alpha > 0$.*

Proof. According to approximation formula (15) and taking into account the decrease estimate (24), we have

$$\Phi(u_\alpha) - \Phi(u) = \delta_1\Phi(u,u_\alpha) + \eta_1 \leq -\alpha\delta_1(u,\alpha) + \eta_1,$$

where

$$\eta_1 = o_\varphi(\|\Delta_\alpha x(t_1)\|) - \int_T o_H(\|\Delta_\alpha x(t)\|)\, dt - \int_T o_H^{(1)}(\|\Delta_\alpha u(t)\|)\, dt.$$

Using the estimates of section 1, we find the order of the error term: $\eta_1 = o(\alpha)$.

Since $\delta_0(u) > 0$, then applying the previous lemma, we get

$$\Phi(u_\alpha) - \Phi(u) \leq -\alpha\delta_0(u) + \alpha(\delta_0(u) - \delta_1(u,\alpha)) + o(\alpha) = -\alpha\delta_0(u) + o_1(\alpha).$$

proving the theorem.

Note that the improvement possibility is higher for the bilinear problem. In this case approximation (15) is exact ($\eta_1 = 0$), therefore

$$\Phi(u_\alpha) - \Phi(u) \leq -\alpha\delta_1(u,\alpha) \leq 0, \quad \alpha \in (0,1].$$

It means that quasigradient procedures (21), (22) in the bilinear problem guarantee nonlocal improvement: for any $\alpha \in (0,1]$, u_α is not worse with respect to the functional than the initial control u.

Now we shall note the second singularity of quasigradient procedures (21), (22). Control $u(t,x,\alpha)$ (formulas (17),(19)) is a piecewise continuous vector-function with respect to x (owing to $v^*(t,x)$). Hence, phase system (20) with this control is also discontinuous with respect to x. Therefore, on discontinuity surfaces there might be more than one solution $x_\alpha(t)$. This is a necessary condition for improvement of the controls that satisfy DMP in problem (1).

In this connection we want to characterize stationary situations from the point of view of quasigradient improvement procedures.

Let $u(t),\ t \in T$ be a stationary control (in the sense of DMP), i.e. $\delta_0(u) = 0$. It means that

$$u(t) = v^*(t, x(t,u)), \quad u_\alpha(t) = u(t), \quad x_\alpha(t) = x(t,u), \quad t \in T, \ \alpha \in (0,1].$$

Thus, stationary process $\{u(t), x(t,u)\}$ is the solution of the Cauchy problem (20) for any $\alpha \in (0,1]$. Here the improvement is possible if (20) has another solution $\{v_\alpha(t), y_\alpha(t)\}$ such that

$$\delta_1(u,\alpha) = \int_T g(t, y_\alpha(t))\, dt > 0.$$

For example, in bilinear problems such situation guarantees improvement of the stationary control.

Let us illustrate this.

Example 2. (improvement of stationary nonsingular control)

$$\Phi(u) = -x(2) + 2\int_0^2 x(2-3u)\, dt \to \min,$$

$$\dot{x} = 2(u-1)t, \quad x(0) = 1, \quad u(t) \in [0,1], \ t \in [0,2].$$

In this case $H = 2\psi(u-1)t - 2x(2-3u)$, conjugate equation $\dot{\psi} = 2(2-3u)$, $\psi(2) = 1$, maximizing control

$$u^*(\psi,x,t) = \begin{cases} 0, & H_u(\psi,x,t) < 0, \\ 1, & H_u(\psi,x,t) > 0, \end{cases}$$

where $H_u = 2\psi t + 6x$.

Let us consider the control

$$u(t) = \begin{cases} 1, & 0 \le t \le 1, \\ 0, & 1 < t \le 2. \end{cases}$$

The corresponding trajectories are

$$x(t,u) = \begin{cases} 1, & 0 \le t \le 1, \\ 2 - t^2, & 1 \le t \le 2, \end{cases} \quad \psi(t,u) = \begin{cases} -2t - 1, & 0 \le t \le 1, \\ 4t - 7, & 1 \le t \le 2. \end{cases}$$

It is not difficult to see that this control strictly satisfies DMP with singular point $t = 1$ (switching point).

Let us apply the improvement procedure (21). First we form the control

$$u(t,x,\alpha) = \begin{cases} 1 + \alpha(v^*(t,x) - 1), & 0 \le t \le 1, \\ \alpha v^*(t,x), & 1 < t \le 2, \end{cases}$$

where $v^*(t,x) = u^*(\psi(t,u), x, t)$. Here

$$H_u(\psi(t,u), x, t) = H_u[t,x] = \begin{cases} -4t^2 - 2t + 6x, & 0 \le t \le 1, \\ 8t^2 - 14t + 6x, & 1 \le t \le 2. \end{cases}$$

Then we solve the phase equation

$$\dot{x} = 2(u(t,x,\alpha) - 1)t, \quad x(0) = 1. \tag{28}$$

For $t \in [0,1)$, $x_\alpha(t)$ is the unique solution and it coincides with $x(t,u)$:

$$x_\alpha(t) = x(t,u), \quad H_u[t, x_\alpha(t)] > 0, \quad v_\alpha(t) = 1, \ u_\alpha(t) = u(t).$$

For $t = 1$ trajectory $x_\alpha(t)$ falls into the discontinuity line of $u(t,x,\alpha)$: $H_u[1, x_\alpha(1)] = 0$, so more than one solution is possible.

Consider two nonsingular cases for $t > 1$:

$$1) \ \ H_u[t, x_\alpha(t)] < 0, \qquad 2) \ \ H_u[t, x_\alpha(t)] > 0.$$

In the first case the initial process is the solution

$$v_\alpha(t) = 0, \quad u_\alpha(t) = u(t), \quad x_\alpha(t) = x(t,u).$$

The second situation leads to an alternative solution

$$v_\alpha(t) = 1, \quad u_\alpha(t) = \alpha, \quad x_\alpha(t) = (\alpha - 1)t^2 + 2 - \alpha.$$

Let us check the condition 2). In this case

$$H_u[t, x_\alpha(t)] = 8t^2 - 14t + 6x_\alpha(t) = (2 + 6\alpha)t^2 - 14t + 12 - 6\alpha,$$

where $H_u[1, x_\alpha(1)] = 0$. The minimum point for this parabola is $t_*(\alpha) = \frac{7}{2+6\alpha}$. To guarantee 2) we demand $t_*(\alpha) \le 1$. Then $\alpha \ge 5/6$, and 2) holds, i.e. the process $\{u_\alpha(t), x_\alpha(t)\}$ is indeed the solution of phase equation (28).

Therefore, for $\alpha \in [5/6, 1]$ (28) has two solutions on [0,2] $\{u(t), x(t,u)\}$ and $\{u_\alpha(t), x_\alpha(t)\}$, where

$$u_\alpha(t) = \begin{cases} 1, & 0 \le t \le 1, \\ \alpha, & 1 < t \le 2, \end{cases} \qquad x_\alpha(t) = \begin{cases} 1, & 0 \le t \le 1, \\ (\alpha - 1)t^2 + 2 - \alpha, & 1 \le t \le 2. \end{cases}$$

Let us check the improvement. By definition

$$g_\alpha(t) = H_u[t, x_\alpha(t)](v_\alpha(t) - u(t)),\ t \in [0,2],$$

where

$$H_u[t, x_\alpha(t)] > 0, \quad v_\alpha(t) - u(t) = \begin{cases} 0, & 0 \le t \le 1, \\ 1, & 1 < t \le 2. \end{cases}$$

Hence,

$$\delta_1(u,\alpha) = \int_0^2 g_\alpha(t)\, dt = \int_1^2 H_u[t, x_\alpha(t)]\, dt > 0.$$

The problem is bilinear, therefore

$$\Phi(u_\alpha) - \Phi(u) = -\alpha\delta_1(u,\alpha) < 0, \quad \alpha \in [5/6, 1].$$

Thus, the modified procedure (21) makes it possible to improve the stationary nonsingular control $u(t)$. The standard conditional gradient method (5) does not have this property.

Example 3. (improvement of singular control)

$$\Phi(u) = -\int_0^1 ux_2\, dt \to \min,$$

$$\dot{x}_1 = u,\ \dot{x}_2 = x_1 + x_2,\ \ x_i(0) = 0,\ i = 1,2,\ \ |u| \le 1.$$

The Pontryagin function: $H = \psi_1 u + \psi_2(x_1 + x_2) + ux_2$.
The Conjugate system: $\dot{\psi}_1 = -\psi_2,\ \ \dot{\psi}_2 = -\psi_2 - u,\ \ \psi_1(1) = 0,\ \psi_2(1) = 0.$
The Maximizing control: $u^* = sign\ (\psi_1 + x_2)$.

Let us consider the control $u(t) = 0$ with trajectories

$$x_i(t,u) = 0,\ i = 1,2, \quad \psi_1(t,u) = 0,\ \psi_2(t,u) = 0.$$

This control is singular since $H_u(\psi(t,u), x(t,u)) = 0$.

We use again the procedure (21). In this case

$$u(t,x,\alpha) = \alpha v^*(t,x), \quad v^*(t,x) = sign\ x_2 \quad (sign\ 0 \in [-1,1]).$$

Now specifying the singular case when $x_2(t) = 0,\ t \in T_0 \subset T$, we get

$$\dot{x}_2(t) = x_1(t) + x_2(t) = 0 \Rightarrow x_1(t) = 0 \Rightarrow \dot{x}_1(t) = u(t) = 0,\ t \in T_0.$$

So, the control $u(t) = 0$ generates a singularity $x_2(t) = 0$, i.e. $sign\ 0 = 0$.

Now we find the solution of the phase system as

$$\dot{x}_1 = \alpha\ sign\ x_2,\ \ \dot{x}_2 = x_1 + x_2,\ \ x_1(0) = x_2(0) = 0.$$

It is clear that there exists a trivial (singular) solution $x_1(t) = x_2(t) = 0$ generated by $u(t) = 0$.

We can check a possibility for the nonsingular solution. If we assume that $x_2(t) > 0$, $t \in (0,1]$, then $x_1(t) = \alpha t$, $\dot{x}_2 = \alpha t + x_2$, $x_2(0) = 0 \Rightarrow x_2(t) = \alpha(e^t - t - 1)$. Here our assumption holds: $x_2(t) > 0$ for $\alpha \in (0,1]$, i.e. the phase system has a nontrivial solution

$$x_1^\alpha(t) = \alpha t, \ \ x_2^\alpha(t) = \alpha(e^t - t - 1), \ t \in [0,1]$$

with the corresponding control $u_\alpha(t) = \alpha, \ \ \alpha \in (0,1]$.

Besides we have an improvement

$$\Phi(u_\alpha) = -\alpha \int_0^1 x_2^\alpha(t)\, dt < 0 = \Phi(0).$$

Therefore, unlike (5) modification (21) can improve the singular controls.

Let us represent the procedures (21), (22) in their iterative forms and discuss the question of convergence of corresponding method in the bilinear problem.

Let $u^k(t)$ be a control and $u_\alpha^k(t)$ be a family of variation controls constructed according to (21) and (22), $k = 0, 1, 2, \ldots$. Then we fix $\alpha \in (0,1]$ and construct the next approximation as $u^{k+1}(t) = u_\alpha^k(t)$, $t \in T$, for which

$$\Phi(u^{k+1}) - \Phi(u^k) \le -\alpha\delta_1(u^k, \alpha)$$

holds. Here, $\delta_1(u^k, \alpha) \ge 0$ is the modified residual of DMP on u^k.

Since the set of admissible pairs $\{u(t), x(t,u)\}$, $u \in V$ is restricted, functional $\Phi(u)$ is bounded from below on V. Therefore, the monotone sequence $\{\Phi(u^k)\}$, $k = 0, 1, 2, \ldots$ is convergent, then $\Phi(u^k) - \Phi(u^{k+1}) \to 0$, $k \to \infty$.

From the following inequality

$$\delta_1(u^k, \alpha) \le \frac{\Phi(u^k) - \Phi(u^{k+1})}{\alpha}$$

we have the convergence of the methods (21), (22) with respect to the modified residual of DMP as $\delta_1(u^k, \alpha) \to 0$, $k \to \infty$.

In conclusion, we represent the modification (22) in an implementable form when a Lagrange multiplier is used as a variation parameter.

First we form λ-parameter family of variation controls ($\lambda > 0, x \in R^n$)

$$u(t,x,\lambda) = \begin{cases} u(t) + \frac{g(t,x)}{\lambda}(v^*(t,x) - u(t)), & g(t,x) < \lambda, \\ v^*(t,x), & g(t,x) \ge \lambda. \end{cases}$$

Then we find the solution $x_\lambda(t)$ of the phase system

$$\dot{x} = f(x, u(t,x,\lambda), t), \ \ x(t_0) = x^0$$

and the control $u_\lambda(t) = u(t, x_\lambda(t), \lambda)$. A problem for searching $\lambda > 0$ is determined by the improvement property: $\Phi(u_\lambda) \le \Phi(u)$.

References

[1] Fedorenko R.P. (1978) *Approximate Solution of Optimal Control Problems*, Science, Moscow.

[2] Tyatushkin A.I. (1992) *Numerical Methods and Software for Optimization of Control Systems*, Science, Novosibirsk.

[3] Vasiliev O.V. (1996) *Optimization Methods*, World Federation Publishers, Atlanta.

[4] Baturin V.A., Urbanovich D.E. (1997) *Approximate Methods of Optimal Control Based on Extension Principle*, Science, Novosibirsk.

[5] Pytlak R. (1999) *Numerical Methods for Optimal Control Problems with State Constraints*, Springer-Verlag, Berlin.

[6] Srochko V.A. (2000) *Iterative Methods for Solving Optimal Control Problems*, Science, Moscow.

[7] Vasiliev F.P. (2002) *Methods of Optimization*, Factorial Press, Moscow.

Optimization and Optimal Control, pp. 287 – 295
P.M. Pardalos, I. Tseveendorj, and R. Enkhbat, Editors

A Simple Model of Improving Survival in Economical Systems

Ulemj Badam (ubadam326@yahoo.com)
Institute of Physics and Technology,
Mongolian Academy of Sciences,
Ulaanbaatar, Mongolia.

Abstract

This article deals with mathematical models of improving survival of business and insurance organizations, as well as with economical interpretation of solutions to formulated problems on the basis of already known approach on survival theory.

Keywords: survival problems, linear system, discrete equation, economic problems, insurance organization, business organization.

The problem of survival is a comparatively new branch of optimization theory. Word "survival" is the ability of any system to cope with unfavorable situations and effects. Formalization of this theory was first given in [1]. Further development of the theory for the linear discrete system was considered in [2]. One approach to optimization problems of survival was proposed in [3]. Applications considered in [1] are related to issues of ecology and economics: namely; stability of an increase of quantities of homogeneous local populations of animals and useful plants, survival of agriculture and banks. This paper is devoted to the development of this methodology to analysis of economic problems.

1 Problem of improving survival of business organizations

In market economy any organization responsible for business and production has to work towards increasing it's profit.
Let us consider a parameter k representing time duration, month, quarter and year.

If the profit of a firm from sales for $(k+1)$ year is written as

$$A^{k+1} = \sum_{i=1}^{n} x_i \tilde{p_i}^{k+1} - \sum_{s=1}^{l} \tilde{q_s}^{k+1} \tag{1}$$

then the target-oriented purpose is treated as an inequality

$$A^{k+1} - A^k \geq \lambda A^k, \quad 0 \leq \lambda \leq 1 \tag{2}$$

where x_i–output of goods,
$\tilde{p_i}^{k+1}$– the market price per unit,
$\sum_{i=1}^{n} x_i \tilde{p_i}^{k+1}$– sum of profit from sale of goods,
$\sum_{s=1}^{l} \tilde{q_s}^{k+1}$– sum of cost of production.

If $\tilde{p_i}^{k+1}$ depends on quantity and quality of goods, public demand and purchasability, the expenses $\tilde{q_s}^{k+1}$ change with the rate of inflation, economic stability and level of taxes. Taking into consideration these reasons, we can propose some permissible limit of changes in $\tilde{p_i}^{k+1}, \tilde{q_s}^{k+1}$ and write

$$\underline{p_i}^{k+1} \leq \tilde{p_i}^{k+1} \leq \bar{p_i}^{k+1}, \quad \underline{q_s}^{k+1} \leq \tilde{q_s}^{k+1} \leq \bar{q_i}^{k+1} \tag{3}$$

We will use the following equalities

$$\tilde{p_i}^{k+1} = \underline{p_i}^{k+1} + (\bar{p_i}^{k+1} - \underline{p_i}^{k+1}) u_i,$$

$$\tilde{q_s}^{k+1} = \underline{q_s}^{k+1} + (\bar{q_s}^{k+1} - \underline{q_s}^{k+1}) v_s \tag{4}$$

introducing parameters of control u_i and perturbation v_s, and the ranges of variables

$$0 \leq u_i \leq 1, \quad 0 \leq v_i \leq 1$$

Substituting (4) in (1) and taking into account (2) and following notations

$$c_s = \bar{q_s}^{k+1} - \underline{q_s}^{k+1}, \qquad b_i = x_i(\bar{p_i}^{k+1} - \underline{p_i}^{k+1})$$

$$a = \sum_{i=1}^{r} x_i \underline{p_i}^{k+1} - \sum_{s=1}^{l} \underline{q_s}^{k+1}, \qquad d = a - (1+\lambda)A^k,$$

we obtain the inequality

$$\sum_{s=1}^{l} c_s v_s \leq d + \sum_{i=1}^{r} b_i u_i \tag{5}$$

The set of safe economic perturbation can be defined as

$$V(u) = \{ \ v \in V = [0,1]^l : \ \sum_{s=1}^{l} c_s v_s \leq d + \sum_{i=1}^{r} b_i u_i \ \} \tag{6}$$

The wider this set is, the higher survival of given organization.
It means that in the case of maximum a value of set, the organization has a possibility of gaining more profit. To reach favorable situation it is necessary to expand the set $V(u)$ by increasing right side of the equality (6). For this purpose, we have to solve following problems:

$$\max_{0\leq u_i\leq 1}\sum_{i=1}^{r} b_i u_i = \sum_{b_i>0} b_i; \quad u_i^* = \begin{cases} 1, & b_i > 0 \\ 0, & b_i \leq 0 \end{cases}$$

$$\max_{0\leq v_s\leq 1}\sum_{s=1}^{l} c_s v_s = \sum_{c_s>0} c_s; \quad v_s^* = \begin{cases} 1, & c_s > 0 \\ 0, & c_s \leq 0 \end{cases}$$

Thus
1^0. If $\sum\limits_{c_s>0} c_s \leq d + \sum\limits_{b_i>0} b_i$, the survival of organization is guaranteed and set V is composed of safe perturbations.
2^0. If $\sum\limits_{c_s>0} c_s > d + \sum\limits_{b_i>0} b_i$ the set $V(u)$ is of the following form

$$V(u^*) = \{\, v \in V : \ \sum_{s=1}^{l} c_s v_s \leq d + \sum_{b_i>0} b_i \,\} \tag{7}$$

In general, choose λ to satisfy following expression

$$d + \sum_{b_i>0} b_i \geq 0$$

that is

$$0 < \lambda \leq \frac{a - A^k + \sum\limits_{b_i>0} b_i}{A^k} = \bar{\lambda}.$$

When the survival of the system is not guaranteed, we have to solve the problem using the following approach.
According to this approach, the problem of inscribing maximum cube in set $V(u)$ is equivalent to the following problem of linear programming:

$$r \to max$$

$$\sum_{s=1}^{l} c_s v_s - \sum_{i=1}^{r} b_i u_i + rN \leq b$$

$$v_s + r \leq 1, \quad -v_s + r \leq 0, \quad s = \overline{1,l}, \quad N = \sum_{s=1}^{l} |c_s| \tag{8}$$

On the basis of solution of (8), we arrive at the conclusion that for a given period a business organization incurs a cost of production equal to

$$\tilde{q}_s^* = \underline{q}_s + (\bar{q}_s - \underline{q}_s)(v_s^* + r^*), \quad s = \overline{1.l}$$

and can hold the selling price at the level

$$\tilde{p}_i^* = \underline{p}_i + (\bar{p}_i - \underline{p}_i)u_i^*, \quad i = \overline{1.r}$$

Thus the organization's profit at the end of a given period does not decrease.

2 Survival of insurance organizations

If an insurance organization proposes i types of insurance with n risk it's profit is defined as

$$\tilde{Q}_i = \tilde{S}_i - \tilde{H}_i, \quad i = 1, 2, \ldots, k \tag{9}$$

where $\tilde{S}_i$–insurance collection for i insurance service, $\tilde{H}_i$–compensation of i–th insuranse services.

If each insurance service has n risk, the corresponding insurance collection and compensation are calculated as

$$\tilde{s}_i = \sum_{j=1}^{n} \tilde{a}_{ij} x_{ij}, \quad \tilde{H}_i = \sum_{j=1}^{n} H_{ij} \tag{10}$$

where x_{ij}–insurance sum of i–th type with j–th risk, $\tilde{a}_{ij}$–corresponding insurance tariff, H_{ij}–compensation.

If the upper and lower limits of parameters $\tilde{a}_{ij}, \tilde{H}_i$ are known, we can write the following equality

$$\underline{a}_{ij} \leq \tilde{a}_{ij} \leq \bar{a}_{ij} \ ; \ \underline{H}_{ij} \leq \tilde{H}_{ij} \leq \bar{H}_{ij}$$

Now we introduce a parameter of control u_{ij} and a parameter of perturbation v_{ij} by formula

$$\tilde{a}_{ij} = \underline{a}_{ij} + u_{ij}(\bar{a}_{ij} - \underline{a}_{ij}) \ , \quad 0 \leq u_{ij} \leq 1 \ ,$$

$$\tilde{H}_{ij} = \underline{H}_{ij} + v_{ij}(\bar{H}_{ij} - \underline{H}_{ij}) \ , \quad 0 \leq v_{ij} \leq 1 \ , \quad (i = \overline{1,k} \ ; \ j = \overline{1,n}).$$

According to (10) and (9)

$$\tilde{Q}_i = \sum_{j=1}^{n} [\underline{a}_{ij} + (\bar{a}_{ij} - \underline{a}_{ij})u_{ij}]x_{ij} - \sum_{j=1}^{n} [\underline{H}_{ij} + (\bar{H}_{ij} - \underline{H}_{ij})v_{ij}].$$

Survival is assumed once the equality

$$\tilde{Q}_i \geq b_i \tag{11}$$

is fulfilled. It means that the profit for the insurance organization is not less than planned b_i. In this case

$$-\sum_{j=1}^{n} (\bar{a}_{ij} - \underline{a}_{ij})x_{ij}u_{ij} + \sum_{j=1}^{n} (\bar{H}_{ij} - \underline{H}_{ij})v_{ij} \leq \sum_{j=1}^{n} \underline{a}_{ij}x_{ij} - \sum_{j=1}^{n} \underline{H}_{ij} - b_i. \tag{12}$$

Using vector notations, we have

$$E_i = (-(\bar{a}_{i1} - \underline{a}_{i1})x_{i1}\ , \ldots , -(\bar{a}_{in} - \underline{a}_{in})x_{in}\ )^T\ ,$$

$$F_i = (\ (\bar{H}_{i1} - \underline{H}_{i1})\ , \ldots , (\bar{H}_{in} - \underline{H}_{in})\ )^T\ ,$$

$$u_i = (u_{i1}\ , \ldots ,\ u_{in})^T\ , \quad v_i = (v_{i1}\ , \ldots ,\ v_{in})^T\ ,$$

$$d_i = \sum_{j=1}^{n} a_{ij}x_{ij} - \sum_{j=1}^{n} \underline{H}_{ij} - b_i$$

and equality (12) is reduced to the form

$$E_i u_i + F_i v_i \leq d_i\ , \quad u_i \in U = [0,1]^k\ ,\ v_i \in V = [0,1]^k. \tag{13}$$

If u_i is fixed, the set of safe perturbation will be defined in the form

$$V(u_i) = \{\ v_i \in V = [0,1]^k :\ E_i u_i + F_i v_i - d_i \leq 0\ \}$$

and the problem of inscribing of the maximum cube

$$B(v_i, r) = \{\ \tilde{v}_i :\ ||\tilde{v}_i - v_i|| = \max_{0 \leq j \leq 1} |\tilde{v}_{ij} - v_{ij}| \leq r\ \}$$

in $V(u_i)$ is reduced to a solution of linear programming:

$$r\ \rightarrow\ max$$

$$E_i u_i + F_i v_i + rN - d_i \leq 0\ ,$$

$$v_i + re \leq e\ , \tag{14}$$

$$-v_i + re \leq 0\ , \quad i = \overline{1,k}$$

for each i. Where $N = (N_1,\ \ldots,\ N_k)'$, $N_j = \sum\limits_{s=1}^{n} |F_{ijs}|$.

Let r^*, u_i^*, $v_i^* = v_i + r^*$ be the solution to (14).

Therefore, we can conclude that when the compensation is maximum $(\underline{H}_{ij} + v_{ij}^*(\bar{H}_{ij} - \underline{H}_{ij}))$ and the insurance tariff varies in the range of $\underline{a}_{ij} + u_{ij}^*(\bar{a}_{ij} - \underline{a}_{ij})$ the effective operation of the organization is ensured.

Our method was tested by using data from "Mongol Daatgal" company.

3 Results of a numerical experiment

In testing the above model data from the Darkhan Branch company "Mongol Daatgal" was used.

Table 1

Type of insurance	1. Owner of transport vehicle insurance and their responsibility	2. citizens accident insurance	3. Insurance of temporary loss of ability to work

Risk (j)	Insurance sum (compen-) sation)	Tariff (mln T)	Insurance sum (compen- sation)	Tariff (mln T)	Insurance sum (compen- sation)	Tariff (mln T)
1. Natural Disasters	1506,89	1,06%	1420,68	1,2%	0	
2. Industrial & Technical accidents	3915,69 (6072,8)	1,32%	2315,70 (1250)	1,4%	0	
3. Auto-vehicle equipment catastrophe and other life mode accidents	7778,6 (3204,4)	1%	3196,53 (2087,5)	1,3%	74,2 (1,48)	0,8%
4. Temporary loss of work ability of insured person	0		0		140,71 (418,0)	5,2%
5. Risks occuring to chauffeur and to the lifes and health of other person	7778,9 (91,7)	1%	1052,48 (400)	0,9%	0	
b_j	230,0	-	80,0	-	180	-

In Table 1, 3 types of insurances and 5 types of risks are represented. Risk sum is a certain conditional sum of money established by contacts between an insurance company and a client which insure objects.
Insurance fee (payment) is sum of money which a client pays to a insurance company for insurance of property.
Insurance tariff is a certain value expressed usually in a percentage and used for calculation of insurance fee.
The following simple formula shows relations between these definitions:

Insurance award=(Insurance tariff)*(Insurance sum)

To eliminate any mistakes in solving the problem (14) let us consider the case where $i = 1$ just for the first case of insurance.
In this case the following values hold:

$$\tilde{s}_1 = \sum_{j=1}^{5} \tilde{a}_{1j} x_{1j} \ , \quad \tilde{H}_1 = \sum_{j=1}^{5} \tilde{H}_{1j}.$$

Then let us determine the lower and upper boundaries of parameters $\tilde{a}_{1j}$, $\tilde{H}_{1j}$ in the following way:

$$\begin{array}{lll}
\tilde{a}_{11} = 0,0106 & \underline{a}_{11} = 0,005 & \bar{a}_{11} = 0,012 \\
\tilde{a}_{12} = 0,0132 & \underline{a}_{12} = 0,005 & \bar{a}_{12} = 0,015 \\
\tilde{a}_{13} = 0,01 & \underline{a}_{13} = 0,005 & \bar{a}_{13} = 0,015 \\
\tilde{a}_{14} = 0 & \underline{a}_{14} = 0 & \bar{a}_{14} = 0 \\
\tilde{a}_{15} = 0,01 & \underline{a}_{15} = 0,005 & \bar{a}_{15} = 0,015 \\
\tilde{h}_{11} = 0 & \underline{h}_{11} = 0 & \bar{h}_{11} = 0 \\
\tilde{h}_{12} = 6072,8 & \underline{h}_{12} = 6000 & \bar{h}_{12} = 6500 \\
\tilde{h}_{13} = 3204,4 & \underline{h}_{13} = 3100 & \bar{h}_{13} = 3500 \\
\tilde{h}_{14} = 0 & \underline{h}_{14} = 0 & \bar{h}_{14} = 0 \\
\tilde{h}_{15} = 91,7 & \underline{h}_{15} = 90 & \bar{h}_{15} = 96
\end{array}$$

Taking into consideration data and numbers from Table 1 we determine the following vectors and scalar values

$$E_1 = (\ -1506.89, \ -3915.69, \ -7778.6, \ 0, \ -7778.9 \ )^T,$$

$$F_1 = (\ 0, \ 500, \ 400, \ 0, \ 6, \ )^T, \ \ d_1 = -21699,95, \ N_1 = 906.$$

Introducing the notations

$$u_1 = (u_{11}, \ u_{12}, \ u_{13}, \ u_{14}, \ u_{15}) = (u_1, \ u_2, \ u_3, \ u_4, \ u_5)^T,$$

$$v_1 = (v_{11}, \ v_{12}, \ v_{13}, \ v_{14}, \ v_{15}) = (v_1, \ v_2, \ v_3, \ v_4, \ v_5)$$

we obtain the following inequalities

$$-1506.89 u_1 - 3915.69 u_2 - 7778.6 u_3 - 7778.9 u_5 + 500 v_2 +$$

$$+400 v_3 + 6 v_5 + 906 r + 21699.95 \leq 0$$

$$v_2 + r \leq 1$$

$$v_3 + r \leq 1$$

$$v_5 + r \leq 1$$

$$-v_2 + r \leq 0$$

$$-v_3 + r \le 0$$
$$-v_5 + r \le 0$$
$$0 \le r \le 1/2$$

For convenience let us introduce the notations for variables as follows:

$$u_1 = x_1 \qquad v_1 = x_5$$
$$u_2 = x_2 \qquad v_2 = x_6$$
$$u_3 = x_3 \qquad v_5 = x_7$$
$$u_5 = x_4 \qquad r = x_8$$

Considering the following linear programming problem,

$$\begin{gathered} x_8 \to \max \\ -1506.89x_1 - 3915.69x_2 - 7778.6x_3 - 7778.9x_4 + 400x_6 + 6x_7 + 906x_8 + 21699.95 \le 0 \\ x_5 + x_8 \le 1 \\ x_6 + x_8 \le 1 \\ x_7 + x_8 \le 1 \\ -x_5 + x_8 \le 1 \\ -x_6 + x_8 \le 0 \\ -x_7 + x_8 \le 0 \\ 0 \le x_8 \le 1/2 \\ 0 \le x_i \le 1\,, \quad i = \overline{1,7} \end{gathered} \tag{15}$$

we obtain

$$u_1^* = x_1 = 1.00 \qquad v_2^* = x_5 = 0.2576$$
$$u_2^* = x_2 = 0 \qquad v_3^* = x_6 = 0.2727$$
$$u_3^* = x_3 = 0 \qquad v_5^* = x_7 = 0.3324$$
$$u_5^* = x_4 = 0 \qquad r^* = x_8 = 0.44115.$$

With the use of these values we find tariff and compensation of insurance as

$$\tilde{a}_{11} = 0.012, \qquad \tilde{h}_{12} = \underline{h}_{12} + (v_{12}^* + r^*)(\bar{h}_{12} - \underline{h}_{12}) = 6350,$$
$$\tilde{a}_{12} = 0.005, \qquad \tilde{h}_{13} = \underline{h}_{13} + (v_{13}^* + r^*)(\bar{h}_{13} - \underline{h}_{13}) = 3784,$$
$$\tilde{a}_{13} = 0.005, \qquad \tilde{h}_{15} = \underline{h}_{15} + (v_{15}^* + r^*)(\bar{h}_{15} - \underline{h}_{15}) = 94.162,$$
$$\tilde{a}_{15} = 0.005.$$

This means that if a change occurs in tariff and compensation in the insurance then the income of the insurance company "Mongol Daatgal" will not decline in the insurance types. Therefore, survival of an insurance company will be understood as the ability to offer maximum compensation with a stable increase of incomes.

References

[1] Ashepkov L.T. (1988), *Models and methods of operational investigations,* Novosibirsk, Nauka.

[2] Dolgii D.V. (1995), *Control of linear systems in conditions of uncertanties,* Vladivostok, University.

[3] Badam.U. (2002), "A necessary condition for optimality in survival problems, Izvestiya VUZov, *Mathematics,* vol.2, pp.18-22.

[4] Badam.U. (2002), "Models of increasing survival of linear discrete control systems", *Journal of optimization, control and intellect,* vol.5(4), pp.36-50.

Optimization and Optimal Control, pp. 297 – 305
P.M. Pardalos, I. Tseveendorj, and R. Enkhbat, Editors

Controlled Systems with Distributed Parameters: optimality conditions and optimization methods in class of smooth restricted controls

O.V. Vasiliev (vasili@math.isu.runnet.ru)
Institute of Mathematics and Economics
Irkutsk State University
1, Karl Marx str.
664003 Irkutsk
RUSSIA

Abstract

In this article we propose a technique for deriving necessary optimality conditions and numerical optimization methods in optimal control problems of systems of partial differential equations with smooth restricted controls.

Keywords: optimal control, numerical methods, systems of partial differential equations

The development of optimal control theory has started from the investigation of controlled processes, the model of which could be described by a finite system of ordinary differential equations. However, in many technical applications the state of the object, as well as its control sometimes, is not possible to describe by giving only its coordinates with respect to time. Control to such objects should be assigned not only at every moment of time, but also at every point of the geometrical region of the physical space that this object occupies. In such case the controlled process is described by a system of partial differential equations (controlled distributed parameter system). It is clear that the constructing theory of optimal control of systems with distributed parameters is a more complex problem than the constructing theory of optimal control of systems of ordinary differential equations. In the later rather effective methods of optimal control search are based on the maximum principle of L.S. Pontryagin. So, the wish of mathematicians to formulate a necessary optimality condition in systems with distributed parameters in analogy with the maximum principle, is natural. For such results the class of admissible controls is extended to one of piecewise-continuous, in general case measurable, controls. It happens because it is in class of piecewise-continuous controls the needle-shaped variation underlying the maximum principle lets us take into account amplitude (point-wise) restrictions on control. However, the existing maximum principle theory is not enough if in the stated mathematical model the class of admissible controls consists of continuously differentiable functions which are also restricted. Moreover, in some real-life problems controlled functions by formulation should be smooth enough. In particular, this is characteristic of mathematical physics problems. Here non-measurable coefficients of a system of partial differential equations with initial-boundary conditions might be considered as controllers. They are to be found by the known solution or a part of it. Note that investigation of optimal control problems in systems of mathematical physics equations with smooth restricted controls lets us use classical theorems of existence, uniqueness, and stability of solution with respect to the control perturbation. We shall note that in class of piecewise continuous controls those theorems do not work.

This article is dedicated to description of a method for deriving necessary optimality condition and construction of optimization methods in controlled systems with distributed parameters for smooth restricted controls. We provide the general scheme of investigation, details of which might be found in [1-4].

1 Problem formulation

Let $t = (t_1, \ldots, t_m)$, $t_i \in T_i = [t_0^i, t_1^i]$, $i = 1, \ldots, m$, be the independent variables, where $T = T_1 \times \cdots \times T_m$ is a given region (parallelepiped). Let also $x = x(t)$, $x(t) \in E^n$, be the state of the process, $u = u(t)$, $u(t) \in E^r$, be the control of the process. Let us note that by $u = u(t)$ we mean not only distributed controls, but lumped ones as well, for example, $u_i = u_i(t_1), u_i = u_i(t_2), \ldots, u_i = u_i(t_1, \ldots, t_{m-1}), \ldots,$ $i \in \overline{1, r}$. The state and the control are linked by a system of partial differential equations:

$$Dx = f(x, Rx, u, t), \ t \in T, \tag{1}$$

with initial-boundary conditions:

$$D_0 x(T_H) = g^0(x, R_0 x, u, T_H). \tag{2}$$

Here D, D_0 are external linear differential operators, R, R_0 are internal ones, $T_H \subset \partial T$ is the given initial part of the boundary of T. The goal functional takes the form:

$$J(u, x) = \varphi(x(T_K)) + \int_T F(x, Rx, u, t)dt, \tag{3}$$

where $T_K \subset \partial T$ is the given terminal part of the boundary of T, $x(T_K) \in C_p^n(T_K)$, $p = 0, 1, \ldots,$ $\varphi \colon C_p^n(T_K) \to E^1$, F is a scalar function which is continuous with respect to its arguments together with its partial derivatives. Vector-functions f and g^0 are also continuous with respect to their arguments together with their partial derivatives.

By admissible controls we mean smooth, i.e. continuously differential functions $u = u(t)$ that satisfy the inequality restrictions: $u \in U$. The form of these restrictions will be given below.

Assumption 1. *For every admissible control $u = u(t)$ there exists a unique smooth solution $x = x(t, u)$ of the initial-boundary problem (1),(2).*

We seek an optimal admissible pair

$$(u^*; x^*): \;\; J(u, x) \to \min, \;\; u \in U. \tag{4}$$

2 Increment formula

It is clear that on two admissible pairs $(u \in U; x = x(t, u)), (\tilde{u} = u + \Delta u \in U; \tilde{x} = x + \Delta x = x(t, \tilde{u}))$ the incremental equations

$$D\Delta x = \Delta f(x, Rx, u, t),$$

$$D_0 \Delta x = \Delta g^0(x, R_0 x, u, T_H)$$

hold, where Δf and Δg^0 are total increments with respect to (x, Rx, u) and $(x, R_0 x, u)$ in comparison with the partial ones that we are going to use later, for example,

$$\Delta_{\tilde{u}} f(x, Rx, u, t) = f(x, Rx, \tilde{u}, t) - f(x, Rx, u, t).$$

Then for any nontrivial smooth vector-function $\psi = \psi(t), \; \psi^0 = \psi^0(T_H)$ the increment of the goal functional (3) on the solutions of (1),(2) might be represented as

$$\begin{aligned} \Delta J(u, x) = J(\tilde{u}, \tilde{x}) - J(u, x) = \Delta\varphi(x(T_K)) + \int_T \Delta F(x, Rx, u, t) dt + \\ + \int_T < \psi(t), D\Delta x(t) - \Delta f(x, Rx, u, t) > dt + \\ + \int_{T_0} < \psi^0(T_H), D_0 \Delta x(T_H) - \Delta g^0(x, R_0 x, u, T_H) > dT_H, \end{aligned} \tag{5}$$

where $< \cdot, \cdot >$ indicates scalar product in Euclidean space.

Remark 1. Further on we shall assume that in problem (1),(2) there are no initial or terminal conditions, i.e. that the initial-boundary conditions (2) are given as

$$D_0 x(T_H) = g^0(T_H). \tag{6}$$

This will simplify the constructions, while the idea of the derivation will remain intact. Taking into account this remark and the following notations:

$$z(t) = (x(t), Rx(t)),$$

$$H(\psi, z, u, t) =< \psi(t), f(z, u, t) > -F(z, u, t),$$

in the increment formula (5) we shall represent $\Delta_{\tilde{z}\tilde{u}}H$ in form

$$\Delta_{\tilde{z}\tilde{u}}H = \Delta_{\tilde{z}}H(\cdot\cdot\tilde{u},t) + \Delta_{\tilde{u}}H(\psi,z,u,t).$$

Next, we shall expand $\Delta_{\tilde{z}}H(\cdot\cdot\tilde{u},t)$, $\Delta\varphi$, $\Delta_{\tilde{u}}H$ into Taylor series up to the first term with the residuals $o_{\varphi}(\|\Delta x(T_K)\|)$, $o_H(\|\Delta z(t)\|)$, $o_H(\|\Delta u(t)\|)$ satisfying $o(\alpha)/\alpha \to 0, \alpha \to 0$. Partial derivative $H_z(\psi,z,\tilde{u},t)$ we shall represent in form

$$H_z(\psi,z,\tilde{u},t) = H_z(\psi,z,u,t) + \Delta_{\tilde{u}}H_z(\psi,z,u,t).$$

Statement 1. *If a function $\psi = \psi(t)$ satisfies the following equality:*

$$\int_T < \psi(t), D\Delta x(t) > dt + < \varphi_x(x(T_K)), \Delta x(T_K) > - \int_T < H_z(\cdot\cdot t), \Delta z(t) > dt = 0 \tag{7}$$

for all Δz with initial conditions $D_0\Delta x(T_0) = 0$, then the increment formula (5) will take form:

$$\Delta J(u,x) = -\int_T < H_u(\cdot\cdot t), \Delta u(t) > dt + \eta(u,\tilde{u}), \tag{8}$$

where

$$\begin{aligned} \eta(u,\tilde{u}) = o_{\varphi}(\|\Delta x(T_K)\|) - \int_T o_H(\|\Delta z(t)\|dt - \\ - \int_T o_H(\|\Delta u(t)\|)dt - \int_T < \Delta_{\tilde{u}}H_z(\cdot\cdot t), \Delta z(t) > dt. \end{aligned} \tag{9}$$

Here (7) defines the conjugate problem. In other words, the form of (8),(9) of the increment formula which is convenient for further investigations, defines the conjugate problem.

Assumption 2. *On an admissible pair $(u;x)$ the smooth solution $\psi = \psi(t,u)$ of the conjugate problem is defined by (7) exists and is unique.*

Finally, the latest assumption comes in connection with stability of the solution of (1),(2) for a small variation of control. Naturally, here it is the best to have pointwise estimations of the following type:

$$\|\Delta x(t)\| \le K \int_T \|\Delta u(t)\|dt,$$

$$\|\Delta Rx(t)\| \le K \int_T \|\Delta u(t)\|dt,$$

but it is sufficient to assume that

$$\eta(u,\tilde{u}) \sim \int_T o(\|\Delta u(t)\|)dt. \tag{10}$$

3 Admissible variations of control

Here for a specific type of U we shall construct such varying control $\tilde{u} = u_\varepsilon(t)$ that will remain admissible for any $\varepsilon \in [0,1]$. So, let us by "A-admissible" controls define smooth vector-functions that satisfy for any $t \in T$ the inclusion-type conditions:

$$u(t) \in U, \tag{11}$$

where $U \subset E^r$ is a compact set. Restriction (11) is equivalent to the following unequality restrictions:

$$G_j(u(t)) \leq 0, \ j = 1, \dots, q, \ t \in T. \tag{12}$$

Next, let us by "B-admissible" controls define smooth vector-functions that satisfy integral-type restrictions:

$$\int_T \Phi_j(u(t))dt \leq L_j, \ j = 1, \dots, q, \tag{13}$$

or integral-amplitude ones (integral with respect to some variables t only), for example,

$$\int_{T_1}\int_{T_2} \Phi_j(u(t))dt_1 dt_2 \leq L_j, \ j = 1, \dots, q. \tag{14}$$

Here (which is essential to the method) scalar functions $\Phi_j(u)$ are positive (power) homogenial:

$$\Phi_j(\lambda u) = \lambda^p \Phi_j(u), \ \lambda(t) > 0, \ p > 0, \ j = 1, \dots, q. \tag{15}$$

Statement 2. *Let $u = u(t)$ be "A-admissible", then*

$$u_\varepsilon(t) = u(t_1 + \varepsilon\delta_1(t_1), \dots, t_m + \varepsilon\delta_m(t_m)) \tag{16}$$

is also "A-admissible" for all $\varepsilon \in [0,1]$ and all smooth scalar functions $\delta_i = \delta_i(t_i)$, $i = 1, \dots, m$, that satisfy

$$t_i + \delta_i(t_i) \in T_i, \ i = 1, \dots, m. \tag{17}$$

Indeed,

$$u_\varepsilon(t) = u(t_1 + \varepsilon\delta_1(t_1), \dots, t_m + \varepsilon\delta_m(t_m)) = u(t_\varepsilon) \in U,$$

since by virtue of (17) $t_\varepsilon \in T$, $\varepsilon \in [0,1]$. The same is true for restrictions (12).

Statement 3. *Let $u = u(t)$ be "B-admissible" in sense of (13) or, for example, (14). Then*

$$u_\varepsilon(t) = \lambda(t)^{1/p} u(t_1 + \varepsilon\delta_1(t_1), \dots, t_m + \varepsilon\delta_m(t_m)), \tag{18}$$

where

$$\lambda(t) = (1 + \varepsilon\dot{\delta}_1(t_1)) \times (1 + \varepsilon\dot{\delta}_2(t_2)) \times \cdots \times (1 + \varepsilon\dot{\delta}_m(t_m)) \tag{19}$$

for restrictions (13) or

$$\lambda(t) = (1 + \varepsilon\dot{\delta}_1(t_1)) \times (1 + \varepsilon\dot{\delta}_2(t_2)) \tag{20}$$

for restrictions (14), is also "B-admissible" for all $\varepsilon \in [0,1]$ *and all smooth vector-functions* $\delta_i = \delta_i(t_i)$, $i = 1, \ldots, m$, *that satisfy (17) and the boundary conditions*

$$\delta_i(t_0^i) = \delta_i(t_1^i) = 0, \ i = 1, \ldots, m. \tag{21}$$

Proof Let $u = u(t)$ be admissible in sense of (13). Taking into account the homogeneity condition (15), in the following equality:

$$\int_T \Phi_j(u_\varepsilon(t))dt = \int_T \lambda(t)^p \Phi_j(u_1 + \varepsilon\delta_1(t_1), \ldots, u_m + \varepsilon\delta_m(t_m))dt_1 dt_2 \ldots dt_m, \ j = 1, \ldots, q,$$

we shall make the change of variables:

$$\tau_i = t_i + \varepsilon\delta_i(t_i), \ d\tau_i = (1 + \varepsilon\dot{\delta}_i(t_i))dt_i,$$

$$dt_i = (1 + \varepsilon\dot{\delta}_i(t_i))^{-1} d\tau_i, \ i = 1, \ldots, m,$$

and set

$$(1 + \varepsilon\dot{\delta}_1(t_1))^{-1} \times \cdots \times (1 + \varepsilon\dot{\delta}_m(t_m))^{-1} \lambda^p(t) = 1, \ \varepsilon \in [0,1],$$

i.e. choose $\lambda = \lambda(t)$ according to (19). Then

$$\int_T \Phi_j(u_\varepsilon(t))dt = \int_T \Phi_j(u(\tau_1, \ldots, \tau_m))d\tau_1 \ldots d\tau_m \leq L_j, \ j = 1, \ldots, q,$$

since by virtue of (17) and (21) $\tau_i \in T_i$, $i = 1, \ldots, m$. If $u = u(t)$ is admissible in sense of (14) then, by making the same change of variables for $i = 1, 2$, we have

$$\int_{T_1}\int_{T_2} \Phi_j(u_\varepsilon(t))dt_1 dt_2 = \int_{T_1}\int_{T_2} \Phi_j(u(\tau_1, \tau_2, t_3 + \varepsilon\delta_3(t_3), \ldots, t_m + \varepsilon\delta_m(t_m))d\tau_1 d\tau_2 \leq L_j,$$
$$j = 1, \ldots, q.$$

Remark 2. While defining the class of the admissible controls, it is possible to combine restrictions of type (13) or (14) with the one-side restriction

$$u(t) \geq 0, \ t \in T.$$

In this case, the admissible varying control $u_\varepsilon(t)$ is constructed by (18),(19) or by (18),(20) with an additional assumption that $\lambda(t)$ is nonnegative. This means that we shall pose an additional restriction on $\delta_i = \delta_i(t_i)$ beside (17) and (21), namely,

$$|\dot{\delta}_i(t_i)| \leq 1, \ i = 1, \ldots, m.$$

4 Necessary optimality conditions, iterative minimization processes.

At this point we shall summarize our conclusions made so far and define our future goals.

So, on any two admissible processes we have constructed a convenient formula of increment of the goal functional (8),(9). Its form defines the conjugate problem (7). For the specific restrictions on smooth controlled functions (controllers), which play an important role in applications, we have constructed such varying control that remain admissible for the changes in the numerical variation parameter $\varepsilon \in [0,1]$. Besides, $\tilde{u} = u_\varepsilon(t) \to u(t)$, $t \in T$, for $\varepsilon \to 0$. The form of such varying control is defined by (16),(17),(18)-(21). Next, on such admissible variations $\Delta u = u_\varepsilon(t) - u(t)$ we have constructed increment formula (8),(9), where the principal term with respect to numerical variation parameter was separated. This principal term determines the necessary optimality condition, and the formula itself serves as a basis for constructing iterative minimization processes convergent to the necessary optimality conditions.

Now we shall proceed with this final part.

Taking into account estimate (10), increment formula (8),(9) for $\Delta u = u_\varepsilon(t) - u(t)$, where $u_\varepsilon = u_\varepsilon(t)$ is defined by (16),(17), takes form:

$$J(u_\varepsilon, x_\varepsilon) - J(u,x) = -\varepsilon \int_T \sum_{i=1}^{m} < H_u(\cdot\cdot t), u_{t_i}(t_1,\dots,t_m) > \delta_i(t_i)dt + o(\varepsilon). \tag{22}$$

Let us inroduce the functions

$$\omega_i([u],t_i) = \int_{T_1} \cdots \int_{T_{i-1}} \int_{T_{i+1}} \cdots \int_{T_m} < H_u(\cdot\cdot t), u_{t_i}(t) > dt_1 \dots dt_{i-1}dt_{i+1}\dots dt_m, \quad i = 1,2,\dots,m. \tag{23}$$

Theorem 1. *If an "A-admissible" process $(u^*;x^*)$ is optimal for problem (4), then*

$$\omega_i([u^*],t_i) = 0, \ t_i \in T_i, \ i = 1,2,\dots,m. \tag{24}$$

This result follows from nonnegativity of increment formula (22),(23) for $u = u^*$, $x = x^*$, $\varepsilon \in [0,1]$ while the choice of the sign of $\delta_i = \delta_i(t_i)$, $i = 1,\dots,m$, is arbitrary. For "B-admissible" controls increment formula (8),(9) takes form:

$$J(u_\varepsilon, x_\varepsilon) - J(u,x) = -\epsilon \int_T \sum_{i=1}^{m} [< H_u(\cdot\cdot t), u_{t_i}(t) > - \frac{1}{p} < H_u(\cdot\cdot t), u(t) >_{t_i}]\delta_i(t_i)dt + o(\varepsilon). \tag{25}$$

In analogy with the previous constructions, we introduce the functions

$$\bar{\omega}_i([u],t_i) = \int_{T_1} \cdots \int_{T_{i-1}} \int_{T_{i+1}} \cdots \int_{T_m} < H_u(\cdot\cdot t), u_{t_i}(t) > - \\ -\tfrac{1}{p} < H_u(\cdot\cdot t), u(t) >_{t_i}]dt_1 \dots dt_{i-1}dt_{i+1}\dots dt_m, \ i = 1,2,\dots,m. \tag{26}$$

Theorem 2. *If a "B-admissible" process $(u^*;x^*)$ is optimal for problem (4), then*

$$\bar{\omega}_i([u^*],t_i)=0,\ t_i\in T_i,\ i=1,2,\ldots,m. \tag{27}$$

This result is justified by the same reasons as for Theorem 1.

Now we start to describe iterational minimization processes for problem (4). Let an initial solution $u^0=u^0(t)$ be chosen from the set of admissible controls, and an admissible $u^k=u^k(t),\ t\in T,\ k=0,1,2,\ldots,$ be calculated by the method. On this control we calculate the solutions $x^k=x(t,u^k),\ \psi^k=\psi(t,u^k)$ of the initial and conjugate systems, construct $\omega_i^k(t_i)=\omega_i([u^k],t_i)$ or $\bar{\omega}_i^k(t_i)=\bar{\omega}_i([u^k],t_i)$ depending on the type of the admissible controls. If $\omega_i^k(t_i)[\bar{\omega}_i^k(t_i)]=0,\ t_i\in T_i,\ i=1,\ldots,m,$ then $u^k(t)$ satisfies the necessary optimality condition and the process of the algorithm terminates. Otherwise, we shall isolate two regions:

$$\Omega_i^{K+}=\{t_i\in T_i:\ \omega_i^k(t_i)[\bar{\omega}_i^k(t_i)]\geq 0\},$$
$$\Omega_i^{K-}=\{t_i\in T_i:\ \omega_i^k(t_i)[\bar{\omega}_i^k(t_i)]<0\}.$$

Next, we shall define the smooth functions

$$\delta_i^k(t_i)=\begin{cases}\alpha_i^k(t_i)\geq 0,\ t_i\in\Omega_i^{K+},\\ \alpha_i^k(t_i)<0,\ t_i\in\Omega_i^{K_i-},\end{cases}$$

that satisfy (17) or (17),(21), construct a one-parameter family of controls

$$u_\varepsilon^k(t)=u^k(t_1+\varepsilon\delta_1^k(t_1),\ldots,t_m+\varepsilon\delta_m^k(t_m))$$

(for "A-admissible" controls) or

$$u_\varepsilon^k(t)=\lambda^k(t)u^k(t_1+\varepsilon\delta_1^k(t_1),\ldots,t_m+\varepsilon\delta_m^k(t_m))$$

(for "B-admissible" controls),
and solve the problem of one-dimensional minimization:

$$\varepsilon_k=\arg\min_{\varepsilon\in[0,1]}J(u_\varepsilon^k,x_\varepsilon^k).$$

The next approximation is found as

$$u^{k+1}=u_{\varepsilon_k}^k(t),\ k=0,1,2,\ldots.$$

The sequence of controls generated by the method is relaxational:

$$J(u^{k+1},x^{k+1})<J(u^k,x^k),\ k=0,1,2,\ldots.$$

If the functional $J(u,x)$ is bounded from below on the admissible processes, and under additional, rather transparent, conditions on the parameters of the initial problem that provide for the estimate of the reminder term of (22),(25) in form:

$$|o(\varepsilon)|\leq K\varepsilon^{1+\nu},\ K>0,\ \nu>0,$$

the sequence $\{u^k\},\ k=0,1,2,\ldots$ is convergent in a weak sense to the necessary optimality conditions (24),(27).

References

[1] Burdukovskaya A.V., Vasiliev O.V. (2000), "Optimization of Systems of Canonical Hyperbolic Equations with Smooth Restricted Controls" (in Russian), *J. of Computational Mathematics and Mathematical Physics*, vol. 40, N1, pp.43-53.

[2] Vasiliev O.V., Arguchintsev A.V., Terletskii V.A. (2001), "Optimization Methods for Systems with Lumped and Distributed Parameters Based on Admissible Variations" (in Russian), *Proc. of XII International Baikalian Conference "Optimization Methods and Their Applications" (Plenary lectures)*, Irkutsk, pp.52-68.

[3] Arguchintsev A.V., Vasiliev O.V. (2000), "Towards Research of Inverse Problems of Mathematical Physics by Optimal Control Methods" *Stability and Control: Theory and Application*, Vol.3, N3, pp.205-211.

[4] Arguchintsev A.V., Vasiliev O.V. (2001), "Iterative Optimization Methods for Systems with Lumped and Distributed Parameters in Class of Smooth Admissible Controls" (in Russian), *Proc. of IM PAS of Belarus*, Minsk, Vol.7, pp.7-16.

Optimization and Optimal Control, pp. 307 – 333
P.M. Pardalos, I. Tseveendorj, and R. Enkhbat, Editors

Duality for the Multiobjective Location Model Involving Sets as Existing Facilities

Gert Wanka (gert.wanka@mathematik.tu-chemnitz.de)
Faculty of Mathematics, Chemnitz University of Technology, D-09107 Chemnitz, Germany

Radu Ioan Boţ (radu.bot@mathematik.tu-chemnitz.de)
Faculty of Mathematics, Chemnitz University of Technology, D-09107 Chemnitz, Germany

Emese Vargyas (emese.vargyas@mathematik.tu-chemnitz.de)
Faculty of Mathematics, Chemnitz University of Technology, D-09107 Chemnitz, Germany

Abstract

In this paper we consider the optimization problem with a multiobjective composed convex function as objective function, namely, being a composite of a convex and componentwise increasing vector function with a convex vector function. By the conjugacy approach, we obtain a dual problem for it. The existence of weak and strong duality is proved.

Using this general result, we introduce the dual problem for the multiobjective location problem in a general normed space, in which the existing facilities are represented by sets of points.

The biobjective Weber-minimax problem, the multiobjective Weber problem and the multiobjective minimax problem with demand sets are studied as particular cases of this problem.

Keywords: multiobjective duality, location problems, optimality conditions, Weber problem, minimax problem

The goal of this paper is to construct, in a general normed space, a dual for a primal multiobjective problem, each component of the multiobjective function being a composite of a convex and componentwise increasing function with a convex vector function. In the past, optimization problems with the objective function being a composed convex function have been considered by different authors. We remind the works [9] and [10] where some results with regard to duality have been given. Recently, optimization problems of this type have found applications in goal programming problems [4] and average distance problems [14].

This article is based on the work of Boţ and Wanka [2], where the authors have examined the case of a single objective function, for which a geometrical characterization of the set of optimal solutions was treated in [13] by Nickel, Puerto and Rodriguez-Chia. Here we study the duality for a multiobjective problem (P). For our original multiobjective problem Pareto-efficient and properly efficient solutions are considered. In order to do this, we consider first the linearly scalarized problem (P_λ) and use a dual problem (D_λ) to derive strong duality and optimality conditions, which later are used to obtain duality assertions for the original and dual multiobjective problem. This dual problem (D_λ) results from a special perturbation of the primal problem, by applying the Fenchel-Rockafellar duality concept based on conjugacy and perturbation (cf. [5]). Therefore, we have the possibility to construct different dual problems to a primal optimization problem, by perturbing it in different ways (see, for instance [22] and [23]).

Among the large number of papers and books dealing with different approaches to multiobjective duality we mention as a representative selection the books [6], [8], [16] and the papers [3], [11], [12], [17], [20], [22], [24] and [25]. Beside presentations in the sense of approaches for general formulated problems there are a lot of contributions devoted to the duality for multiobjective programming problems of special type, as for example linear problems [7], location and approximation problems [18], [20], portfolio optimization problems [19], [21], etc. In this paper, applying the conjugacy approach and using an appropiate perturbation as the authors in [2], we construct a dual for (P_λ) and we prove the existence of strong duality between them. The dual problem is given in terms of conjugate functions and has the advantage that its structure gives an idea how to formulate the multiobjective dual problem to the original problem. By means of strong duality for (P_λ) and its dual, we derive some optimality conditions for the primal optimization problem and construct a dual for the multiobjective primal problem (P).

Using the general result, we introduce the dual problem and study the weak and strong duality for the multiobjective location problem in a general normed space in which the existing facilities are represented by sets. Afterwards, as particular cases of this problem, the multiobjective Weber and minimax location problems are studied.

1 The optimization problem with a multiobjective composed convex function as objective function

Let $(X, \|\cdot\|)$ be a normed space, $g_j : X \to \mathbb{R},\ j = 1, ..., m$, convex and continuous functions and $f = (f_1, \dots, f_l)^T,\ f_i : \mathbb{R}^m \to \mathbb{R},\ i = 1, \dots, l$, convex and componentwise increasing functions, i.e. for $y = (y_1, \dots, y_m),\ z = (z_1, \dots, z_m) \in \mathbb{R}^m$, such that $y_j \geq z_j, j = 1, ..., m$, holds $f_i(y) \geq f_i(z)$ for $i = 1, \dots, l$.

We consider the following multiobjective optimization problem

$$(P)\quad \text{v-}\min_{x \in X} f(g(x))$$

with $g(x) = (g_1(x), \dots, g_m(x))$.

The problem (P) is a multiobjective optimization problem in the form of a vector minimum problem and for such kind of problems different notions of solutions are known. We will use in our paper the so-called efficient and properly efficient solutions. Let us recall the two solution concepts.

Definition 1. *An element $\bar{x} \in X$ is said to be efficient (or Pareto - efficient) with respect to (P) if from*

$$f(g(x)) \leqq_{\mathbb{R}^l_+} f(g(\bar{x})) \text{ for } x \in X, \text{ follows } f(g(x)) = f(g(\bar{x})).$$

Remark 1. Here we consider the partial ordering in $\mathbb{R}^l$ given by the cone $\mathbb{R}^l_+ = \{y = (y_1, \dots, y_l)^T \in \mathbb{R}^l : y_i \geq 0, i = 1, \dots, l\}$ by $y^1 \leqq_{\mathbb{R}^l_+} y^2$ iff $y^2 - y^1 \in \mathbb{R}^l_+$.

Definition 2. *An element $\bar{x} \in X$ is said to be properly efficient with respect to (P) if there exists $\lambda = (\lambda_1, \dots, \lambda_l)^T \in int\mathbb{R}^l_+$ (i.e. $\lambda_i > 0,\ i = 1, \dots, l$), such that $\sum_{i=1}^{l} \lambda_i f_i(g(\bar{x})) \leq \sum_{i=1}^{l} \lambda_i f_i(g(x)), \forall x \in X$.*

2 Duality for the scalarized problem

In order to study the duality for the multiobjective problem (P), first we will study the duality for the scalarized problem (cf. Definition 2).

$$(P_\lambda)\quad \inf_{x \in X} \sum_{i=1}^{l} \lambda_i f_i(g(x)),$$

where $\lambda = (\lambda_1, \dots, \lambda_l)^T \in int\mathbb{R}^l_+$ is a fixed vector. For (P_λ) we will derive a dual (D_λ) by means of the conjugacy approach, which permits us to construct different dual problems to an original primal problem depending on the kind of perturbation.

We introduce the following perturbation function $\Psi : \underbrace{X \times \ldots \times X}_{m+1} \times \mathbb{R}^m \to \mathbb{R}$,

$$\Psi(x, q, d) = \sum_{i=1}^{l} \lambda_i f_i((g_1(x + q_1), \ldots, g_m(x + q_m)) + d),$$

with the perturbation variables $q = (q_1, \ldots, q_m) \in X \times \ldots \times X$ and $d = (d_1, \ldots, d_m) \in \mathbb{R}^m$. A dual problem to (P_λ), obtained by using the perturbation function Ψ, is then

$$(D_\lambda) \qquad \sup_{\substack{q^* = (q_1^*, \ldots, q_m^*), \\ q_j^* \in X^*, j = 1, \ldots, m, \\ d^* \in \mathbb{R}^m}} \{-\Psi^*(0, q^*, d^*)\},$$

where $\Psi^* : \underbrace{X^* \times \ldots \times X^*}_{m+1} \times \mathbb{R}^m \to \mathbb{R} \cup \{+\infty\}$ is the conjugate function of Ψ, $q_j^* \in X^*$, $j = 1, \ldots, m$, and $d^* = (d_1^*, \ldots, d_m^*)^T \in \mathbb{R}^m$ are the dual variables. X^* denotes the topological dual space to X. The conjugate function of Ψ is by definition

$$\begin{aligned}\Psi^*(x^*, q^*, d^*) &= \sup_{\substack{x \in X, d \in \mathbb{R}^m, q = (q_1, \ldots, q_m), \\ q_j \in X, j = 1, \ldots, m}} \left\{ \langle x^*, x \rangle + \sum_{j=1}^{m} \langle q_j^*, q_j \rangle + \langle d^*, d \rangle - \Psi(x, q, d) \right\} \\ &= \sup_{\substack{x \in X, d \in \mathbb{R}^m, q = (q_1, \ldots, q_m), \\ q_j \in X, j = 1, \ldots, m}} \left\{ \langle x^*, x \rangle + \sum_{j=1}^{m} \langle q_j^*, q_j \rangle + \langle d^*, d \rangle \right. \\ &\qquad \left. - \sum_{i=1}^{l} \lambda_i f_i((g_1(x + q_1), \ldots, g_m(x + q_m)) + d) \right\},\end{aligned}$$

where $\langle \cdot, \cdot \rangle$ denotes the bilinear pairing between X^* and X (for $\langle x^*, x \rangle$ and $\langle q_j^*, q_j \rangle$, $j = 1, \ldots, m$) and, respectively, the scalar product in $\mathbb{R}^m$ (for $\langle d^*, d \rangle$). To calculate this expression we introduce first the new variables r_j instead of q_j and then t instead of d by

$$r_j = x + q_j \in X, \; j = 1, \ldots, m,$$

and

$$t = (g_1(x + q_1), \ldots, g_m(x + q_m)) + d \; \in \mathbb{R}^m.$$

This implies

$$\Psi^*(x^*,q^*,d^*) = \sup_{\substack{x\in X,\, t\in\mathbb{R}^m,\\ r_j\in X,\, j=1,\dots,m}} \left\{ \langle x^*,x\rangle + \sum_{j=1}^m \langle q_j^*, r_j - x\rangle + \langle d^*,t\rangle - \langle d^*, (g_1(r_1),\dots,g_m(r_m))\rangle \right.$$

$$\left. - \sum_{i=1}^l \lambda_i f_i(t) \right\} = \sup_{\substack{x\in X,\, t\in\mathbb{R}^m,\\ r_j\in X,\, j=1,\dots,m}} \left\{ \langle x^*,x\rangle + \sum_{j=1}^m \langle q_j^*, r_j\rangle - \sum_{j=1}^m \langle q_j^*, x\rangle + \langle d^*,t\rangle \right.$$

$$\left. - \langle d^*, (g_1(r_1),\dots,g_m(r_m))\rangle - \sum_{i=1}^l \lambda_i f_i(t) \right\} = \sup_{t\in\mathbb{R}^m} \left\{ \langle d^*,t\rangle - \sum_{i=1}^l \lambda_i f_i(t) \right\}$$

$$+ \sup_{\substack{r_j\in X,\\ j=1,\dots,m}} \left\{ \sum_{j=1}^m \langle q_j^*, r_j\rangle - \langle d^*, (g_1(r_1),\dots,g_m(r_m))\rangle \right\} + \sup_{x\in X} \left\langle x^* - \sum_{j=1}^m q_j^*, x \right\rangle$$

$$= \left(\sum_{i=1}^l \lambda_i f_i\right)^* (d^*) + \sum_{j=1}^m \sup_{r_j\in X} \left(\langle q_j^*, r_j\rangle - d_j^* g_j(r_j)\right) + \sup_{x\in X} \left\langle x^* - \sum_{j=1}^m q_j^*, x \right\rangle$$

$$= \left(\sum_{i=1}^l \lambda_i f_i\right)^* (d^*) + \sum_{j=1}^m (d_j^* g_j)^*(q_j^*) + \sup_{x\in X} \left\langle x^* - \sum_{j=1}^m q_j^*, x \right\rangle.$$

Setting $x^* = 0$, the dual problem of (P_λ) takes the form

$$(D_\lambda) \qquad \sup_{\substack{q_j^*\in X^*,\, j=1,\dots,m,\\ d^*\in\mathbb{R}^m}} \left\{ -\left(\sum_{i=1}^l \lambda_i f_i\right)^* (d^*) - \sum_{j=1}^m (d_j^* g_j)^*(q_j^*) + \inf_{x\in X} \left\langle \sum_{j=1}^m q_j^*, x \right\rangle \right\}.$$

In the objective function of (D_λ), if $\sum_{j=1}^m q_j^* \neq 0$, then it holds $\inf_{x\in X} \left\langle \sum_{j=1}^m q_j^*, x \right\rangle = -\infty$. Thus, for the calculation of (D_λ) only $\sum_{j=1}^m q_j^* = 0$ is relevant. On the other hand (cf. [15]),

$$\left(\sum_{i=1}^l \lambda_i f_i\right)^* (d^*) = \inf \left\{ \sum_{i=1}^l (\lambda_i f_i)^*(d^{i*}) \ : \ \sum_{i=1}^l d^{i*} = d^* \right\}.$$

Further, because $\lambda_i > 0$, we get $(\lambda_i f_i)^*(d^{i*}) = \lambda_i f_i^*(\frac{d^{i*}}{\lambda_i})$, for each $i = 1,...,l$. We can make the substitutions $a^i = (a_1^i,...,a_m^i) := \frac{1}{\lambda_i} d^{i*}$, $i = 1,\dots,l$, and then the dual (D_λ) becomes

$$(D_\lambda) \qquad \sup_{\substack{a^i\in\mathbb{R}^m,\ \sum_{i=1}^l \lambda_i a^i = d^*,\ d^*\in\mathbb{R}^m,\\ q_j^*\in X^*,\ \sum_{j=1}^m q_j^* = 0}} \left\{ -\sum_{i=1}^l \lambda_i f_i^*(a^i) - \sum_{j=1}^m (d_j^* g_j)^*(q_j^*) \right\}.$$

Let us point out that, by construction, between (P_λ) and (D_λ) weak duality holds (cf. [5]), i.e. $\inf(P_\lambda) \geq \sup(D_\lambda)$. But, we are interested in the existence of strong duality, i.e. $\inf(P_\lambda) = \sup(D_\lambda)$, or even $\inf(P_\lambda) = \max(D_\lambda)$, meaning the existence of the solution of the dual problem. This can be shown by proving that the problem (P_λ) is stable (cf. [5]).

Proposition 1. *The function* $\Psi : \underbrace{X \times \ldots \times X}_{m+1} \times \mathbb{R}^m \to \mathbb{R}$,

$$\Psi(x, q, d) = \sum_{i=1}^{l} \lambda_i f_i((g_1(x + q_1), \ldots, g_m(x + q_m)) + d)$$

is convex.

The convexity of Ψ follows from the convexity of the functions f_i, $i = 1, \ldots, l$, and g_j, $j = 1, ...m$, and the fact that f_i, $i = 1, \ldots, l$, are componentwise increasing functions.

Proposition 2. *If* $\inf(P_\lambda) > -\infty$, *then the dual problem has a solution and strong duality holds, i.e.*

$$\inf(P_\lambda) = \max(D_\lambda).$$

Proof. See Theorem 1 in [2]. □

To investigate later the multiobjective duality for (P) we need the optimality conditions regarding to the scalar problem (P_λ) and its dual (D_λ). These are formulated in the following theorem.

Theorem 1.

(1) Let $\bar{x} \in X$ be a solution to (P_λ). Then there exists a tupel $(\bar{a}, \bar{q}^*, \bar{d}^*)$, such that the following optimality conditions are satisfied

(i) $f_i(g(\bar{x})) + f_i^*(\bar{a}^i) - \langle \bar{a}^i, g(\bar{x}) \rangle = 0, \; i = 1, \ldots, l,$

(ii) $(\bar{d}_j^* g_j)^*(\bar{q}_j^*) + \bar{d}_j^* g_j(\bar{x}) - \langle \bar{q}_j^*, \bar{x} \rangle = 0, \; j = 1, \ldots, m,$

(iii) $\sum\limits_{j=1}^{m} \bar{q}_j^* = 0,$

(iv) $\sum\limits_{i=1}^{l} \lambda_i \bar{a}^i = \bar{d}^*.$

(2) If $\bar{x} \in X$ and $(\bar{a}, \bar{q}^*, \bar{d}^*)$ satisfy (i)-(iv), then $\bar{x}$ is a solution to (P_λ), $(\bar{a}, \bar{q}^*, \bar{d}^*)$ is a solution to (D_λ) and strong duality holds, i.e.

$$\sum_{i=1}^{l} \lambda_i f_i(g(\bar{x})) = -\sum_{i=1}^{l} \lambda_i f_i^*(\bar{a}^i) - \sum_{j=1}^{m} (\bar{d}_j^* g_j)^*(\bar{q}_j^*). \tag{1}$$

Remark 2. Obviously, the tupel $(\bar{a}, \bar{q}^*, \bar{d}^*)$ in the part (1) of Theorem 1 is a solution of (D_λ) (cf. the proof).

Proof.

(1) By Proposition 2, it follows that there exist $\bar{a} = (\bar{a}^1, ..., \bar{a}^l)$, $\bar{a}^i \in \mathbb{R}^m$, $i = 1, ..., l$, $\bar{q}^* = (\bar{q}_1^*, ..., \bar{q}_m^*)$, $\bar{q}_j^* \in X^*$, $j = 1, ..., m$, and $\bar{d}^* \in \mathbb{R}^m$, such that $(\bar{a}, \bar{q}^*, \bar{d}^*)$ is a solution to (D_λ) and $\inf(P_\lambda) = \max(D_\lambda)$. This means that $\sum\limits_{j=1}^{m} \bar{q}_j^* = 0$, $\sum\limits_{i=1}^{l} \lambda_i \bar{a}^i = \bar{d}^*$, i.e. (iii) and (iv) are true, and

$$\sum_{i=1}^{l} \lambda_i f_i(g(\bar{x})) = -\sum_{i=1}^{l} \lambda_i f_i^*(\bar{a}^i) - \sum_{j=1}^{m} (\bar{d}_j^* g_j)^*(\bar{q}_j^*).$$

This equality is equivalent to

$$\sum_{i=1}^{l} \lambda_i f_i(g(\bar{x})) + \sum_{i=1}^{l} \lambda_i f_i^*(\bar{a}^i) - \sum_{i=1}^{l} \lambda_i \langle \bar{a}^i, g(\bar{x}) \rangle + \sum_{j=1}^{m} (\bar{d}_j^* g_j)^*(\bar{q}_j^*)$$
$$+ \sum_{j=1}^{m} \bar{d}_j^* g_j(\bar{x}) - \sum_{j=1}^{m} \langle \bar{q}_j^*, \bar{x} \rangle + \sum_{i=1}^{l} \lambda_i \langle \bar{a}^i, g(\bar{x}) \rangle - \sum_{j=1}^{m} \bar{d}_j^* g_j(\bar{x}) = 0.$$

From here follows

$$\sum_{i=1}^{l} \lambda_i \{ f_i^*(\bar{a}^i) + f_i(g(\bar{x})) - \langle \bar{a}^i, g(\bar{x}) \rangle \} + \sum_{j=1}^{m} \{ (\bar{d}_j^* g_j)^*(\bar{q}_j^*) + \bar{d}_j^* g_j(\bar{x}) - \langle \bar{q}_j^*, \bar{x} \rangle \}$$
$$+ \sum_{i=1}^{l} \lambda_i \langle \bar{a}^i, g(\bar{x}) \rangle - \sum_{j=1}^{m} \bar{d}_j^* g_j(\bar{x}) = 0.$$

But $\langle \sum\limits_{i=1}^{m} \lambda_i \bar{a}^i, g(\bar{x}) \rangle - \langle \bar{d}^*, g(\bar{x}) \rangle = 0$, which implies that

$$\sum_{i=1}^{l} \lambda_i \{ f_i^*(\bar{a}^i) + f_i(g(\bar{x})) - \langle \bar{a}^i, g(\bar{x}) \rangle \} + \sum_{j=1}^{m} \{ (\bar{d}_j^* g_j)^*(\bar{q}_j^*) + \bar{d}_j^* g_j(\bar{x}) - \langle \bar{q}_j^*, \bar{x} \rangle \} = 0. \tag{2}$$

Because of the Young-Fenchel inequality which is expressing that for a function f and its conjugate f^*, $f(x) + f^*(x^*) \geqq \langle x^*, x \rangle$ is fulfilled, obviously all terms of the sum in (2) are non-negative and therefore they must be even equal to zero. This gives the optimality conditions (i) and (ii).

(2) All the calculations and transformations done within part (1) may be carried out in the reverse direction starting from the conditions (i), (ii), (iii) and (iv). Thus the equality (1) results, which is the strong duality, and shows that $\bar{x}$ solves (P_λ) and $(\bar{a}, \bar{q}^*, \bar{d}^*)$ solves (D_λ). □

3 The multiobjective dual problem

A dual multiobjective optimization problem (D) to (P) is introduced by

$$(D) \quad \underset{(a,q,d,\lambda,t)\in Y}{\text{v-max}} \; h(a,q,d,\lambda,t),$$

with

$$h(a,q,d,\lambda,t) = \begin{pmatrix} h_1(a,q,d,\lambda,t) \\ h_2(a,q,d,\lambda,t) \\ \vdots \\ h_l(a,q,d,\lambda,t) \end{pmatrix},$$

and

$$h_i(a,q,d,\lambda,t) = -f_i^*(a^i) - \frac{1}{l\lambda_i}\sum_{j=1}^{m}(d_j g_j)^*(q_j) + t_i, \; i = 1,\dots,l.$$

The dual variables are

$$a = (a^1,\dots,a^l), \; a^i \in \mathbb{R}^m, \; i = 1,\dots,l, \; q = (q_1,\dots,q_m), \; q_j \in X^*, \; j = 1,\dots,m,$$

$$d = (d_1,\dots,d_m)^T \in \mathbb{R}^m, \; \lambda = (\lambda_1,\dots,\lambda_l)^T \in \mathbb{R}^l, \; t = (t_1,\dots,t_l)^T \in \mathbb{R}^l,$$

and the set of constraints is

$$Y = \left\{ (a,q,d,\lambda,t) : \lambda \in int\mathbb{R}^l_+, \sum_{i=1}^{l}\lambda_i a^i = d, \; \sum_{j=1}^{m} q_j = 0, \; \sum_{i=1}^{l}\lambda_i t_i = 0 \right\}.$$

Remark 3. For the sake of simplicity of the denotation of the dual variables we write here and in the following q_j and q instead of q_j^* and q^* and d instead of d^*.

Definition 3. *An element $(\bar{a},\bar{q},\bar{d},\bar{\lambda},\bar{t}) \in Y$ is said to be efficient (or Pareto-efficient) to (D) if from $h(a,q,d,\lambda,t) \underset{\mathbb{R}^l_+}{\geqq} h(\bar{a},\bar{q},\bar{d},\bar{\lambda},\bar{t})$ for $(a,q,d,\lambda,t) \in Y$ follows $h(a,q,d,\lambda,t) = h(\bar{a},\bar{q},\bar{d},\bar{\lambda},\bar{t})$.*

The following theorem states the weak duality assertion.

Theorem 2. *There is no $x \in X$ and no $(a,q,d,\lambda,t) \in Y$ fulfilling $h(a,q,d,\lambda,t) \underset{\mathbb{R}^l_+}{\geqq} f(g(x))$ and $h(a,q,d,\lambda,t) \neq f(g(x))$.*

Proof. Let us assume that there exist $x \in X$ and $(a,q,d,\lambda,t) \in Y$, such that $h(a,q,d,\lambda,t) \underset{\mathbb{R}^l_+}{\geqq} f(g(x))$ and $h(a,q,d,\lambda,t) \neq f(g(x))$, i.e. $h_i(a,q,d,\lambda,t) \geq f_i(g(x))$,

$\forall i = 1, \ldots, l$, and $h_k(a, q, d, \lambda, t) > f_k(g(x))$ for at least one $k \in \{1, \ldots, l\}$. This means that

$$\sum_{i=1}^{l} \lambda_i h_i(a, q, d, \lambda, t) > \sum_{i=1}^{l} \lambda_i f_i(g(x)). \tag{3}$$

On the other hand, we have

$$\sum_{i=1}^{l} \lambda_i h_i(a, q, d, \lambda, t) = -\sum_{i=1}^{l} \lambda_i f_i^*(a^i) - \sum_{i=1}^{l} \lambda_i \frac{1}{l\lambda_i} \sum_{j=1}^{m} (d_j g_j)^*(q_j) + \sum_{i=1}^{l} \lambda_i t_i$$

$$= -\sum_{i=1}^{l} \lambda_i f_i^*(a^i) - \sum_{j=1}^{m} (d_j g_j)^*(q_j).$$

By the Young-Fenchel inequality

$$-f_i^*(a^i) \leq f_i(g(x)) - \langle a^i, g(x) \rangle, \; i = 1, \ldots, l$$

and

$$-(d_j g_j)^*(q_j) \leq d_j g_j(x) - \langle q_j, x \rangle, \; j = 1, ..., m,$$

we obtain

$$\begin{aligned} \sum_{i=1}^{l} \lambda_i h_i(a, q, d, \lambda, t) \leq \; & \sum_{i=1}^{l} \lambda_i f_i(g(x)) - \sum_{i=1}^{l} \lambda_i \langle a^i, g(x) \rangle + \sum_{j=1}^{m} d_j g_j(x) - \sum_{j=1}^{m} \langle q_j, x \rangle \\ = \; & \sum_{i=1}^{l} \lambda_i f_i(g(x)) + \left\langle d - \sum_{i=1}^{l} \lambda_i a^i, g(x) \right\rangle - \left\langle \sum_{j=1}^{m} q_j, x \right\rangle \\ = \; & \sum_{i=1}^{l} \lambda_i f_i(g(x)), \end{aligned}$$

and, therefore,

$$\sum_{i=1}^{l} \lambda_i h_i(a, q, d, \lambda, t) \leq \sum_{i=1}^{l} \lambda_i f_i(g(x)).$$

But this inequality contradicts relation (3). □

The following theorem expresses the so-called strong duality between the two multiobjective problems (P) and (D).

Theorem 3. *Let $\bar{x}$ be a properly efficient element to (P). Then an efficient solution $(\bar{a}, \bar{q}, \bar{d}, \bar{\lambda}, \bar{t}) \in Y$ to (D) exists and strong duality $f(g(\bar{x})) = h(\bar{a}, \bar{q}, \bar{d}, \bar{\lambda}, \bar{t})$ holds.*

Proof. Assume $\bar{x}$ to be properly efficient to (P). From Definition 2 the existence of a corresponding vector $\tilde{\lambda} = (\tilde{\lambda}_1, \dots, \tilde{\lambda}_l)^T \in int\mathbb{R}^l_+$ follows such that $\bar{x}$ solves the scalar problem

$$(P_{\tilde{\lambda}}) \quad \inf_{x \in X} \sum_{i=1}^{l} \tilde{\lambda}_i f_i(g(x)).$$

Because of $\inf(P_{\tilde{\lambda}}) > -\infty$, Proposition 2 ensures the existence of a solution $(\tilde{a}, \tilde{q}, \tilde{d})$ of the dual $(D_{\tilde{\lambda}})$ of $(P_{\tilde{\lambda}})$. The optimality conditions $(i) - (iv)$ are satisfied because of Theorem 1. Now we construct by means of $\bar{x}$ and $(\tilde{a}, \tilde{q}, \tilde{d})$ the efficient solution $(\bar{a}, \bar{q}, \bar{d}, \bar{\lambda}, \bar{t})$ of (D). We consider $\bar{\lambda} := \tilde{\lambda}$, $\bar{a} := \tilde{a}$, $\bar{q} := \tilde{q}$, $\bar{d} := \tilde{d}$. It remains to introduce $\bar{t} = (\bar{t}_1, \dots, \bar{t}_l)^T$. Let for $i = 1, \dots, l$,

$$\bar{t}_i := \frac{1}{l\bar{\lambda}_i} \sum_{j=1}^{m} (\bar{d}_j g_j)^*(\bar{q}_j) + \langle \bar{a}^i, g(\bar{x}) \rangle \in \mathbb{R}.$$

By the optimality conditions $(i) - (iv)$, for this tupel $(\bar{a}, \bar{q}, \bar{d}, \bar{\lambda}, \bar{t})$ it holds

$$\bar{\lambda} \in int\mathbb{R}^l_+, \ \sum_{i=1}^{l} \bar{\lambda}_i \bar{a}^i = \bar{d}, \ \sum_{j=1}^{m} \bar{q}_j = 0$$

and

$$\begin{aligned}
\sum_{i=1}^{l} \bar{\lambda}_i \bar{t}_i &= \sum_{i=1}^{l} \bar{\lambda}_i \tfrac{1}{l\bar{\lambda}_i} \sum_{j=1}^{m} (\bar{d}_j g_j)^*(\bar{q}_j) + \sum_{i=1}^{l} \bar{\lambda}_i \langle \bar{a}^i, g(\bar{x}) \rangle \\
&= \sum_{j=1}^{m} (\bar{d}_j g_j)^*(\bar{q}_j) + \left\langle \sum_{i=1}^{l} \bar{\lambda}_i \bar{a}^i, g(\bar{x}) \right\rangle \\
&= \sum_{j=1}^{m} (\bar{d}_j g_j)^*(\bar{q}_j) + \sum_{j=1}^{m} \bar{d}_j g_j(\bar{x}) \\
&= \left\langle \sum_{j=1}^{m} \bar{q}_j, \bar{x} \right\rangle = 0.
\end{aligned}$$

This means that the element $(\bar{a}, \bar{q}, \bar{d}, \bar{\lambda}, \bar{t})$ is feasible to (D). It remains to show that the values of the objective functions are equal, i.e. $f(g(\bar{x})) = h(\bar{a}, \bar{q}, \bar{d}, \bar{\lambda}, \bar{t})$. Therefore we will prove that $f_i(g(\bar{x})) = h_i(\bar{a}, \bar{q}, \bar{d}, \bar{\lambda}, \bar{t})$ for each $i = 1, \dots, l$. For this we use the relation (i) from Theorem 1 and obtain the following equalities

$$\begin{aligned}
h_i(\bar{a}, \bar{q}, \bar{d}, \bar{\lambda}, \bar{t}) &= -f_i^*(\bar{a}^i) - \tfrac{1}{l\bar{\lambda}_i} \sum_{j=1}^{m} (\bar{d}_j g_j)^*(\bar{q}_j) + \bar{t}_i \\
&= -f_i^*(\bar{a}^i) - \tfrac{1}{l\bar{\lambda}_i} \sum_{j=1}^{m} (\bar{d}_j g_j)^*(\bar{q}_j) + \tfrac{1}{l\bar{\lambda}_i} \sum_{j=1}^{m} (\bar{d}_j g_j)^*(\bar{q}_j) + \langle \bar{a}^i, g(\bar{x}) \rangle \\
&= -f_i^*(\bar{a}^i) + \langle \bar{a}^i, g(\bar{x}) \rangle \\
&= f_i(g(\bar{x})), \quad i = 1, \dots, l.
\end{aligned}$$

In conclusion,

$$f(g(\bar{x})) = h(\bar{a}, \bar{q}, \bar{d}, \bar{\lambda}, \bar{t}).$$

□

4 The case of monotonic norms

In this section we particularize the multiobjective problem presented in the previous section. Therefore, let be $\Phi_i : \mathbb{R}^m \to \mathbb{R}$, $i = 1, \ldots, l$, monotonic norms on $\mathbb{R}^m$, i.e. (cf. [1])

$$\forall\ u,\ v \in \mathbb{R}^m,\ |u_j| \leq |v_j|,\ j = 1, ..., m,\ \text{it holds}\ \Phi_i(u) \leq \Phi_i(v).$$

Let us introduce now the following multiobjective problem

$$(P_\Phi) \quad \underset{x \in X}{\text{v-min}} \begin{pmatrix} \Phi_1^+(g(x)) \\ \vdots \\ \Phi_l^+(g(x)) \end{pmatrix},$$

where $\Phi_i^+(t) := \Phi_i(t^+)$, $i = 1, ..., l$, with $t^+ = (t_1^+, \ldots, t_m^+)$ and $t_j^+ = \max\{0, t_j\}$, $j = 1, ..., m$.

Proposition 3. *The functions $\Phi_i^+ : \mathbb{R}^m \to \mathbb{R}$, $i = 1, \ldots, l$, are convex and componentwise increasing.*

Proof. See Proposition 2 in [2]. □

In order to study the duality for the problem (P_Φ) we will study, like in section 2, the duality for the scalarized problem

$$(P_{\Phi\lambda}) \quad \inf_{x \in X} \sum_{i=1}^{l} \lambda_i \Phi_i^+(g(x)),$$

where $\lambda = (\lambda_1, \ldots, \lambda_l)^T \in int\mathbb{R}^l_+$ is a fixed vector. By the approach described in section 2, a dual problem to $(P_{\Phi\lambda})$ is

$$(D_{\Phi\lambda}) \quad \sup_{\substack{a^i \in \mathbb{R}^m,\ \sum_{i=1}^{l} \lambda_i a^i = d^*,\ d^* \in \mathbb{R}^m, \\ q_j^* \in X^*,\ \sum_{j=1}^{m} q_j^* = 0}} \left\{ -\sum_{i=1}^{l} \lambda_i (\Phi_i^+)^*(a^i) - \sum_{j=1}^{m} (d_j^* g_j)^*(q_j^*) \right\}.$$

Proposition 4. *The conjugate functions $(\Phi_i^+)^* : \mathbb{R}^m \to \mathbb{R} \cup \{+\infty\}$, $i = 1, \ldots, l$, of Φ_i^+ are*

$$(\Phi_i^+)^*(a^i) = \begin{cases} 0, & \text{if } a^i \underset{\mathbb{R}^m_+}{\geqq} 0 \text{ and } \Phi_i^0(a^i) \leq 1, \\ +\infty, & \text{otherwise,} \end{cases}$$

where Φ_i^0 is the dual norm of Φ_i in $\mathbb{R}^m$.

Proof. See Proposition 3 in [2]. □

By Proposition 4, the dual of $(P_{\Phi\lambda})$ has the following form

$$(D_{\Phi\lambda}) \qquad \sup_{\substack{a^i \in \mathbb{R}^m_+,\, \Phi_i^0(a^i)\le 1,\, i=1,\dots,l,\\ \sum_{i=1}^{l} \lambda_i a^i = d^*,\, d^*\in\mathbb{R}^m_+,\; q_j^*\in X^*,\, \sum_{j=1}^{m} q_j^*=0}} \left\{ -\sum_{j=1}^{m} (d_j^* g_j)^*(q_j^*) \right\}.$$

Let us consider now the set-valued variable $I \subseteq \{1,\dots,m\}$. In the objective function of $(D_{\Phi\lambda})$ let us separate the terms for which $d_j^* > 0$ (i.e. $j \in I$) from the terms for which $d_j^* = 0$ (i.e. $j \notin I$). Then it follows

$$(D_{\Phi\lambda}) \qquad \sup_{\substack{a^i \in \mathbb{R}^m_+,\, \Phi_i^0(a^i)\le 1,\, i=1,\dots,l,\\ \sum_{i=1}^{l} \lambda_i a^i = d^*,\, I\subseteq\{1,\dots,m\},\, d_j^*>0,\,(j\in I),\\ d_j^*=0,\,(j\notin I),\, q_j^*\in X^*,\, \sum_{j=1}^{m} q_j^*=0}} \left\{ -\sum_{j\in I} (d_j^* g_j)^*(q_j^*) - \sum_{j\notin I} (d_j^* g_j)^*(q_j^*) \right\}.$$

Let us notice that, in the case $d_j^* > 0$ $(j \in I)$ there must exist at least one $i \in \{1,\dots,l\}$, such that $a_j^i > 0$.

Because of $0^*(q_j^*) = \sup\limits_{x\in X}\{\langle q_j^*, x\rangle - 0\} = \sup\limits_{x\in X}\langle q_j^*, x\rangle = \begin{cases} 0, & \text{if } q_j^* = 0,\\ +\infty, & \text{otherwise}\end{cases}$, in order to have supremum in $(D_{\Phi\lambda})$, we must take $q_j^* = 0$ for all $j \notin I$.

The problem $(D_{\Phi\lambda})$ can then be written as

$$(D_{\Phi\lambda}) \qquad \sup_{\substack{a^i \in \mathbb{R}^m_+,\, \Phi_i^0(a^i)\le 1,\, i=1,\dots,l,\\ \sum_{i=1}^{l} \lambda_i a^i = d^*,\, I\subseteq\{1,\dots,m\},\, d_j^*>0,\,(j\in I),\\ d_j^*=0,\,(j\notin I),\, q_j^*\in X^*,\, j\in J,\, \sum_{j\in I} q_j^*=0}} \left\{ -\sum_{j\in I} (d_j^* g_j)^*(q_j^*) \right\}.$$

For $d_j^* > 0$, i.e. $j \in I$, we apply again the formula $(d_j^* g_j)^*(q_j^*) = d_j^* g_j^*(\frac{1}{d_j^*} q_j^*)$ (cf. [5]). Denoting $q_j^* := \frac{q_j^*}{d_j^*}$ for $j \in I$ we get

$$(D_{\Phi\lambda}) \qquad \sup_{(I,a,d^*,q^*)\in Y_{\Phi\lambda}} \left\{ -\sum_{j\in I} d_j^* g_j^*(q_j^*) \right\},$$

with

$$\begin{aligned} Y_{\Phi\lambda} = \Big\{ (I,a,d^*,q^*) : \; & I\subseteq\{1,\dots,m\},\, a=(a^1,\dots,a^l),\, a^i\in\mathbb{R}^m_+,\, \Phi_i^0(a^i)\le 1,\, i=1,\dots,l,\\ & d^*=(d_1^*,\dots,d_m^*)\in\mathbb{R}^m,\; q^*=(q_1^*,\dots,q_m^*),\; q_j^*\in X^*,\; j=1,\dots,m,\\ & \sum_{i=1}^{l} \lambda_i a^i = d^*,\; d_j^*>0,\, j\in I,\; d_j^*=0,\, j\notin I,\; \sum_{j\in I} d_j^* q_j^* = 0 \Big\}. \end{aligned}$$

Now we can eliminate the variable d^*, observing that $\sum_{i=1}^{l} \lambda_i a_j^i = d_j^*$, for $j = 1, ..., m$. Then the dual becomes (setting $q := q^*$)

$$(D_{\Phi\lambda}) \qquad \sup_{(I,a,q)\in Y_{\Phi\lambda}} \left\{ -\sum_{j\in I} \left(\sum_{i=1}^{l} \lambda_i a_j^i \right) g_j^*(q_j) \right\}$$

with

$$\begin{aligned} Y_{\Phi\lambda} = \Big\{ (I,\, a,\, q) : I \subseteq \{1, \dots, m\}, a = (a^1, \dots, a^l), a^i \in \mathbb{R}_+^m, \Phi_i^0(a^i) \leq 1, \\ q = (q_1, \dots, q_m),\, q_j \in X^*,\, j = 1, \dots, m, \sum_{j\in I} \left(\sum_{i=1}^{l} \lambda_i a_j^i \right) q_j = 0, \\ \sum_{i=1}^{l} \lambda_i a_j^i > 0,\, j \in I,\; a_j^i = 0,\, j \notin I,\; i = 1, \dots, l \Big\}. \end{aligned}$$

Because of the functions Φ_i^+, $i = 1, ..., l$, are convex and componentwise increasing, it follows that $\sum_{i=1}^{l} \lambda_i \Phi_i^+$ is also convex and componentwise increasing. One can notice that $\inf(P_{\Phi\lambda})$ is finite, being greater than or equal to zero. This observation, together with Proposition 2, permits us to formulate the following strong duality theorem for the problems $(P_{\Phi\lambda})$ and $(D_{\Phi\lambda})$.

Theorem 4. *The dual problem $(D_{\Phi\lambda})$ has a solution and strong duality holds, i.e.*

$$\inf(P_{\Phi\lambda}) = \max(D_{\Phi\lambda}).$$

Analogously to problem (P_λ) we can derive now the optimality conditions for $(P_{\Phi\lambda})$.

Theorem 5.

(1) Let $\bar{x} \in X$ be a solution to $(P_{\Phi\lambda})$. Then there exists $(\bar{I}, \bar{a}, \bar{q}) \in Y_{\Phi\lambda}$, solution to $(D_{\Phi\lambda})$, such that the following optimality conditions are satisfied

(i) $\bar{I} \subseteq \{1, \dots, m\}$, $\bar{a}^i \in \mathbb{R}_+^m$, $\sum_{i=1}^{l} \lambda_i \bar{a}_j^i > 0$, $j \in \bar{I}$, $\bar{a}_j^i = 0$, $j \notin \bar{I}$, $i = 1, \dots, l$,

(ii) $\Phi_i^0(\bar{a}^i) \leq 1$, $i = 1, \dots, l$, $\sum_{j\in\bar{I}} \left(\sum_{i=1}^{l} \lambda_i \bar{a}_j^i \right) \bar{q}_j = 0$,

(iii) $\Phi_i^+(g(\bar{x})) = \langle \bar{a}^i, g(\bar{x}) \rangle$, $i = 1, \dots, l$,

(iv) $g_j(\bar{x}) + g_j^*(\bar{q}_j) = \langle \bar{q}_j, \bar{x} \rangle$, $j \in \bar{I}$.

(2) If $\bar{x} \in X$, $(\bar{I}, \bar{a}, \bar{q})$ is feasible to $(D_{\Phi\lambda})$ and (i)-(iv) are fulfilled, then $\bar{x}$ is a solution to $(P_{\Phi\lambda})$, $(\bar{I}, \bar{a}, \bar{q})$ is a solution to $(D_{\Phi\lambda})$ and strong duality holds, i.e.

$$\sum_{i=1}^{l} \lambda_i \Phi_i^{+}(g(\bar{x})) = -\sum_{j\in\bar{I}} \left(\sum_{i=1}^{l} \lambda_i \bar{a}_j^i \right) g_j^*(\bar{q}_j).$$

Proof.

(1) By Theorem 4, it follows that there exists $(\bar{I}, \bar{a}, \bar{q}) \in Y_{\Phi\lambda}$, a solution to $(D_{\Phi\lambda})$, such that $(i) - (ii)$ are fulfilled and

$$\sum_{i=1}^{l} \lambda_i \Phi_i^{+}(g(\bar{x})) = -\sum_{j\in\bar{I}} \Big(\sum_{i=1}^{l} \lambda_i \bar{a}_j^i \Big) g_j^*(\bar{q}_j).$$

This is equivalent to

$$\sum_{i=1}^{l} \lambda_i \Phi_i^{+}(g(\bar{x})) + \sum_{i=1}^{l} \lambda_i (\Phi_i^{+})^*(\bar{a}^i) - \sum_{i=1}^{l} \lambda_i \langle \bar{a}^i, g(\bar{x}) \rangle +$$

$$\sum_{i=1}^{l} \lambda_i \langle \bar{a}^i, g(\bar{x}) \rangle + \sum_{j\in\bar{I}} \left(\sum_{i=1}^{l} \lambda_i \bar{a}_j^i \right) g_j^*(\bar{q}_j) = 0.$$

Hence,

$$\sum_{i=1}^{l} \lambda_i [\Phi_i^{+}(g(\bar{x})) + (\Phi_i^{+})^*(\bar{a}^i) - \langle \bar{a}^i, g(\bar{x}) \rangle] +$$

$$\sum_{i=1}^{l} \lambda_i \langle \bar{a}^i, g(\bar{x}) \rangle + \sum_{j\in\bar{I}} \left(\sum_{i=1}^{l} \lambda_i \bar{a}_j^i \right) g_j^*(\bar{q}_j) = 0.$$

Because of (i) and (ii), it holds

$$\sum_{i=1}^{l} \lambda_i \langle \bar{a}^i, g(\bar{x}) \rangle + \sum_{j\in\bar{I}} \left(\sum_{i=1}^{l} \lambda_i \bar{a}_j^i \right) g_j^*(\bar{q}_j) = \sum_{i=1}^{l} \lambda_i \langle \bar{a}^i, g(\bar{x}) \rangle$$

$$+ \sum_{j\in\bar{I}} \left(\sum_{i=1}^{l} \lambda_i \bar{a}_j^i \right) g_j^*(\bar{q}_j) - \left\langle \sum_{j\in\bar{I}} \Big(\sum_{i=1}^{l} \lambda_i \bar{a}_j^i \Big) \bar{q}_j, \bar{x} \right\rangle = \sum_{i=1}^{l} \lambda_i \left(\sum_{j\in\bar{I}} \bar{a}_j^i g_j(\bar{x}) \right)$$

$$+ \sum_{j\in\bar{I}} \left(\sum_{i=1}^{l} \lambda_i \bar{a}_j^i \right) \left(g_j^*(\bar{q}_j) - \langle \bar{q}_j, \bar{x} \rangle \right)$$

$$= \sum_{j \in \bar{I}} \left(\sum_{i=1}^{l} \lambda_i \bar{a}_j^i \right) \left(g_j(\bar{x}) + g_j^*(\bar{q}_j) - \langle \bar{q}_j, \bar{x} \rangle \right),$$

implying

$$\sum_{i=1}^{l} \lambda_i \left[\Phi_i^+(g(\bar{x})) + (\Phi_i^+)^*(\bar{a}^i) - \langle \bar{a}^i, g(\bar{x}) \rangle \right] +$$

$$\sum_{j \in \bar{I}} \left(\sum_{i=1}^{l} \lambda_i \bar{a}_j^i \right) \left(g_j(\bar{x}) + g_j^*(\bar{q}_j) - \langle \bar{q}_j, \bar{x} \rangle \right) = 0.$$

In conclusion, using again the Young-Fenchel inequality, we obtain

$$\Phi_i^+(g(\bar{x})) + (\Phi_i^+)^*(\bar{a}^i) - \langle \bar{a}^i, g(\bar{x}) \rangle = 0, \; i = 1, \dots, l,$$

and

$$g_j(\bar{x}) + g_j^*(\bar{q}_j) - \langle \bar{q}_j, \bar{x} \rangle = 0, \; j \in \bar{I},$$

and, so, the equalities in *(iii)* and *(iv)* must hold. □

Further, like in the general case, we can construct a multiobjective dual problem to the primal problem (P_Φ)

$$(D_\Phi) \qquad \underset{(I,a,q,\lambda,t) \in Y_\Phi}{\text{v-max}} \begin{pmatrix} h_1(I,a,q,\lambda,t) \\ \vdots \\ h_l(I,a,q,\lambda,t) \end{pmatrix},$$

with

$$h_k(I,a,q,\lambda,t) = -\frac{1}{l\lambda_k} \sum_{j \in I} \left(\sum_{i=1}^{l} \lambda_i a_j^i \right) g_j^*(q_j) + t_k, \; k = 1, \dots, l,$$

the dual variables

$$a = (a^1, \dots, a^l), \; a^i \in \mathbb{R}^m, \; i = 1, \dots, l, \; q = (q_1, \dots, q_m), \; q_j \in X^*, j = 1, \dots, m,$$

$$\lambda = (\lambda_1, \dots, \lambda_l)^T \in \mathbb{R}^l, \; t = (t_1, \dots, t_l)^T \in \mathbb{R}^l,$$

and the set of constraints

$$Y_\Phi = \Bigg\{ (I,a,q,\lambda,t) : I \subseteq \{1, ..., m\}, a^i \in \mathbb{R}_+^m, \Phi_i^0(a^i) \le 1, \sum_{j \in I} \left(\sum_{i=1}^{l} \lambda_i a_j^i \right) q_j = 0,$$
$$\left(\sum_{i=1}^{l} \lambda_i a_j^i \right) > 0, j \in I, a_j^i = 0, j \notin I, i = 1, \dots, l, \lambda \in int\mathbb{R}_+^l, \sum_{k=1}^{l} \lambda_k t_k = 0 \Bigg\}.$$

Let us present now the weak and strong duality theorems for these problems.

Theorem 6. *There is no $x \in X$ and no $(I, a, q, \lambda, t) \in Y_\Phi$, such that $\Phi_i^+(g(x)) \le h_i(I, a, q, \lambda, t)$, $i = 1, \dots, l$, and $\Phi_k^+(g(x)) < h_k(I, a, q, \lambda, t)$ for at least one $k \in \{1, \dots, l\}$.*

Theorem 7. *Let $\bar{x}$ be a properly efficient element to (P_Φ)). Then an efficient solution $(\bar{I}, \bar{a}, \bar{q}, \bar{\lambda}, \bar{t}) \in Y_\Phi$ to (D_Φ) exists and strong duality holds, i.e.*

$$\Phi_k^+(g(\bar{x})) = -\frac{1}{l\bar{\lambda}_k} \sum_{j \in \bar{I}} \left(\sum_{i=1}^{l} \bar{\lambda}_i \bar{a}_j^i \right) g_j^*(\bar{q}_j) + \bar{t}_k, \ k = 1, \dots, l.$$

5 The multiobjective location model involving sets as existing facilities

Let $\mathcal{A} = \{A_1, \dots, A_m\}$ be a family of convex sets in X such that $\bigcap_{j=1}^{m} \bar{A}_j = \emptyset$. We consider the same vector function $d : X \to \mathbb{R}^m$ as in [2], i.e.

$$d(x) := (d_1(x, A_1), \dots, d_m(x, A_m)),$$

where

$$d_j(x, A_j) = \inf\{\gamma_j(x - y_j) : y_j \in A_j\}, \ j = 1, ..., m,$$

and γ_j, $j = 1, ..., m$, are continuous norms on X. For $j = 1, ..., m$, we consider the functions $g_j : X \to \mathbb{R}$, $g_j(x) = d_j(x, A_j)$. This means that the functions g_j, $j = 1, \dots, m$, are convex and continuous on X. The multiobjective location problem with sets as existing facilities is

$$(P_\Phi(\mathcal{A})) \quad \underset{x \in X}{\text{v-min}} \begin{pmatrix} \Phi_1(d(x)) \\ \vdots \\ \Phi_l(d(x)) \end{pmatrix},$$

with $\Phi_i : \mathbb{R}^m \to \mathbb{R}$, $i = 1, ..., l$, monotonic norms on $\mathbb{R}^m$.

Because of

$$\Phi_i^+(d(x)) = \Phi_i((d(x))^+) = \Phi_i(d(x)), \ \forall x \in X, \ i = 1, ..., l,$$

where $(d(x))^+ = ((d_1(x))^+, \dots, (d_m(x))^+)$ with $(d_i(x))^+ = \max\{0, d_i(x)\}$, for $i = 1 \dots, m$, we can write $(P_\Phi(\mathcal{A}))$ in the equivalent form

$$(P_\Phi(\mathcal{A})) \quad \underset{x \in X}{\text{v-min}} \begin{pmatrix} \Phi_1^+(d(x)) \\ \vdots \\ \Phi_l^+(d(x)) \end{pmatrix}.$$

This problem is a particular case of the problem studied in section 4. In order to study the duality for this problem, we will study again, at first, the duality for the scalarized problem

$$(P_{\Phi\lambda}(\mathcal{A})) \quad \inf_{x\in X} \sum_{i=1}^{l} \lambda_i \Phi_i^+(d(x)),$$

with $\lambda = (\lambda_1, \ldots, \lambda_l)^T \in int\mathbb{R}_+^l$ fixed.

Then the dual of $(P_{\Phi\lambda}(\mathcal{A}))$ is

$$(D_{\Phi\lambda}(\mathcal{A})) \quad \sup_{(I,a,q)\in Y_{\Phi\lambda}(\mathcal{A})} \left\{ -\sum_{j\in I} \left(\sum_{i=1}^{l} \lambda_i a_j^i \right) d_j^*(q_j) \right\},$$

where $d_j^*(q_j)$ is the conjugate function to $d_j(x, A_j), j = 1, ..., m$, and

$$\begin{aligned} Y_{\Phi\lambda}(\mathcal{A}) = \Big\{(I,a,q):\ & I \subseteq \{1,\ldots,m\}, a = (a^1,\ldots,a^l), a^i \in \mathbb{R}_+^m, \Phi_i^0(a^i) \le 1, \\ & q = (q_1,\ldots,q_m), q_j \in X^*, \left(\sum_{i=1}^{l} \lambda_i a_j^i\right) > 0, j \in I, \\ & a_j^i = 0, j \notin I, i = 1, ..., l, \sum_{j\in I} \left(\sum_{i=1}^{l} \lambda_i a_j^i\right) q_j = 0 \Big\}. \end{aligned}$$

Using the Theorems 4 and 5, we can present for $(P_{\Phi\lambda}(\mathcal{A}))$ and $(D_{\Phi\lambda}(\mathcal{A}))$ the strong duality theorem and the optimality conditions.

Theorem 8. *The dual problem $(D_{\Phi\lambda}(\mathcal{A}))$ has a solution and strong duality holds, i.e.*

$$\inf(P_{\Phi\lambda}(\mathcal{A})) = \max(D_{\Phi\lambda}(\mathcal{A})).$$

Theorem 9.

(1) Let $\bar{x} \in X$ be a solution to $(P_{\Phi\lambda}(\mathcal{A}))$. Then there exists $(\bar{I}, \bar{a}, \bar{q}) \in Y_{\Phi\lambda}(\mathcal{A})$, solution to $(D_{\Phi\lambda}(\mathcal{A}))$, such that the following optimality conditions are satisfied

(i) $\bar{I} \subseteq \{1, \ldots, m\}, \bar{I} \neq \emptyset, \left(\sum_{i=1}^{l} \lambda_i \bar{a}_j^i\right) > 0, j \in \bar{I}, \bar{a}_j^i = 0,\ j \notin \bar{I}, i = 1, \ldots, l,$

(ii) $\bar{a}^i \in \mathbb{R}_+^m,\ \Phi_i^0(\bar{a}^i) = 1,\ i = 1, \ldots, l,\ \sum_{j\in\bar{I}} \left(\sum_{i=1}^{l} \lambda_i \bar{a}_j^i\right) \bar{q}_j = 0,$

(iii) $\Phi_i(d(\bar{x})) = \langle \bar{a}^i, d(\bar{x}) \rangle,\ i = 1, \ldots, l,$

(iv) $\bar{x} \in \partial d_j^*(\bar{q}_j),\ j \in \bar{I}.$

(2) If $\bar{x} \in X$, $(\bar{I}, \bar{a}, \bar{q})$ is feasible to $(D_{\Phi\lambda}(\mathcal{A}))$ and (i)-(iv) are fulfilled, then $\bar{x}$ is a solution to $(P_{\Phi\lambda}(\mathcal{A}))$, $(\bar{I}, \bar{a}, \bar{q})$ is a solution to $(D_{\Phi\lambda}(\mathcal{A}))$ and strong duality holds, i.e.

$$\sum_{i=1}^{l} \lambda_i \Phi_i^+(d(\bar{x})) = -\sum_{j\in\bar{I}} \left(\sum_{i=1}^{l} \lambda_i \bar{a}_j^i\right) d_j^*(\bar{q}_j).$$

Proof.

(1) By Theorem 5 follows that there exists $(\bar{I}, \bar{a}, \bar{q}) \in Y_{\Phi\lambda}(\mathcal{A})$, solution to $(D_{\Phi\lambda}(\mathcal{A}))$, such that

$$\begin{aligned}
&(i') && \bar{I} \subseteq \{1,\dots,m\}, \bar{I} \neq \emptyset, \left(\textstyle\sum_{i=1}^{l} \lambda_i \bar{a}_j^i\right) > 0, j \in \bar{I}, \bar{a}_j^i = 0, j \notin \bar{I}, i = 1,\dots,l,\\
&(ii') && \bar{a}^i \in \mathbb{R}_+^m,\ \Phi_i^0(\bar{a}^i) \leq 1,\ i = 1,\dots,l,\ \textstyle\sum_{j\in\bar{I}} \left(\sum_{i=1}^{l} \lambda_i \bar{a}_j^i\right) \bar{q}_j = 0,\\
&(iii') && \Phi_i^+(d(\bar{x})) = \langle \bar{a}^i, d(\bar{x})\rangle,\ i = 1,\dots,l,\\
&(iv') && d_j(\bar{x}, A_j) + d_j^*(\bar{q}_j) = \langle \bar{q}_j, \bar{x}\rangle,\ j \in \bar{I}.
\end{aligned}$$

We will prove that $(\bar{I}, \bar{a}, \bar{q})$ verifies the relations $(i) - (iv)$. If $\bar{I}$ would be empty, then it would follow by (i') that $\bar{a}_j^i = 0, j = 1,\dots,m, i = 1,\dots,l$. From (iii') it holds then $\Phi_i((d(\bar{x}))^+) = \Phi_i^+(d(\bar{x})) = 0$ which actually means that $d(\bar{x}) = (d(\bar{x}))^+ = 0$, i.e.

$$g_j(\bar{x}) = d_j(\bar{x}, A_j) = 0,\ j = 1, ..., m.$$

But, this would imply that $\bar{x} \in \bigcap_{j=1}^{m} \bar{A}_j$. This is a contradiction to the hypothesis $\bigcap_{j=1}^{m} \bar{A}_j = \emptyset$. By this, the relation (i) is proved.

From (iii'), we have that

$$\Phi_i^+(d(\bar{x})) = \Phi_i(d(\bar{x})) = \langle \bar{a}^i, d(\bar{x})\rangle,\ i = 1,\dots,l,$$

and, so, (iii) is also proved. From (iv'), we have that $\bar{q}_j \in \partial d_j(\bar{x}, A_j)$ for $j \in \bar{I}$ (cf. [5]). On the other hand, d_j being a convex and continuous function, verifies (cf. [5])

$$\bar{q}_j \in \partial d_j(\bar{x}, A_j) \Leftrightarrow \bar{x} \in \partial d_j^*(\bar{q}_j),\ j \in \bar{I},$$

which proves (iv).

Now, it remains us to show that $\Phi_i^0(\bar{a}^i) = 1,\ i = 1, ..., l$. By the definition of the dual norm, we have

$$\Phi_i^0(\bar{a}^i) = \sup_{\substack{\Phi_i(v)\leq 1\\ v\in\mathbb{R}^m}} \{|\langle \bar{a}^i, v\rangle|\},\ i = 1, ..., l.$$

Because of $\bigcap_{j=1}^{m} \bar{A}_j = \emptyset$, it holds $\Phi_i(d(\bar{x})) > 0$, for $i = 1, ..., l$. Let be $\bar{v}_i = \frac{1}{\Phi_i(d(\bar{x}))} d(\bar{x}) \in \mathbb{R}^m$. We have $\Phi_i(\bar{v}_i) = 1,\ i = 1, ..., l$, and then, by (iii),

$$\Phi_i^0(\bar{a}^i) = \Phi_i(\bar{v}_i)\Phi_i^0(\bar{a}^i) \geq \langle \bar{a}^i, \bar{v}_i \rangle = \frac{\sum_{k=1}^{m} \bar{a}_k^i d_k(\bar{x}, A_k)}{\Phi_i(d(\bar{x}))} = \frac{\langle \bar{a}^i, d(\bar{x}) \rangle}{\Phi_i(d(\bar{x}))} = 1.$$

In conclusion, by (ii'), $\Phi_i^0(\bar{a}^i) = 1,\ i = 1, \ldots, l$. □

As a multiobjective dual problem of the primal problem $(P_\Phi(\mathcal{A}))$ we can introduce

$$(D_\Phi(\mathcal{A})) \qquad \underset{(I,a,q,\lambda,t) \in Y_\Phi(\mathcal{A})}{\text{v-max}} \begin{pmatrix} h_1^d(I,a,q,\lambda,t) \\ \vdots \\ h_l^d(I,a,q,\lambda,t) \end{pmatrix},$$

with

$$h_k^d(I,a,q,\lambda,t) = -\frac{1}{l\lambda_k} \sum_{j \in I} \left(\sum_{i=1}^{l} \lambda_i a_j^i \right) d_j^*(q_j) + t_k,\ k = 1, \ldots, l,$$

the dual variables

$$I \subseteq \{1, \ldots, m\},\ a = (a^1, \ldots, a^l),\ a^i \in \mathbb{R}^m,\ q = (q_1, \ldots, q_m),\ q_j \in X^*,$$
$$\lambda = (\lambda_1, \ldots, \lambda_l)^T \in \mathbb{R}^l,\ t = (t_1, \ldots, t_l)^T \in \mathbb{R}^l,$$

and the set of constraints

$$\begin{aligned} Y_\Phi(\mathcal{A}) = \Big\{ &(I,a,q,\lambda,t) : I \subseteq \{1, ..., m\},\ a^i \in \mathbb{R}_+^m,\ \Phi_i^0(a^i) \leq 1,\ i = 1, \ldots, l, \\ &q_j \in X^*,\ j = 1, \ldots, m,\ \left(\sum_{i=1}^{l} \lambda_i a_j^i \right) > 0,\ j \in I,\ a_j^i = 0,\ j \notin I, \\ &i = 1, \ldots, l, \sum_{j \in I} \left(\sum_{i=1}^{l} \lambda_i a_j^i \right) q_j = 0, \lambda \in int\mathbb{R}_+^l, \sum_{k=1}^{l} \lambda_k t_k = 0 \Big\}. \end{aligned}$$

The following theorems state the weak and strong duality assertions applying Theorem 6 and Theorem 7.

Theorem 10. *There is no $x \in X$ and no $(I,a,q,\lambda,t) \in Y_\Phi(\mathcal{A})$, such that $\Phi_i(d(x)) \leq h_i^d(I,a,q,\lambda,t)$, $i = 1, \ldots, l$, and $\Phi_k(d(x)) < h_k^d(I,a,q,\lambda,t)$ for at least one $k \in \{1, \ldots, l\}$.*

Theorem 11. *Let $\bar{x}$ be a properly efficient element to $(P_\Phi(\mathcal{A}))$. Then there exists an efficient solution $(\bar{I}, \bar{a}, \bar{q}, \bar{\lambda}, \bar{t}) \in Y_\Phi(\mathcal{A})$ to $(D_\Phi(\mathcal{A}))$ and strong duality*

$$\Phi_k(d(\bar{x})) = -\frac{1}{l\bar{\lambda}_k} \sum_{j \in \bar{I}} \left(\sum_{i=1}^{l} \bar{\lambda}_i \bar{a}_j^i \right) d_j^*(\bar{q}_j) + \bar{t}_k,\ k = 1, ..., l,$$

holds.

6 The biobjective Weber-minimax problem with infimal distances

In this section, for the same data set $\mathcal{A} = \{A_1, \dots, A_m\}$ as in the previous one, we consider a multiobjective minimization problem with a two-dimensional objective function, its first component being given by the Weber location problem and the second one by the minimax location problem with infimal distances. Thus, the primal problem is

$$(P_{WM}(\mathcal{A})) \quad \underset{x \in X}{\text{v-min}} \begin{pmatrix} \sum\limits_{j=1}^{m} w_j d_j(x, A_j) \\ \max\limits_{j=1,\dots,m} w_j d_j(x, A_j) \end{pmatrix},$$

where $d_j(x, A_j) = \inf\limits_{y_j \in A_j} \gamma_j(x - y_j)$, $j = 1, ..., m$, and $w_j > 0$, $j = 1, \dots, m$, are positive weights. Let be, for $j = 1, \dots, m$, the continuous norms $\gamma'_j : X \to \mathbb{R}$, $\gamma'_j = w_j \gamma_j$ and the corresponding distance functions $d'_j(\cdot, A_j) : X \to \mathbb{R}$, $d'_j(x, A_j) = \inf\limits_{y_j \in A_j} \gamma'_j(x - y_j) = w_j d_j(x, A_j)$. This means that the primal problem $(P_{WM}(\mathcal{A}))$, as a special case of $(P_\Phi(\mathcal{A}))$ in section 5 with $\Phi_1 = l_1$ and $\Phi_2 = l_2$, becomes

$$(P_{WM}(\mathcal{A})) \quad \underset{x \in X}{\text{v-min}} \begin{pmatrix} l_1(d'(x)) \\ l_\infty(d'(x)) \end{pmatrix},$$

with $d'(x) = (d'_1(x, A_1), ..., d'_m(x, A_m))$ and the norms $l_1, l_\infty : \mathbb{R}^m \to \mathbb{R}$, $l_1(z) = \sum\limits_{j=1}^{m} |z_j|$, $l_\infty(z) = \max\limits_{j=1,\dots,m} |z_j|$, for $z \in \mathbb{R}^m$. We remark that $l_1^0(z) = l_\infty(z)$ and $l_\infty^0(z) = l_1(z)$. Obviously, l_1 and l_∞ are monotonic norms.

Taking into consideration the form of $D_\Phi(\mathcal{A})$ in section 5, observing $d'^*_j(q_j) = (w_j d_j)^*(q_j) = w_j d_j^*(\frac{1}{w_j} q_j)$, and, denoting by $q_j := \frac{1}{w_j} q_j$, we construct the multiobjective dual to the primal problem $(P_{WM}(\mathcal{A}))$. This becomes

$$(D_{WM}(\mathcal{A})) \quad \underset{(I,a,q,\lambda,t) \in Y_{WM}(\mathcal{A})}{\text{v-max}} \begin{pmatrix} h_1(I, a, q, \lambda, t) \\ h_2(I, a, q, \lambda, t) \end{pmatrix},$$

with

$$\begin{aligned} h_1(I, a, q, \lambda, t) &= -\tfrac{1}{2\lambda_1} \sum_{j \in I} \left(\sum_{i=1}^{2} \lambda_i a_j^i \right) w_j d_j^*(q_j) + t_1, \\ h_2(I, a, q, \lambda, t) &= -\tfrac{1}{2\lambda_2} \sum_{j \in I} \left(\sum_{i=1}^{2} \lambda_i a_j^i \right) w_j d_j^*(q_j) + t_2, \end{aligned}$$

the dual variables

$$I \subseteq \{1, \dots, m\},\ a = (a^1, a^2),\ a^1,\ a^2 \in \mathbb{R}^m,\ q = (q_1, \dots, q_m),\ q_j \in X^*,$$

$$\lambda = (\lambda_1, \lambda_2)^T \in \mathbb{R}^2,\ t = (t_1, t_2)^T \in \mathbb{R}^2,$$

and the set of constraints

$$Y_{WM}(\mathcal{A})=\Big\{(I,a,q,\lambda,t): I\subseteq\{1,...,m\}, a^1,a^2\in\mathbb{R}^m_+, q_j\in X^*, j=1,\dots,m,$$
$$\max_{j=1,\dots,m}|a^1_j|\le 1, \sum_{i=1}^{m}|a^2_j|\le 1, \left(\sum_{i=1}^{2}\lambda_i a^i_j\right)>0, j\in I, a^i_j=0, j\notin I, i=1,2,$$
$$\sum_{j\in I}\left(\sum_{i=1}^{2}\lambda_i a^i_j\right)w_j q_j=0, \lambda\in int\mathbb{R}^2_+, \sum_{k=1}^{2}\lambda_k t_k=0\Big\},$$

Let us give also for this problem the weak and strong duality theorems.

Theorem 12. *There is no $x\in X$ and no $(I,a,q,\lambda,t)\in Y_{WM}(\mathcal{A})$ such that*

$$\sum_{j=1}^{m}w_j d_j(x,A_j)\le h_1(I,a,q,\lambda,t), \max_{j=1,\dots,m}w_j d_j(x,A_j)\le h_2(I,a,q,\lambda,t)$$

and

$$\sum_{j=1}^{m}w_j d_j(x,A_j)< h_1(I,a,q,\lambda,t) \text{ or } \max_{j=1,\dots,m}w_j d_j(x,A_j)<h_2(I,a,q,\lambda,t).$$

Theorem 13. *Let $\bar{x}$ be a properly efficient element to $(P_{WM}(\mathcal{A}))$. Then there exists an efficient solution $(\bar{I},\bar{a},\bar{q},\bar{\lambda},\bar{t})\in Y_{WM}(\mathcal{A})$ to $(D_{WM}(\mathcal{A}))$ and the strong duality holds, i.e.*

$$\sum_{j=1}^{m}w_j d_j(\bar{x},A_j)=-\frac{1}{2\bar{\lambda}_1}\sum_{j\in\bar{I}}\left(\sum_{i=1}^{2}\bar{\lambda}_i\bar{a}^i_j\right)w_j d^*_j(\bar{q}_j)+\bar{t}_1$$

and

$$\max_{j=1,\dots,m}w_j d_j(\bar{x},A_j)=-\frac{1}{2\bar{\lambda}_2}\sum_{j\in\bar{I}}\left(\sum_{i=1}^{2}\bar{\lambda}_i\bar{a}^i_j\right)w_j d^*_j(\bar{q}_j)+\bar{t}_2.$$

7 The multiobjective Weber problem with infimal distances

We consider, as another application of the multiobjective duality results in section 5, the multiobjective Weber problem with infimal distances for the data $\mathcal{A}$

$$(P_W(\mathcal{A}))\quad \underset{x\in X}{\text{v-min}}\begin{pmatrix}\sum_{j=1}^{m}w^1_j d_j(x,A_j)\\ \vdots\\ \sum_{j=1}^{m}w^l_j d_j(x,A_j)\end{pmatrix},$$

where $d_j(x, A_j) = \inf\limits_{y_j \in A_j} \gamma_j(x - y_j),\ j = 1, ..., m,$ and $w_j^i,\ j = 1, \ldots, m,\ i = 1, ..., l,$ are positive weights. Again, the norms $\gamma_j,\ j = 1, \ldots, m$, are assumed to be continuous. Considering the norms $\Phi_i^W : \mathbb{R}^m \to \mathbb{R},\ i = 1, ..., l$, defined by

$$\Phi_i^W(x) := \sum_{j=1}^{m} w_j^i |x_j|,$$

we have

$$\Phi_i^W(d(x)) = \sum_{j=1}^{m} w_j^i d_j(x, A_j).$$

We notice that $\Phi_i^W, i = 1, ..., l$, are monotonic norms, with the dual norm $(\Phi_i^W)^0(x) = \max\limits_{j=1,...,m} \frac{|x_j|}{w_j^i}$.
So, the primal problem $(P_W(\mathcal{A}))$ becomes

$$(P_W(\mathcal{A})) \quad \underset{x \in X}{\text{v-min}} \begin{pmatrix} \Phi_1^W(d(x)) \\ \vdots \\ \Phi_l^W(d(x)) \end{pmatrix}.$$

Due to section 5, a multiobjective dual problem to $(P_W(\mathcal{A}))$ is

$$(DP_W(\mathcal{A})) \quad \underset{(I,a,q,\lambda,t) \in Y_W(\mathcal{A})}{\text{v-max}} \begin{pmatrix} h_1^W(I, a, q, \lambda, t) \\ \vdots \\ h_l^W(I, a, q, \lambda, t) \end{pmatrix},$$

with

$$h_k^W(I, a, q, \lambda, t) = -\frac{1}{l\lambda_k} \sum_{j \in I} \left(\sum_{i=1}^{l} \lambda_i a_j^i \right) d_j^*(q_j) + t_k,\ k = 1, ..., l,$$

the dual variables

$$I \subseteq \{1, \ldots, m\},\ a = (a^1, ..., a^l),\ a^i \in \mathbb{R}^m,\ q = (q_1, ..., q_m),\ q_j \in X^*,$$

$$\lambda = (\lambda_1, ..., \lambda_l)^T \in \mathbb{R}^l,\ t = (t_1, ..., t_l)^T \in \mathbb{R}^l,$$

and the set of constraints

$$\begin{aligned} Y_W(\mathcal{A}) = \Big\{ & (I, a, q, \lambda, t) : I \subseteq \{1, ..., m\},\ a^i \in \mathbb{R}_+^m,\ q_j \in X^*, ,j = 1, \ldots, m, \\ & \lambda \in int\mathbb{R}_+^l,\ \max_{j=1,\ldots,m} \frac{a_j^i}{w_j^i} \le 1,\ \left(\sum_{i=1}^{l} \lambda_i a_j^i \right) > 0, j \in I, a_j^i = 0, j \notin I, i = 1, \ldots, l, \\ & \sum_{j \in I} \left(\sum_{i=1}^{l} \lambda_i a_j^i \right) q_j = 0,\ \sum_{k=1}^{l} \lambda_k t_k = 0 \Big\}, \end{aligned}$$

which can be written equivalently as (setting $a^i_j := \frac{a^i_j}{w^i_j}$)

$$(DP_W(\mathcal{A})) \qquad \underset{(I,a,q,\lambda,t)\in Y_W(\mathcal{A})}{\text{v-max}} \begin{pmatrix} h^W_1(I,a,q,\lambda,t) \\ \vdots \\ h^W_l(I,a,q,\lambda,t) \end{pmatrix},$$

with

$$h^W_k(I,a,q,\lambda,t) = -\frac{1}{l\lambda_k}\sum_{j\in I}\left(\sum_{i=1}^{l}\lambda_i w^i_j a^i_j\right) d^*_j(q_j) + t_k,\ k = 1,...,l,$$

the dual variables

$$I \subseteq \{1,\dots,m\},\ a = (a^1,...,a^l),\ a^i \in \mathbb{R}^m,\ q = (q_1,...,q_m),\ q_j \in X^*,$$

$$\lambda = (\lambda_1,...,\lambda_l)^T \in \mathbb{R}^l,\ t = (t_1,...,t_l)^T \in \mathbb{R}^l,$$

and the set of constraints

$$\begin{aligned} Y_W(\mathcal{A}) = \Big\{ & (I,a,q,\lambda,t) : I \subseteq \{1,...,m\}, a^i \in \mathbb{R}^m_+,\ q_j \in X^*,\ j = 1,\dots,m, \\ & \lambda \in int\mathbb{R}^l_+,\ \max_{j=1,\dots,m} a^i_j \leq 1,\ \left(\sum_{i=1}^{l}\lambda_i w^i_j a^i_j\right) > 0, j \in I, a^i_j = 0, j \notin I, i = 1,\dots,l, \\ & \sum_{j\in I}\left(\sum_{i=1}^{l}\lambda_i w^i_j a^i_j\right) q_j = 0, \sum_{k=1}^{l}\lambda_k t_k = 0\Big\}. \end{aligned}$$

Using the Theorems 10 and 11 we can formulate the following duality results.

Theorem 14. *There is no $x \in X$ and no (I,a,q,λ,t) in $Y_W(\mathcal{A})$ such that $\sum_{j=1}^{m} w^i_j d_j(x,A_j) \leq h^W_i(I,a,q,\lambda,t)$, $i = 1,...,l$, and $\sum_{j=1}^{m} w^k_j d_j(x,A_j) < h^W_k(I,a,q,\lambda,t)$ for at least one $k \in \{1,...,l\}$.*

Theorem 15. *Let $\bar{x}$ be a properly efficient element to $(P_W(\mathcal{A}))$. Then there exists an efficient solution $(\bar{I},\bar{a},\bar{q},\bar{\lambda},\bar{t}) \in Y_W(\mathcal{A})$ to $(DP_W(\mathcal{A}))$ and strong duality, i.e.*

$$\sum_{j=1}^{m} w^k_j d_j(\bar{x},A_j) = -\frac{1}{l\bar{\lambda}_k}\sum_{j\in\bar{I}}\left(\sum_{i=1}^{l}\bar{\lambda}_i w^i_j \bar{a}^i_j\right) d^*_j(\bar{q}_j) + \bar{t}_k,\ k = 1,...,l,$$

holds.

8 The multiobjective minimax location problem with infimal distances

The last optimization problem we are going to consider in this paper is the multiobjective minimax location problem with infimal distances for the data $\mathcal{A}$

$$(P_M(\mathcal{A})) \quad \text{v-}\min_{x\in X} \begin{pmatrix} \max\limits_{j=1,\dots,m} w_j^1 d_j(x,A_j) \\ \vdots \\ \max\limits_{j=1,\dots,m} w_j^l d_j(x,A_j) \end{pmatrix},$$

where $d_j(x,A_j) = \inf\limits_{y_j\in A_j} \gamma_j(x-y_j),\ j=1,...,m$, and w_j^i, $j=1,\dots,m$, $i=1,...,l$, are positive weights. Considering the norms $\Phi_i^M : \mathbb{R}^m \to \mathbb{R},\ i=1,...,l$, defined by

$$\Phi_i^M(x) = \max_{j=1,\dots,m} w_j^i |x_j|,$$

we have that

$$\Phi_i^M(d(x)) = \max_{j=1,\dots,m} w_j^i d_j(x,A_j).$$

We notice that $\Phi_i^M, i = 1,...,l$, are monotonic norms, with the dual norm $(\Phi_i^M)^0(x) = \sum\limits_{j=1}^m \frac{|x_j|}{w_j^i}$.

Thus, the primal problem $(P_M(\mathcal{A}))$ becomes

$$(P_M(\mathcal{A})) \quad \text{v-}\min_{x\in X} \begin{pmatrix} \Phi_1^M(d(x)) \\ \vdots \\ \Phi_l^M(d(x)) \end{pmatrix}.$$

Its multiobjective dual problem is (cf. section 5)

$$(DP_M(\mathcal{A})) \quad \text{v-}\max_{(I,a,q,\lambda,t)\in Y_M(\mathcal{A})} \begin{pmatrix} h_1^M(I,a,q,\lambda,t) \\ \vdots \\ h_l^M(I,a,q,\lambda,t) \end{pmatrix},$$

with

$$h_k^M(I,a,q,\lambda,t) = -\frac{1}{l\lambda_k}\sum_{j\in I}\left(\sum_{i=1}^{l}\lambda_i a_j^i\right) d_j^*(q_j) + t_k,\ k=1,...,l,$$

the dual variables

$$I \subseteq \{1,\dots,m\},\ a=(a^1,...,a^l),\ a^i\in\mathbb{R}^m,\ q=(q_1,...,q_m),\ q_j\in X^*,$$

$$\lambda = (\lambda_1,...,\lambda_l)^T \in \mathbb{R}^l,\ t=(t_1,...,t_l)^T\in\mathbb{R}^l,$$

and the set of constraints

$$Y_M(\mathcal{A}) = \Big\{ (I,a,q,\lambda,t) : I \subseteq \{1,...,m\}, a^i \in \mathbb{R}^m_+,\ q_j \in X^*,\ j = 1,\dots,m,$$
$$\sum_{j=1}^m \frac{a_j^i}{w_j^i} \le 1, \left(\sum_{i=1}^l \lambda_i a_j^i\right) > 0, j \in I,\ a_j^i = 0,\ j \notin I,\ i = 1,\dots,l,$$
$$\sum_{j\in I} (\sum_{i=1}^l \lambda_i a_j^i) q_j = 0, \lambda \in int\mathbb{R}^l_+, \sum_{k=1}^l \lambda_k t_k = 0 \Big\},$$

which can be written equivalently as (setting $a_j^i := \frac{a_j^i}{w_j^i}$)

$$(DP_M(\mathcal{A})) \qquad \underset{(I,a,q,\lambda,t)\in Y_M(\mathcal{A})}{\text{v-max}} \begin{pmatrix} h_1^M(I,a,q,\lambda,t) \\ \vdots \\ h_l^M(I,a,q,\lambda,t) \end{pmatrix},$$

with

$$h_k^M(I,a,q,\lambda,t) = -\frac{1}{l\lambda_k} \sum_{j\in I} \left(\sum_{i=1}^l \lambda_i w_j^i a_j^i \right) d_j^*(q_j) + t_k,\ k = 1,...,l,$$

the dual variables

$$I \subseteq \{1,\dots,m\},\ a = (a^1,...,a^l),\ a^i \in \mathbb{R}^m,\ q = (q_1,...,q_m),\ q_j \in X^*,$$
$$\lambda = (\lambda_1,...,\lambda_l)^T, \in \mathbb{R}^l,\ t = (t_1,...,t_l)^T \in \mathbb{R}^l,$$

and the set of constraints

$$Y_M(\mathcal{A}) = \Big\{ (I,a,q,\lambda,t) : I \subseteq \{1,...,m\}, a^i \in \mathbb{R}^m_+,\ q_j \in X^*,\ j = 1,\dots,m,$$
$$\sum_{j=1}^m a_j^i \le 1, \left(\sum_{i=1}^l \lambda_i w_j^i a_j^i\right) > 0, j \in I,\ a_j^i = 0,\ j \notin I,\ i = 1,\dots,l,$$
$$\sum_{j\in I} \left(\sum_{i=1}^l \lambda_i w_j^i a_j^i\right) q_j = 0,\ \lambda \in int\mathbb{R}^l_+, \sum_{k=1}^l \lambda_k t_k = 0 \Big\}.$$

Remark 4. We emphasize the interesting observation that both dual problems $(DP_W(\mathcal{A}))$ and $(DP_M(\mathcal{A}))$ differ only in the constraints $\max\limits_{i=1,\dots,m} a_j^i \le 1$ and $\sum\limits_{i=1}^m a_j^i \le 1$, respectively.

The corresponding duality results for $(DP_M(\mathcal{A}))$ are the following.

Theorem 16. *There is no $x \in X$ and no $(I,a,q,\lambda,t) \in Y_M(\mathcal{A})$ such that $\max\limits_{j=1,...,m} w_j^i d_j(x,A_j) \le h_i^M(I,a,q,\lambda,t)$, $i = 1,...,l$, and $\max\limits_{j=1,...,m} w_j^k d_j(x,A_j) < h_k^M(I,a,q,\lambda,t)$ for at least one $k \in \{1,...,l\}$.*

Theorem 17. *Let $\bar{x}$ be a properly efficient element to $(P_M(\mathcal{A}))$. Then there exists an efficient solution $(\bar{I}, \bar{a}, \bar{q}, \bar{\lambda}, \bar{t}) \in Y_M(\mathcal{A})$ to $(DP_M(\mathcal{A}))$ and strong duality, i.e.*

$$\max_{j=1,\dots,m} w_j^k d_j(\bar{x}, A_j) = -\frac{1}{l\bar{\lambda}_k} \sum_{j \in \bar{I}} \left(\sum_{i=1}^{l} \bar{\lambda}_i w_j^i \bar{a}_j^i \right) d_j^*(\bar{q}_j) + \bar{t}_k, k = 1, \dots, l,$$

holds.

References

[1] Bauer, F.L., Stoer, J., Witzgall, C. (1961), "Absolute and monotonic norms," *Numerische Mathematik*, 3, 257-264.

[2] Boţ, R.I., Wanka, G. (2002), "Duality for composed convex functions with applications in location theory," *(to appear)*.

[3] Breckner, W.W., Kolumbán, I. (1968), "Dualität bei Optimierungsaufgaben in topologischen Vektorräumen," *Mathematica*, 10, 229-244.

[4] Carrizosa, E., Fliege, J. (1999), "Generalized goal programming: Polynomial methods and applications," *(submitted to Mathematical Programming)*.

[5] Ekeland, I., Temam, R. (1976), *Convex analysis and variational problems*, North-Holland Publishing Company, Amsterdam.

[6] Göpfert, A., Nehse, R. (1990), *Vektoroptimierung, Theorie, Verfahren und Anwendungen*, Teubner, Leipzig.

[7] Iserman, H. (1978), "On some relations between a dual pair of multiple objective linear programs," *ZOR-Methods and Models of Operations Research*, 22, 33-41.

[8] Jahn, J. (1986), *Mathematical vector optimization in partially ordered linear spaces*, Verlag Peter Lang, Frankfurt am Main.

[9] Lemaire, B. (1984), "Subdifferential of a convex composite functional: Application to optimal control in variational inequalities," *Nondifferentiable Optimization*, Sopron, Hungary.

[10] Levin, V.L. (1970), "Sur le sous-différentiels de fonctions composées," *Doklady Akademia Nauk, Belorussian Academy of Sciences*, 194, 28-29.

[11] Nakayama, H. (1984), "Geometric consideration of duality in vector optimization," *Journal of Optimization Theory and Applications*, 44(4), 625-655.

[12] Nakayama, H. (1996), "Some remarks on dualization in vector optimization," *Journal of Multi-Criteria Decision Analysis*, 5, 218-255.

[13] Nickel, S., Puerto, J., Rodriguez-Chia, M. (2001), "An approach to location models involving sets as existing facilities," *(to appear)*.

[14] Puerto, J. (2002), "Average distance problems with respect to infinite sets," *(to appear)*.

[15] Rockafellar, R.T. (1970), *Convex Analysis*, Princton University Press, Princeton.

[16] Sawaragi, Y., Nakayama, H., Tanino, T. (1985), *Theory of multiobjective optimization*, Academic Press, New York.

[17] Tanino, T. (1992), "Conjugate duality in vector optimization," *Journal of Mathematical Analysis and Applications*, 167, 84-97.

[18] Wanka, G. (1991), "Duality in vectorial control approximation problems with inequality restrictions," *Optimization*, 22, 755-764.

[19] Wanka, G. (1999), "Multiobjective duality for the Markowitz portfolio optimization problem," *Control and Cybernetics*, 28(4), 691-702.

[20] Wanka, G. (2000), "Multiobjective control approximation problems: duality and optimality," *Journal of Optimization Theory and Applications*, 105, 457-475.

[21] Wanka, G., Göhler, L. (2001), "Duality for portfolio optimization with short sales," *Mathematical Methods of Operations Research*, 53, 247-263.

[22] Wanka, G., Boţ, R.I. (2002), "A new duality approach for multiobjective convex optimization problems," *Journal of Nonlinear and Convex Analysis*, 3 (1), 41-57.

[23] Wanka, G., Boţ, R.I. (2002), "On the relations between different dual problems in convex mathematical programming," *In Chamoni, P., Leisten, R., Martin, A., Minnemann, J., Stadtler, H., (eds.), Operation Research Proceedings 2001*, Springer, Heidelberg, 255-262.

[24] Weir, T. (1987), "Proper efficiency and duality for vector valued optimization problems," *Journal of Australian Mathematical Society*, 43, 21-34.

[25] Weir, T., Mond, B. (1991), "Multiple objective programming duality without a constraint qualification," *Utilitas Mathematica*, 39, 41-55.

Optimization and Optimal Control, pp. 335 – 345
P.M. Pardalos, I. Tseveendorj, and R. Enkhbat, Editors

The Points of A Linear Manifold Nearest to a Given Vector[1]

Zorkaltsev V.I. (zork@isem.sei.irk.ru)
Energy Systems Institute, SB RAS,
13, Lermontov street,
Irkutsk, Russia

Abstract

The properties of the solution sets of the classical problem of determining the point of a linear manifold nearest the origin of coordinates are analysed in three formulations: maximization of the set of zero components, Pareto-minimization of the absolute values of the components of the unknown vector, and minimization of special penalty functions with an absolute minimum at the origin of coordinates. It is proved that the solution sets are bounded and that in all three formulations solutions can be obtained to any required accuracy by the method of least squares by varying of the weights, together with other properties. Applications and generalizations of the results are discussed.

Keywords: linear manifold, the method of least squares.

Introduction and statement of a problem

Many applied problems are represented as a geometric problem of searching for a point from set $\tilde{Y} \subset R^n$ maximum close to a given vector $x \in R^n$ a point from set $\tilde{Y} \subset R^n$

$$\tilde{y} = \arg\min_{y \in Y} \rho(\ x, y\ )$$

There can be different ways of the representation

1. Set Y. We consider cases when Y is a linear manifold.

[1]The work is supported by RFBR grant N 00-01-00530

2. Measure of proximity $\rho(x, y)$. Three measures will be considered. In all cases

$$\rho(x, y) = \rho(\dot{0}, y - x).$$

Therefore the problem is represented as searching for the point, the nearest to the origin of coordinates

$$\tilde{y} = \arg\min_{y \in Y} \rho(0, y) = \tilde{y} - x$$

from the given set

$$Y = \tilde{Y} - x.$$

The problems which can be reduced to the geometric problem of searching for the points of a set, least removed from the origin of coordinates.

1. State identification of systems (including power energy and pipeline system) based on inexact, surplus observations.

2. Determination of feasible solutions which are maximum close to the given infeasible vector, least norm solutions with purposes of minimization of linearization errors.

3. Coordination of independent prognosises and solutions on separate parameters of a system.

4. Approximation, construction of the regression equations, detection of time series components.

5. Determination of pseudo-solutions of models with inconsistent conditions.

6. Model with minimization of production deficiency.

The given article develops the investigations published in the authors papers [1-5]. These results were generalized in the book [6].

1 Initial definitions

The initial object here is a non-empty m-dimensional linear manifold Y in n-dimensional real space R^n. For vector $y \in R^n$ sets

$$J_0(y) = \{j : \ y_j = 0\}, \ J_+(y) = \{j : \ y_j > 0\},$$

$$J_-(y) = \{j : \ y_j < 0\}, \ J\ (y) = \{j : \ y_j \neq 0\}$$

denote the sets of indices of its components with zero, positive,negative and non-zero values. The set $J(y)$ will be called the carrier of the vector y. This paper presents some conceptually different problems (approximation, search for the pseudo-solutions to models with contradictory conditions, determination of the feasible solution close to the given unfeasible vector), which are reduced in the simplest cases to the general geometrical problem - determination of the nearest to the origin point of the linear manifold.

We consider the following formulations of the problem of determining the point of Y which is nearest the origin of coordinates.

1. *Minimization of the set of zero components of the required vector* Following Rockafellar [7], vectors of the set Y with a non-expandable set of zero components will be called the support vectors of that manifold. The set of support vectors, each of which is a solution of the problem, is denoted by

$$B = \{b \in Y : \neg \exists y \in Y, y \neq b, J_0(b) \subset J_0(y)\}.$$

This statement for the approximation problems reflects a naive, but to some extent reasonable approach. In the linear approximation the components of vector from Y are equal to the discrepancies between the observed values of the resulting indicators and its representation in the form of a linear function of the values of indicators-factors.

This statement has a somewhat more valid base in the models of interests coordination. Let the values y_j, $j = \overline{1,n}$, characterize discrepancies between the desired and factual states of the participants (sharing of limited wealths, whose deficit and excess cause discomfort). Determination of the vector with the maximum support will mean formation of the maximum wide coalition of all participants which completely achieve their goals.

2. *To find the Pareto-optimum solution of the multicriterion problem*

$$|y_j| \to \min_{y \in Y}, \quad j = 1, 2, ..., n.$$

In other words, it is required to obtain a vector of the set

$$Q = y \in Y : \nexists \tilde{y} \in Y, \tilde{y} \neq y, |\tilde{y}_j| \leq |y_j|, \; j = 1, 2, ..., n \tag{1.1}$$

of vectors of the linear manifold with Pareto-minimal absolute values of the components.

This statement is an extension of the notion of optimality for the multi-objective problems, which was introduced by the well-known economist Pareto. According to Pareto the solution which cannot be improved with respect to any applied parameter without deterioration with respect to others, is optimal. They are called the Pareto-optimal solutions.

3. *Minimization the penalty function*

$$f(y) \to \min_{y \in Y} \tag{1.2}$$

The function f can have a conceptual interpretation. For example, in the economic models, in which the components of the vector y are equal to imbalances for individual types of resources, the function f can measure the cost of losses due to the product deficit or excess.

We will confine ourselves in this paper to considering strict psevdoconvex (hence differential) penalty functions which, for any $y \in R^n$, satisfy the condition

$$\text{sign}\nabla_j f(y) = \text{sign } y_j, \ j = 1, 2, ..., n. \tag{1.3}$$

This relation shows that the value of the penalty function increases when any component of the vector of variables increases if that value is positive, and when the value of the component decreases, if it is negative.

We will denote the set of strict pseudoconvex functions satisfying condition (1.3) by F.

Problem (1.2) has a unique solution for $f \in F$. Let this solution be denoted by

$$y(f) = \arg\min_{y \in Y} f(y).$$

Let also

$$PF = \{y(f) : f \in F\}$$

denote the set of solutions of problem (1.2) for different penalty functions from F.

The Hölder norms comprising the class of functions

$$\varphi_p^h\,(y) = [\,\sum_{j=i}^{n} h_j\,(\,|y_j|\,)^p\,]^{1/p},$$

where $p > 1$ is the power factor, $h \in R^n_\oplus$ is the vector of weighting factors, belong to the set F. Here $R^n_\oplus$ is the set of vectors of R^n with positive components.

The vector $y(\varphi_p^h)$ will be called the Hölder projection of coordinates on to the linear manifold Y. We will introduce sets of projection for fixed p and for all possible power factors

$$P_p = \{y\,(\varphi_p^h) : \ h \in R^n_\oplus\}, \ \ p > 1, \ \ P = \bigcup_{p>1} P_p\,.$$

Notice that when $p = 2$ the function φ_p^h is the Euclidean norm. The use of a penalty function of this kind means that the problem can be solved by the method of least squares. The set P_2 will consist of the Euclidean projections of the origin of coordinates on to the linear manifold Y.

The following two methods of defining a linear manifold are often used in specific problems which reduce to the problem considered here.

Method 1. The set of solutions of the system of linear equations

$$Y = \{y \in R^n : \ Ay = b\}, \tag{1.4}$$

where A is a $r \times n$ matrix for some integer r, and b is a vector of R^n.

Method 2. The translation by a given vector of the linear subspace defined as the set of linear combinations of some vector set:

$$Y = \{y = y_0 - G\lambda : \ \lambda \in R^m\}, \tag{1.5}$$

where G is an $n \times m$ matrix consisting of column vectors g^i, $i = 1, 2, ..., m$.

From now on conditions (1.4) and (1.5) will be assumed to define the same manifold, and the vectors g^i will be assumed to be linearly independent. Hence

$$\text{rank } A = n - m \tag{1.6}$$

A subspace parallel to the manifold Y will be denoted by

$$Z = \{z = G\lambda : \ \lambda \in R^m\}.$$

Essentially different problems that can be reduced to this geometric problem include: estimating the parameters of linear regression equations, looking for pseudosolutions in models with incompatible conditions, finding admissible solutions which are the nearest to given inadmissible solutions, etc.

These problems are often formulated in a more complicated way: with inequality constraints or non-linear dependences applied to the solution sets of the optimization models. In practice, models given in the form of linear equations are frequently used and so the study of the geometric problem considered here is of interest itself. At the time, it can also be regarded as a necessary basis for the analysis of more difficult cases.

The first attempts of a strict formulation of the problem, as we know, were made in connection with the processing of astronomical observations. According to G.V. *Bagratun'*, writing in the preface to [8], the first clear formulation of the problem as a minimization of the sum of moduli of the required vector was given by the Italian mathematician and astronomer Boscovich. Laplace and Gauss and, of course, other mathematicians at the beginning of the 19th century, knew that the vectors of a linear manifold with the maximum set of zero components can be worked out in a formulation of that kind.

Formulations of the second type had been employed intuitively even earlier, although they were strictly only feasible after the Italian mathematician and economist Pareto had introduced the optimality criterion that bears his name for problems with several objective functions, in 1906.

The method of least squares was discovered by Gauss in 1794 and, independently, by Lagrange a few years later. It is worth noting that Gauss gave special importance to an "intuitive" proof of the method of least squares, as a special case of the minimization of penalty functions, especially in his second work on the method [8]. Possible penalty functions that Gauss considered included what would now be called Hölder norms with arbitrary even coefficients. He commented that the second degree is to be preferred from a computational point of view.

The solutions obtained by the method of least squares depend on the choice of weighting factors for the squares of the deviations. There is often some arbitrariness in the designation of these factors. It is useful to know what will happen if the numerical values of the weighting factors are varied, how widely the solutions can vary as a result, and what type of structure the sets of solutions calculated by the method of least squares and those obtained in other formulations have.

2 The Properties of solution sets

We will now give two auxiliary lemmas which follow from property (1.3) of penalty functions and definition (1.1) of the set R.

Lemma 1. Let $f \in F$. The vector $y \in Y$ will be the same as the vector $y(f)$ if, and only if, for any $z \in Z$

$$\sum_{j=1}^{n} \nabla_j f(y)\, z_j = 0. \tag{2.1}$$

Lemma 2. The vector $y \in Y$ will not be in set Q if,and only if, there is a vector $z \in Z$ such that

$$J(z) \neq \oslash, \;\; J_+(z) \subset J_+(y), \;\; J_-(z) \subset J_-(y). \tag{2.2}$$

The following theorems can be proved with these lemmas.

Theorem 1. The sets *PF*, it P and P_p are equal for any $p > 1$.

Theorem 2. The points of a minimum on the set Y of differentiable penalty functions satisfying condition (1.3) are vectors with Pareto-minimal absolute values of the components:

$$PF \subset Q.$$

Theorem 3. The support vectors of a linear manifold have Pareto-minimal absolute values of the components:

$$B \subset Q.$$

Theorem 4. All vectors with Pareto-minimal absolute values of the components belong to the convex hull of the set of support vectors:

$$Q \subset \text{co}B.$$

Theorem 5. The set B contains a finite number of vectors, which is not larger than

$$C_n^m = \frac{n!}{m!(n-m)!}.$$

Proof of Theorem 1. Since the Hölder norms with fixed power coefficient form the subset of all Hölder norms which, in their turn, belong to the set of functions F, then for any $p > 1$

$$P_p \subset P \subset PF.$$

We must prove the relation

$$PF \subset P_p. \tag{2.3}$$

Let $f \in F$, if $y_j(f) \neq 0$, put

$$h_j = \mid \nabla_j f(y(f)) \mid / \mid y_j(f) \mid^{p-1}.$$

If $y_j(f) = 0$, then h_j can have any positive value, $h_j = 1$ say.

It follows from the definition of the vector h that

$$\nabla \varphi_p^h(y(f)) = [\varphi_p^h(y(f))]^{1-p} \nabla f(y(f)). \tag{2.4}$$

Since relation (2.1) is satisfied for the function f at the point $y(f)$, according to (2.4) this condition will also hold at that point for the function φ_p^h. Hence, by Lemma 1,

$$y(\varphi_p^h) = \varphi(f),$$

from which the relation (2.3) is obtained.

Proof of Theorem 2. Suppose that $y(f) \notin Q$ for some function $f \in F$. From Lemma 2 and condition (1.5), for some $z \in Z$, $z \neq 0$

$$J_+(z) \subset J_+(\nabla f(y(f))), \quad J_-(z) \subset J_-(\nabla f(y(f))).$$

Hence, for $y = y(f)$ Eq.(2.1) will not hold for the given vector z. This contradicts Lemma 1. Hence the assumption is false.

Proof of Theorem 3. Let $y \in B$. We will assume that $y \notin Q$. Then, by Lemma 2, there is a vector $z \in Z$ which possesses properties (2.2). It follows from (2.2) that for any real λ

$$J(y - \lambda z) \subset J(y). \tag{2.5}$$

For $j \in J(z)$, put

$$\lambda = y_j / z_j.$$

This index j will not be in the set $J(y - \lambda z)$. Thus relation (2.5) will be satisfied in its strict form, contradicting the condition $y \in B$. The assumption is false.

Proof of Theorem 4. Let $\bar{y} \in Q$. Assume that $y \neq \mathbf{0}$. Otherwise the linear manifold Y will be a subspace and,therefore, will have one y one support vector $\mathbf{0}$, in which case the theorem is true.

Let H be a finite set of vectors of the manifold Y whose convex hull contains the vector $\bar{y}$, and for any $y \in H$

$$J_+(y) \subset J_-(\bar{y}), \quad J_-(z) \subset J_-(\bar{y}). \tag{2.6}$$

It can be assumed initially that H is the one-element set containing the vector $\bar{y}$.

Suppose there is a vector $\tilde{y} \in H$ which is not a support vector of the manifold Y. Then there is a vector $b \in B$ such that

$$J(b) \subset J(\tilde{y}). \tag{2.7}$$

The vector $z = \tilde{y} - b$ will be non-zero and, according to (2.7), $J(z) \subset J(\tilde{y})$. Since relations (2.6) are satisfied for the vector $\tilde{y}$, from Lemma 2 we know that both sets

$$J_1 = \{j : \ z_j\tilde{y}_j > 0\}, \quad J_2 = \{j : \ z_j\tilde{y}_j < 0\},$$

will be non-empty. Hence, there are quantities

$$\alpha_1 = \min_{y \in Y_1} \frac{-\tilde{y}_j}{z_j}, \quad \alpha_2 = \min_{y \in Y_2} \frac{\tilde{y}_j}{z_j}. \tag{2.8}$$

The vector $\tilde{y}$ is a convex combination of the vectors

$$y^1 = \tilde{y} + \alpha_1 z, \quad y^2 = \tilde{y} - \alpha_2 z.$$

Also, according to (2.8)

$$\begin{aligned} J_+(y^1) \subset J_+(\tilde{y}), \quad J_-(y^1) \subset J_-(\tilde{y}), \\ J_+(y^2) \subset J_+(\tilde{y}), \quad J_-(y^2) \subset J_-(\tilde{y}). \end{aligned} \tag{2.9}$$

Thus relations (2.6) will also hold for the vectors y_1 and y_2.

We remove the vector $\tilde{y}$ from H and introduce y_1 and y_2 into it. The set H will then satisfy both of the required properties, as before.

Note that the relations

$$J(y^1) \subset J(\tilde{y}), \quad J(y^2) \subset J(\tilde{y}).$$

which follow from (2.9), are satisfied in their strict form. The set $J(y^1)$ does not contain an index for which the value α_1 is realized in (2.8). And the set $J(y^2)$ does not contain an index for which the value α_2 is realized. Thus, after a finite number of such replacements, the set H will contain only vectors with a minimum carrier. That is, the vector $\bar{y}$ will be represented by a convex combination of support vectors, as it was required to prove.

Proof of Theorem 5. It follows from the description of the linear manifold (1.5) and condition (1.6) that the carrier of any support vector will contain not fewer than $n - m$ indices. The carriers of two different support vectors must differ. Hence, the number of support vectors cannot be greater than the number of different selections of $n-m$ elements from the total number n. This also proves that the number of support vectors if finite and that the upper bound of their number is that given in the theorem.

3 Some applications and generalizations

1. *Boundedness of the solution sets. The use of this fact for proving the convergence of interior point algorithms to prove convergence* It follows from the above theorems that the solutions obtained for finding the point of a linear manifold nearest the origin of coordinates in any of the formulations of the problem in Section 1 will lie in a bounded set, a polygon formed by convex combinations of support vectors of the manifold. For any $p > 1$

$$P_p = P = PF \subset Q \subset \text{co } B. \tag{3.1}$$

We have proved special cases of these relations before. Thus, in [1] it was established that $P \subset \text{co } B$.

The boundedness of the set of Euclidean projections of the origin of coordinates on to a linear manifold was used in [1] and later works to prove the convergence of interior point algorithms. We will describe these methods for the linear programming problem:

$$\alpha \to \min, \quad Ax + \alpha b = d, \quad x \geq 0. \tag{3.2}$$

Here the $m \times n$ matrix A and vectors b, $d \in R^n$, are given, and the vector $x \in R^n$ and the value of α are unknown.

Suppose there is a vector $x^0 \in R^n_\oplus$ and a quantity α^0 which comprise an admissible solution of problem (3.2). The solution can be improved using the formulae

$$x^{k+1} = x^k + \lambda_k s^k, \tag{3.3}$$

$$\alpha^{k+1} = \alpha^k - \lambda_k, \quad k = 0, 1, ..., \tag{3.4}$$

where

$$s^k = \arg\min \, [\sum_{j=1}^{n} h_j^k (s_j^k)^2 : \; As = b], \tag{3.5}$$

$$\lambda_k = \gamma^k \min \, [\frac{-x_j^k}{s_j^k} : \; j \in J_-(s^k)] \tag{3.6}$$

for given $\gamma^k \in (0,1)$ and vector $h^k \in R^n_\oplus$. In particular, this vector can be defined by the rule

$$h_j^k = 1/(x_j^k)^p \tag{3.7}$$

for given $p \geq 1$. Algorithms with $p = 2$ were considered in [9] and subsequently reconsidered as modifications of Karmarkar's method [10-13].

It is relatively easy to prove that the process (3.3)-(3.6) converges whatever the method of choosing the vectors $h^k \in R^n_\oplus$, using the fact that the Euclidean norms of the projections of the origin of coordinates on to a linear manifold are bounded.

We will assume that the system $As = b$ is consistent (otherwise the solution (x^0,α^0) is optimal) and on all iterations the set $J_-(s^k)$ is non-empty (otherwise the objective function of problem (3.2) does not have a lower bound in the region of admissible solutions). The solutions (x^k,α^k) will be admissible on all iterations. If $\sum_{k=0}^{\infty}\lambda_k = \infty$, problem (3.2) does not have a solution (the objective function is not bounded below in the admissible region). But if $\sum_{k=0}^{\infty}\lambda_k < \infty$, it follows from this fact and the boundedness of the set of vectors s^k that the vectors x^k and values α^k will converge to

$$\bar{x} = x^0 + \sum_{k=0}^{\infty}\lambda_k s^k, \quad \bar{\alpha} = \alpha^0 - \sum_{k=0}^{\infty}\lambda_k,$$

which is an admissible solution of problem (3.2).

It was proved in [1], under certain conditions on the choice of the vector h^k, satisfied in particular by rule (3.7), that $(\bar{x},\bar{\alpha})$ will be optimal for non-degenerate problems. For algorithms with weighting factors (3.7), a proof for $p = 1$ is given in [1] which does not assume non-degeneracy.

Improvement of the relations and properties of sets. The theorems proved in Section 2 were announced in [3], together with a number of refinements. In particular, it is easy to show that the only difference between the set PF and the set Q might be that the former is not closed, whereas Q is: then

$$\mathrm{cl}PF = Q. \tag{3.8}$$

It can also be proved that, although they are not always convex, the set P, PF and Q prossess some of the properties of convex sets. Thus, they all have relatively interior points (points that are interior relative to the minimum linear manifold that contains those sets) and are connected.

It should be noted that it follows from Theorem 3 and relations (3.1) and (3.8) that the set cl P_2 contains B, Q and PF. This means that in the formulations considered in Section 1, any solution can be obtained to any accuracy by the method of lest squares and a choice of the weighting factors.

Approximation by a polygon and estimation of the ranges of variation of the components of solution vectors It follows from Theorem 3 and relations (3.1) and (3.8) that

$$\mathrm{co\ cl}\ P_p = \mathrm{co\ cl}PF = \mathrm{co}Q = \mathrm{co}B \tag{3.9}$$

for any $p > 1$. The polygon formed by convex combinations of the support vectors is the minimum polygon containing any of the sets under consideration.

On the basis of (3.9) one can, in particular, estimate the range of possible variations of the values of the components of the vectors obtained in the different formulations of the problem. This could be done by scanning a finite number of vectors: for $j = 1, 2, ..., n$, for any $p > 1$

$$\inf_{y\in P_p} y_j = \inf_{y\in PF} y_j = \min_{y\in Q} y_j = \min_{b\in B} b_j,$$

$$\sup_{y \in P_p} y_j = \sup_{y \in PF} y_i = \max_{y \in Q} y_i = \max_{b \in B} b_j$$

References

[1] Zorkaltsev V.I.,(1984) *A relatively interior point of optimal solutions*, Komi SC Akad. Nauk S.S.S.R., Syktyvkar.

[2] Zorkaltsev V.I., (1989) "Properties of Hölder projections", *In collected materials Methods of mathematical programming and software,* JMM UD RAS, Sverdlovsk.

[3] Batoeva I.V., Bedenkov A.R., Zorkaltsev V.I., Sadov S.L.,(1989) *Coordination of private forecasts in balance models.* Komi NTs UrO Akad. Nauk S.S.S.R., Syktyvkar.

[4] Zorkalsev V.I.,(1993) *The Method Liast Squares and Its Competitors.* Sibirskii Energetich. Inst. SB Rus. Akad. Sciences.

[5] Zorkaltsev V.I.,(1995) "The points of a linear manifold, nearest the origin of coordinates," *Comp. Maths. Math. Phys.,* v. 5.

[6] Zorkaltsev V.I.,(1995) *The method of least squares: geometrical properties, alternative approaches, applications* - "Nauka", Novosibirsk.

[7] Rockafellar R.T. (1970) *Convex Analysis.* Princeton University Press, Princeton.

[8] Gauss K.F.,(1957) *Selected Geophysical Essays.* Izd. Geof. Lit., Moscow.

[9] Dikin I.I., Zorkaltsev V.I.,(1980) *The Iterative Solution of Mathematical Programming Problems (Interior Point Algorithms).* Nauka, Novosibirsk.

[10] Adler I., Karmarkar N., Resende M.G.S and Veiga G.,(1989) "An implementation of Karmarkar's algorithm for linear programming" *Mathematical programming,* 44, 297-335.

[11] Bernes E.B.,(1986) "Variation of Karmarkar's algorithm for solving linear problems", *Reports of the IBM Boston Research Center* May, 1986.

[12] Monma C.L. and Morton A.J., (1987) *Computational experience with a dual affine variant of Karmarkar's methods for linear programming,* Bell Commun. Res. Morristown, NJ.

[13] Wei Zi Luan, (1987) "An interior point method for linear programming," *J. Comp. Math.* **10,** 342-350.

Optimization and Optimal Control, pp. 347 – 367
P.M. Pardalos, I. Tseveendorj, and R. Enkhbat, Editors

Some Remarks on D.C. Programming[1]

Alexander Strekalovsky (strekal@icc.ru)
Institute of System Dynamics and Control Theory SB of RAS,
Lermontov Str., 134, Irkutsk-33, 664033, Russia

Keywords: d.c. programming, global optimization, optimality conditions, global search algorithm.

1 Introduction

At the end of past millenium the needs of practice make turn specialists on optimization to consider a large field of nonconvex problems, in which there exist a lot of local extrema or even stationary (critical) points that are different from global solutions.

It is commonly known that in such problems the classical algorithms of convex programming have the ability of being trapped at a critical point, which may be very far from a global solution.

That is probably why, the Global Optimization took the way of Branch & Bounds methods, cutting method idea and so on [1–5].

Recall that these tools were initially invented into the bosom of Discrete Optimization and for Discrete Optimization (DO). Only after, the apparatus was transferred to the area of Continuous Non-Convex Optimization where it dominates the situation till now. These tools stay evidently the most applicable in the field, although the so called "dimension's curse" becomes more and more influential in practical applications. To our opinion this feature of the approach was visible from the very beginning.

Regardless the deficiency, almost all specialists in so-called Global Optimization (GO) were educated in the ideology of B&B. As a consequence, it is very difficult, often impossible to explain to the adepts the importance of Extremum Theory, the

[1]The investigation was carried out under financial support of RFBR (the grant number 01-01-00048)

theory of Convex Optimization Methods (COM), for instance, and the value of its Convergence Theorems in Optimization as a whole.

At the same time, the following four classes of non-convex are regarded by all scientists working in the field as principal objects of investigations of today [1,2].

a) d.c. minimization:

$$F(x) := g(x) - f(x) \downarrow \min, \quad x \in D, \tag{1.1}$$

where $g(\cdot)$ and $f(\cdot)$ are convex functions;

b) d.c. constraint problems:

$$\begin{aligned} h(x) \downarrow \min, \quad x \in S, \\ F(x) := g(x) - f(x) \geq 0, \end{aligned} \tag{1.2}$$

where $g(\cdot)$ and $f(\cdot)$ are as above and $h(\cdot)$ is continuous and S is convex in $\mathbb{R}^n$.

Two following problems arising in various applications turn out to be particular cases of (1.1) and (1.2) correspondingly:

c) convex maximization (concave minimization)

$$f(x) \uparrow \max, \quad x \in D, \tag{1.3}$$

where $f(\cdot)$ is convex;

d) reverse convex problem:

$$\begin{aligned} h(x) \min, \quad x \in S, \\ g(x) \geq 0, \end{aligned} \tag{1.4}$$

where $g(\cdot)$ is convex function.

This classification is obviously (almost) exhausting for Quadratic Optimization (excluding the cases when in (1.2) and (1.4) one has several inequality-constraints and its combinations).

Only this fact is sufficient to attract the attention of GO specialists to the problems (1.1)–(1.4). If we remember that the space $\mathcal{DC}$(C) of d.c. functions on a compact $C \subset R^n$ is dence in the space $\mathcal{C}$(C) of all continuous functions (in the uniform convergence topology), we see the possibility for applying the classical optimization ideology under condition that three following question have the positive answers.

First, is it possible to characterize a global solution to problems (1.1) and (1.2)? Second, may one construct a Global Search Algorithm (GSA) based on the characterization (Global Optimality Condition)? Third, does there exist a possibility to solve computationally a well-known problem and to get satisfactory results for problems of rather high size by means of so developed GSA?

At present, these positive answers have been got for all three questions [20].

2 Minimizimg sequences

Consider a d.c. minimization problem:

$$F(x) = g(x) - f(x) \downarrow \min, \quad x \in D, \tag{P}$$

where $g(\cdot)$ and $f(\cdot)$ are convex differentiable functions over an open domain Ω containing a convex set $D \subset R^n$.

First of all, let us remind of Global Optimality Conditions [11,15] in Problem (P).

Theorem 1 *If a feasible point $z \in D$ is a global solution to Problem (P), then*

$$\left.\begin{array}{c} \forall (y, \beta) : \; y \in D, \;\; \beta - f(y) = \zeta \triangleq F(z), \\ g(y) \le \beta \le \sup(g, D), \\ g(x) - \beta \ge \langle \nabla f(y), x - y \rangle \;\; \forall x \in D. \end{array}\right\} \tag{E}$$

If in addition the following condition takes place

$$\exists v \in D : F(v) > F(z) \triangleq \zeta, \tag{H}$$

then the condition (E) becomes sufficient for a feasible point z to be a global solution to Problem (P).#

The fact, that z is a global solution to (P) will be denoted below $z \in Sol(P)$.

Remarks. 3) Let us underline the connection of the condition (E) with Classical Extremum Theory. Actually, when $y = z$, from (E) it follows $\beta = g(z)$ and

$$g(x) - g(z) \ge \langle \nabla f(z), x - z \rangle \;\; \forall x \in D.$$

When $D = R^n$ and $g(\cdot)$ is differentiable it results the famous Fermat rule $\nabla f(z) = \nabla g(z)$ or $\nabla F(z) = 0$.

On the same way one can obtain Kuhn-Tucker Theorem (cf. [11]).

4) On the other hand, the conditions (E) inherit the classical Linearization idea. Actually, for each pair (y, β), $\beta = f(y) + \zeta$, $\zeta \triangleq F(z)$, in (P) one implements the linearization w.r.t. the function $f(\cdot)$ (as it was in Problem (5.1)), which generates the basic nonconvexity in (P). This linearization leads us to the problem:

$$g(x) - \langle \nabla f(y), x \rangle \downarrow \min \quad x \in D. \tag{2.1}$$

Since D is supposed to be convex, Problem (2.1) turns out to be convex, and one can apply standard convex programming methods for solving (2.1).

5) If $\overline{x} = x(y) \in D$ is a solution to (2.1), according to Theorem 2 one has to verify the inequality:

$$g(\overline{x}) - \langle \nabla f(y), \overline{x} \rangle \ge \beta - \langle \nabla f(y), y \rangle .$$

If the latter inequality is violated, then the feasible point $\overline{x} \in D$ is better than the point z under study because of convexity of $f(\cdot)$

$$g(\overline{x}) < \beta + \langle \nabla f(y), \overline{x} - y \rangle \leq \beta - f(y) + f(\overline{x}) = F(z) + f(\overline{x})$$

$$\text{or } F(\overline{x}) < F(z).$$

Thus, GOC (E) succeeds to so-called Algorithmic Property (AP) of Classical Optimality Conditions. It means that, if an optimality condition is violated, then there is a procedure to construct a better feasible point.

6) It is important to note that when $g \equiv 0$ from GOC(E) it follows well-known GOC for Convex Maximization Problem (5.8):

$$\left.\begin{array}{c} \forall y \in D, \quad f(y) = f(z), \\ \langle \nabla f(y), x - y \rangle \leq 0 \;\; \forall x \in D, \end{array}\right\} \tag{2.2}$$

what completely corresponds to the relation between the statements of the Problems (P) and (5.8).#

Now consider the following function:

$$\phi(z) = \sup_{x,y,\beta}\{\langle \nabla f(y), x - y \rangle + \beta - g(x) /$$

$$x, y \in D, \beta = f(y) + \zeta, \; g(y) \leq \beta \leq \sup(g, D)\} \tag{2.3}$$

Since $\beta = \beta_0 \triangleq g(z)$ with $y = z$, one has

$$0 = \beta_0 + \langle \nabla f(z), z - z \rangle - g(z) \leq \phi(z).$$

Therefore

$$\phi(z) \geq 0 \;\; \forall z \in D. \tag{2.4}$$

It can be readily seen, that GOC(E) may be expressed in the form:
if $z \in Sol(P)$, then $\phi(z) = 0$.

Theorem 2 *If a sequence $\{z^k\} \subset D$ is minimizing to Problem (P), then*

$$\lim_{k \to \infty} \phi(z^k) = 0. \tag{2.5}$$

If the following condition takes place

$$\exists v \in D, \;\; \exists \chi > 0 : F(v) \geq F(z^k) + \chi, \;\; k = 0, 1, 2, ... \tag{2.6}$$

then the optimality condition (2.5) becomes sufficient for the sequence $\{z^k\}$ to be minimizing.

Proof. 1) In the sequel we denote the fact, that $\{z^k\}$ is a minimizing sequence to (P) by the inclusion $\{z^k\} \in \mathcal{M}$. If $\{z^k\} \in \mathcal{M}$, i.e.

$$\lim F(z^k) = \inf(F, D),$$

then due to convexity of $f(\cdot)$ one has

$$\forall (y, \beta) : \ \beta - f(y) = F(z^k), \ \forall x \in D,$$

$$\langle \nabla f(y), x - y \rangle + \beta - g(x) \leq F(z^k) - F(x).$$

Therefore, according to (2.3) and (2.4) one obtains

$$0 \leq \phi(z^k) \leq \sup_{x \in D} [F(z^k) - F(x)] = F(z^k) - \inf(F, D).$$

Whence, it follows (2.5).

2) Now suppose, (2.5) and (2.6) are verified, but the sequence $\{z^k\}$ is not minimizing, i.e. for some $\gamma > 0$ we have

$$F(z^k) \geq \inf(F, D) + \gamma + \varepsilon \ \ (\varepsilon > 0).$$

There then exists a feasible point $u \in D$, such that

$$F(z^k) \geq F(u) + \gamma, \quad k = 0, 1, 2, ... \tag{2.7}$$

Denote $\zeta_k := F(z^k)$. Then (2.7) is equivalent to the fact, that

$$(u, g(u)) \notin epi(f(\cdot) + \zeta_k).$$

On the other hand, (2.6) means that

$$(v, g(v)) \in int \ epi(f(\cdot) + \zeta_k).$$

Therefore, one can find a number $\alpha_k, \ \ 0 < \alpha_k < 1$, such that the convex combination

$$(y^k, \beta_k) = \alpha_k (u, g(u)) + (1 - \alpha_k)(v, g(v))$$

belongs to the boundary of the set $epi(f(\cdot) + \zeta_k)$. In other words

$$\left.\begin{aligned} y^k &= \alpha_k u + (1 - \alpha_k) v, \\ \beta_k &= \alpha_k g(u) + (1 - \alpha_k) g(v) = f(y^k) + \zeta_k. \end{aligned}\right\} \tag{2.8}$$

Moreover, due to convexity of $g(\cdot)$ we have

$$g(y^k) \leq \beta_k \leq \sup(g, D).$$

Besides, it follows from (2.8)

$$\left.\begin{aligned} u &= \alpha_k^{-1}[y^k - (1-\alpha_k)v], \\ g(u) &= \alpha_k^{-1}[\beta_k - (1-\alpha_k)g(v)]. \end{aligned}\right\} \tag{2.9}$$

Further, it results from the definition (2.3) of the function $\phi(\cdot)$

$$\phi(z^k) \geq \left\langle \nabla f(y^k), u - y^k \right\rangle + \beta_k - g(u).$$

Taking into account (2.9) and convexity of $f(\cdot)$ the latter inequality takes the form

$$\phi(z^k) \geq \frac{1-\alpha_k}{\alpha_k}[g(v) - \beta_k] + \frac{\alpha_k - 1}{\alpha_k}\left\langle \nabla f(y^k), v - y^k \right\rangle \geq \frac{1-\alpha_k}{\alpha_k}[F(v) - \zeta_k].$$

Whence, due to the hypothesis (2.6) one has

$$\phi(z^k) \geq \frac{1-\alpha_k}{\alpha_k}\chi > 0.$$

Then, Optimality Condition (2.5) makes us conclude that, first, $\lim \alpha_k = 1$ and, second, in virtue of (2.8)

$$\lim y^k = u, \quad \lim \beta_k = g(u).$$

Therefore, one gets due to continuity of the function $f(\cdot)$

$$\lim f(y^k) = f(u).$$

On the other hand, in virtue of the equality $\zeta_k = \beta_k - f(y^k)$ we obtain

$$\lim \zeta_k = \lim F(z^k) = g(u) - f(u) = F(u).$$

The latter contradicts (2.7). So, under conditions (2.5) and (2.6) the sequence $\{z^k\}$ must be minimizing.

3 Theoretical minimization scheme

In this section we consider a theoretical scheme of global search in Problem (P), based upon Optimality Condition (2.5). Theorem 3 suggests that in order to decide whether a current point $z^k \in D$, $k = 0, 1, 2, ...$ is a solution to (P) or not, one has to calculate the function $\phi(z^k)$. Therefore, the following procedure seems to be too natural. Given a feasible point $z^k \in D$, the following point $z^{k+1} \in D$ is constructed according the rule:

$$\left.\begin{aligned} &\left\langle \nabla f(y^k), z^{k+1} - y^k \right\rangle + \beta_k - g(z^{k+1}) \geq \\ &\qquad \geq \Theta_k \phi(z^k) - \varepsilon_k; \end{aligned}\right\} \tag{3.1}$$

where

$$\left.\begin{array}{c}\beta_k = f(y^k) + \zeta_k, \;\; \zeta_k := F(z^k), \\ 0 < \Theta_- \le \Theta_k \le \Theta_+ \le 1, \\ \varepsilon_k > 0, \;\; k = 0,1,2,..., \; \sum\limits_0^\infty \varepsilon_k < +\infty. \end{array}\right\} \qquad (3.2)$$

So, the rule (3.1) means that we may calculate the function $\phi(z^k)$ (by means z^{k+1} and y^k), first, approximately and, second, partially (Θ_k may be strictly less than 1). Suppose, in addition, that the initial point z^0 verifies the condition:

$$\left.\begin{array}{c}\exists v \in D, \;\; \exists \chi > 0 : \\ F(z^0) \le F(v) - \chi - \sum\limits_0^\infty \varepsilon_k. \end{array}\right\} \qquad (3.3)$$

It is obvious, that the assumption (3.3) is not very burden or severe accordingly to the statement of Problem (P). It turns out, that the rule (3.1)–(3.2) allows to construct a minimizing sequence to Problem (P).

Theorem 3 *Assume, the objective function $F(\cdot)$ is bounded from below over the feasible set.*

Then any sequence $\{z^k\}$ constructed accordingly the rule (3.1)–(3.2) verifies Optimality Condition (2.5).

If in addition the assumption (3.3) holds, then $\{z^k\}$ turns out to be minimizing to Problem (P).

Besides, every cluster point of the sequence $\{z^k\}$ furnishes the infimum of $F(\cdot)$ over the feasible set D.

Finally, in the case, when D is closed, this cluster point turns out to be a global solution to Problem (P).

Proof. Due to the property (2.4), the rule (3.1) and the convexity of $f(\cdot)$ one has

$$-\varepsilon_k \le \Theta_k \phi(z^k) - \varepsilon_k \le \left\langle \nabla f(y^k), z^{k+1} - y^k \right\rangle + \beta_k - g(z^{k+1}) \le$$

$$\le f(z^{k+1}) - f(y^k) + \beta_k - g(z^{k+1}) = F(z^k) - F(z^{k+1}). \qquad (3.4)$$

In particular, from (3.4) it follows

$$F(z^{k+1}) \le F(z^k) + \varepsilon_k, \qquad (3.5)$$

i.e. the sequence $\{F(z^k)\}$ is "almost monotonically decreasing". Therefore (cf. [5], lemma 2.3) there exists a finite limit

$$\lim_{k\to\infty} F(z^k) \ge \inf(F, D) > -\infty,$$

and then (2.5) immediately follows from (3.4).

Now let show that (3.3) implies the regularity condition (3.6).

Then according to Theorem 3 Optimality Condition (3.5) becomes sufficient for the sequence $\{z^k\}$ being minimizing.

Actually, in virtue of (3.3) and (3.5) one has

$$F(v) - \chi - \sum_{s=0}^{\infty} \varepsilon_s \geq F(z^0) \geq F(z^1) - \varepsilon_0 \geq F(z^k) - \sum_{s=0}^{k-1} \varepsilon_s \geq F(z^k) - \sum_{s=0}^{\infty} \varepsilon_s,$$

and (2.6) is proven. The latter assertions of Theorem 4 are obvious. ♯

4 Global search strategy

We have advanced Global Search Approach based on GOC of Theorem 2 for the first time in [15]. Let now discuss some features of the proposed strategy taking into account Theorems 3 and 4.

The crucial question for an optimization problem is that whether a given feasible point z is global solution or not, and if not, how to construct another feasible point, which is better than this one of interest? In order to do it Theorem 2 and 3 suggest the considering the following problem:

$$\left.\begin{aligned} &\Psi(x,y,\beta) \triangleq \langle \nabla f(y), x-y\rangle + \beta - g(y) \uparrow \max_{(x,y,\beta)}, \\ &x \in D, \quad (y,\beta) \in R^{n+1} : \beta - f(y) = \zeta \triangleq F(z), \\ &g(y) \leq \beta \leq \sup(g,D). \end{aligned}\right\} \tag{4.1}$$

Theorem 4 and the rule (3.1)–(3.2) precise that at each iteration Problem (4.1) may be solved not only approximately, but in addition partially. In other words, one is not obliged to solve (4.1) as whole at every step of a Global Search Algorithm (GSA). One may "bite off" a part $\Theta_k (0 < \Theta_k < 1)$ of the problem. Nevertheless, the problem (4.1) seems to be rather difficult, may-be, comparable, to the initial problem (P) as to the hardness .

That is why we propose to decompose Problem (4.1) into a few simpler problems.

a) Given a number β, such that

$$\beta_- := \inf(g,D) \leq \beta \leq \beta_+ := \sup(g,D). \tag{4.2}$$

Then for the level surface $(\zeta \triangleq F(z))$

$$Y(\zeta,\beta) = \{y \in R^n : \ f(y) = \beta - \zeta\}$$

of the function $f(\cdot)$ construct some approximation

$$\mathcal{A}(\zeta,\beta) = \{y^1, ..., y^N / f(y^i) = \beta - \zeta, \ \ i = 1, ..., N\}$$

b) For each $y^i \in \mathcal{A}(\zeta, \beta)$ solve the partially Linearized Problem:

$$g(x) - \left\langle \nabla f(y^i), x \right\rangle \downarrow \min, \quad x \in D. \tag{4.3}$$

Let u^i be a global solution to (4.3).

c) Solve the so-called "Level" Problem for each $i = 1, ..., N$

$$\left\langle \nabla f(v), u^i - v \right\rangle \uparrow \max_v, \quad f(v) = \beta - \zeta. \tag{4.4}$$

Let w^i be a global solution to (4.4).

Notice, that in the quadratic case the problem (4.4) can be solved analytically [7,13,15].

c1) Sometimes, instead of solving the problem (4.4) one operates directly with the values $F(u^i)$ of the goal function of (P) at the solutions of Linearized Problems (4.3). In order to do it, one has to estimate apriori the volume and the hardness of computations and to choose a corresponding althernative.

d) As a result of the solvings of three basic subproblems one obtains the number

$$\begin{array}{c} \eta(\beta, \zeta) = \langle \nabla f(w^j), u^j - w^j \rangle + \beta - g(u^j) = \\ \max_i \{ \langle \nabla f(w^i), u^i - w^i \rangle + \beta - g(u^i) \}, \end{array} \tag{4.5}$$

where the number j is any (for instance, the first) index at which the maximum is attained.

Recall, if $\eta(\beta, \zeta) > 0$, then the feasible point u^j turns out to be better, than z:

$$F(u^j) < F(z).$$

Having $\eta(\beta, \zeta)$ for all $\beta \in [\beta_-, \beta_+]$ one is able to undertake the unidimentional global maximization over $[\beta_-, \beta_+]$.

Besides notice, it suffices to get $\eta(\beta, \zeta) > 0$ only for a one $\beta \in [\beta_-, \beta_+]$ in order to move to a better feasible point.

Now let us put all above suggestions in a more algorithmic form.

Let given a feasible point $x^0 \in D$ and a number sequence $\{\varepsilon_k\}$, $\varepsilon_k > 0, \varepsilon_k \downarrow 0 \ (k \to \infty)$.

4.1. Global Search Strategy.

Step 0. Set $k := 0, \quad x^k := x^0$.

Step 1. Starting at $x^k \in D$ by means of some Local Search Method obtain ε_k-stationary point $z^k \in D$. Set $\zeta_k := F(z^k)$.

Step 2. Choose some $\beta \in [\beta_-, \beta_+]$. In particular one can start the unidimentional search w.r.t. β with $\beta_0 = g(z^k)$.

Step 3. Construct some approximation

$$\mathcal{A}_k(\beta) = \left\{y^1, ..., y^{N_k} / \; f\left(y^i\right) = \beta - \zeta_k, \;\; i = 1: \;\; N_k = N_k(\beta)\right\}.$$

Step 4. Introduce the family of indices

$$I_k = I_k(\beta) = \{i \in \{1, ..., N_k\} / \; g(y^i) \leq \beta\}$$

Step 5. For every $i \in I_k$ find a global ε_k- solution u^i of Linearized Problem (4.3)

$$g(u^i) - \left\langle \nabla f(y^i), u^i \right\rangle - \varepsilon_k \leq \inf_x \{g(x) - \left\langle \nabla f(y^i), x \right\rangle / x \in D\}. \tag{4.6}$$

Step 6. For every $i \in I_k$ find an ε_k-solution w^i, $\;f(w^i) = \beta - \zeta_k$, of Level Problem (4.4) corresponding to u^i:

$$\left\langle \nabla f(w^i), u^i - w^i \right\rangle + \varepsilon_k \geq \sup_y \{\left\langle \nabla f(y), u^i - y \right\rangle / f(w^i) = \beta - \zeta_k\}. \tag{4.7}$$

Step 7. Set $\eta_k(\beta) := \eta(\beta, \zeta_k)$ according to (4.5).

Step 8. If $\eta_k(\beta) > 0$, then set $\;k := k + 1, \;\; x^{k+1} := u^j\;$ and loop to Step 1.

Step 9. If $\eta_k(\beta) \leq 0$, set $\beta := \beta + \triangle\beta$ and go to Step 3.

(A number $\triangle\beta$ can be found by one of the standart methods of unidimentional maximization.)

Step 10. If $\eta_k(\beta) \leq 0$ for all $\beta \in [\beta_-, \beta_+]$ (i.e unidimentional search w.r.t. β over the segment $[\beta_-, \beta_+]$ is finished) and $\varepsilon_k \leq \delta$, (where δ is a given tolerence), then STOP.

If $\varepsilon_k > \delta$, set $k := k + 1, \;\; x^k := z^{k-1}$ and loop to Step 1.

#

Remarks. 7) In order to be, in a certain extend, theoretically based let us do the following assumptions.

(HL). *On Step 5 it is possible to solve approximately, but globally the Linearized Problem (4.3). More precisely, for every numbers ε, β and ζ such that $\varepsilon > 0$, $\beta \in [\beta_-, \beta_+]$, $\zeta \in F(D)$, and for every y^i, $\;f(y^i) = \beta - \zeta$, one can find a feasible point u^i verifying (4.6) with $\varepsilon_k = \varepsilon$.*

(HY). *On Step 6 the Level Problem (4.4) is approximately, but globally solvable, that is under the same assumptions as in (HL) and for any $u^i \in D$ one can find w^i, $\;f(w^i) = \beta - \zeta$, verifying (4.7) with $\varepsilon_k = \varepsilon$, $\;\zeta_k = \zeta$.*

8) Due to the describing of Step 1 it can be readily seen that the sequence $\{z^k\}$ generated by the algorithm above is a sequence of ε_k-stationary points.

9) On Steps 1 and 5 one suppose to apply standard methods of convex optimization. This property of the algorithm above can be viewed as advantageous, since it makes us to use standard software libraries. Note only, that they must be rather fast, because one uses them several times at each iteration.

10) It is easy to see, that the above method is not an algorithm in a common sense, because, for instance, one don't precise what local search algorithm one use on Step 1 or what method we have to use for solving the Linearized Problem or the Level Problem.

This process can be viewed as a principal algorithm [4] or as a strategy [13-18]. However, this method, it seems to us, describes in a fully adequate manner the strategy of Global Search based upon GOC of Theorem 2 and following from Theorems 3 and 4. #

Clearly, the way of solving the subproblems b) and c), or what is the same, the way of choosing the solution methods for the Linearized Problem (4.3) and the Level Problem (4.4) are very influent to the success of Global Search, but in a sense standard, "already seen" [2,4,5].

On the other hand, the subproblem a) or a constructing of a "good" approximation on Step 3 of Principal Algorithm is apparently of the paramount importance from the view point of real global search.

Together with the unidimentional maximization w.r.t. β the approximation problem seems to be very difficult, taught, and, of cause, unprecedented, "never seen before".

As a consequence, it gives rise to unforeseen difficulties and novel problems.

In the same time, it is obvious, that even a satisfactory solving of these new problems has a deciding impact on the Global Search viewed as a whole [10, 13-19].

Actually, as our computational experience [7-16] shows, a choice of a such approximation turns out to be a crucial moment to escape out of a stationary point, provided that all other components of Global Search are made out in a skilled manner. Therefore, this is the most important element determining the unified success of Global Search.

5 Local search method

Suppose also, that $g(\cdot)$ is bounded from below and from above on D. Let investigate the following procedure which generates a sequence $\{x^s\} \subset D$ of feasible points. Suppose, it is known an initial feasible point $x^0 \in D$. If some feasible point $x^s \in D$ is given, then the following element $x^{s+1} \in D$ is constructed as an approximate solution of (partially) Linearized (at the point x^s) Problem:

$$g(x) - \langle \nabla f(x^s), x \rangle \downarrow \min, \quad x \in D. \tag{5.1}$$

Due to convexity of Problem (5.1), one can set that at the point x^{s+1} the classical necessary and sufficient optimality condition takes place:

$$\inf_{x\in D}\left\langle \nabla g(x^{s+1}) - \nabla f(x^s), x - x^{s+1}\right\rangle \geq -\delta_s, \tag{5.2}$$

where $\{\delta_s\}$ is a number sequence, such that

$$\delta_s > 0, \quad s = 0, 1, 2, \ldots \ \sum_{s=0}^{\infty} \delta_s < +\infty. \tag{5.3}$$

Notice, that the linearization was performed as above w.r.t. to the function $f(\cdot)$ which generates the nonconvexity in Problem (P). The point x^s verifying (5.2) with $x^{s+1} = x^s$ will be called below approximately (δ_s) stationary or critical.

Theorem 4 *Let the goal function $F(\cdot) = g(\cdot) - f(\cdot)$ in Problem (P) is bounded from below on D, $g(\cdot)$ and $f(\cdot)$ are continuously differentiable over an open set $\Omega \subset R^n$, containing D. Then the sequence $\{x^s\}$ generated by the method (5.2) verifies the following condition*

$$\lim_{s\to\infty}[\inf_{x\in D}\left\langle \nabla g(x^{s+1}) - \nabla f(x^s), x - x^{s+1}\right\rangle] = 0. \tag{5.4}$$

Besides, for any cluster point x_* of the sequence $\{x^s\}$ the classical necessary optimality condition of local minimum in Problem (P) takes place:

$$\langle \nabla g(x_*) - \nabla f(x_*), x - x_*\rangle \geq 0 \ \ \forall x \in D. \tag{5.5}$$

Finally, in the case, when the feasible set is closed, this cluster point x_* turns out to be stationary (or critical) to Problem (P), i.e. x_* verify (5.5).

Proof. Due to convexity of $g(\cdot)$ and $f(\cdot)$ it follows from (5.2)

$$-\delta_s \leq \inf_{x\in D}\left\langle \nabla g(x^{s+1}) - \nabla f(x^s), x - x^{s+1}\right\rangle \leq$$

$$\leq \left\langle \nabla g(x^{s+1}), x^s - x^{s+1}\right\rangle + \left\langle \nabla f(x^s), x^{s+1} - x^s\right\rangle \leq$$

$$\leq g(x^s) - g(x^{s+1}) + f(x^{s+1}) - f(x^s) = F(x^s) - F(x^{s+1}). \tag{5.6}$$

So one has the number sequence $\{F(x^s)\}$ such that

$$F(x^{s+1}) \leq F(x^s) + \delta_s. \tag{5.7}$$

Since $F(\cdot)$ is bounded from below and (5.3) is verified, there exists a finite limit:

$$\lim_{s\to\infty} F(x^s) = F_*$$

due to Lemma 2.3 from [5].

Then the condition (5.4) follows from (5.6). If now $x^s \to x_*$, then (5.4) implies (5.5) because $\nabla g(\cdot)$ and $\nabla f(\cdot)$ are continuous. The latter assertion is obvious. #

Remarks. 1) It can be readily seen that the method (5.1)–(5.2) is not an algorithm in a common sense, because it is not clear how to solve Linearized Problem (5.1), which is convex. So this method (5.1)–(5.2) can be viewed as a principal algorithm [4] or a strategy, or conseptual algorithm, like the method (strategy) of feasible directions. On the other hand, this property can be estimated as advantageous because one can choose a suitable algorithm from the standard library of programs. As our first computational experiments certify, the effectiveness of proposed method vigorously depends on the choice of algorithm for solving the (partially) Linearized Problem [7,8,13-19].

2) It is easy to see that when $g(x) \equiv 0$ Problem (P) transforms itself into well-known problem of convex maximization (or concave minimization)

$$f(x) \uparrow \max, \quad x \in D. \tag{5.8}$$

Respectively, the method (5.1)–(5.2) is transmuted into the already known local search method [15] for (5.8). Given $x^s \in D$, x^{s+1} is sought as a global (may be, approximative) solution of the following linearized problem:

$$x^{s+1} \in D: \quad \langle \nabla f(x^s), x \rangle \uparrow \max, \quad x \in D. \tag{5.9}$$

Since the method (5.9) is a particular case of Method (5.1)–(5.2), then the convergence theorem for Method (5.9) follows from this one for Method (5.1)–(5.2) that completely corresponds the relation between the statements of the Problem (P) and (5.8).

3) One can find several publications (see, for example [22-24] and references in [24]) devoted to special local search methods using linearization w.r.t. the basic nonconvexity for convex maximization (concave minimization) and reverse convex problems.

But not any of the methods gives the reponces to the following questions.

i) Is it possible to avoid the compactness assumption for feasible sets?

ii) What happens to the methods if the auxiliary problems are solved approximately?

iii) Are the new methods converging and what are the properties of a cluster point, if one exist?

iv) What is the stopping criterion of the modified method, and can one derive it from the convergence theorem?

v) How the solving accuracy of the original problem must be accorded with this one of auxiliary problem?

vi) What do we get when the stopping criterion is satisfied approximately?

It can be readily seen that the proof of Theorem 4 gives the positive answer to the first question and characterize the convergence (see (5.4)) and the cluster point (see (5.5)) of the modified method.

In addition, answering to iv) (5.6) gives two stopping criterions:

$$\left\langle \nabla g(x^{s+1}) - \nabla f(x^s), x^s - x^{s+1} \right\rangle \leq \tau, \tag{5.10}$$

or

$$F(x^s) - F(x^{s+1}) \leq \tau, \tag{5.11}$$

where τ is a given solving accuracy.

If the solving accuracy for Linearized Problem δ_s (see (5.2)) verifies $\delta_s \leq \varepsilon$, then due to (5.6) one gets

$$| \inf_{x \in D} \left\langle \nabla g(x^{s+1}) - \nabla f(x^s), x - x^{s+1} \right\rangle | \leq \varepsilon, \tag{5.12}$$

The latter inequality with $\delta_s \leq \varepsilon$ gives the answers to fifth and sixth questions. Another variant of Linearization method for d.c. minimization problem (P) is considered in [21].

6 An overview on d.c. constraint's optimization

Let us consider the problem (1.2)

$$\left.\begin{array}{r} h(x) \downarrow \min, \quad x \in S, \\ F(x) \triangleq g(x) - f(x) \geq 0; \end{array}\right\} \tag{$P1$}$$

where h, g and f are convex functions and $S \subset R^n$ is a closed convex set. It was noted above that well-known reverse convex problem:

$$h(x) \downarrow \min, \quad x \in S, \quad g(x) \geq 0; \tag{PR}$$

is the particular case of $(P1)$, when $f \equiv 0$.

A global solution to $(P1)$ can be characterized by the GOC [19] which have a relation to Classical Extremum Theory.

On the other hand, these GOC use the linearization w.r.t. the basic nonconvexity (namely the function $g(\cdot)$ in $(P1)$.

Finally, as all GOC [6, 7, 9, 11, 12, 19–21], these ones for $(P1)$ possess so-called Algorithmic Property (AP) the development of which can lead us to a Global Search Strategy.

But the difference of GOC for (P) and these ones for $(P1)$ consists in the fact that for the latter's the distance between the necessary and sufficient OC is considerably more large than in the case of d.c. minimization. Let us show it.

Suppose, the basic constraint is "essential" [1–3],i.e.

$$(G): \qquad Sol(P1) \cap \{x \mid F(x) > 0\} = \emptyset. \tag{6.1}$$

Then the necessary OC take following form.

Theorem 5 *Let assumption (6.1) take place, the point z be a global solution to $(P1)$ $(z \in Sol(P1))$ and $F(z) = 0$. Then*

$$(E_0): \qquad \left.\begin{array}{c} \forall (y,\beta): \ \beta = g(y), \quad \forall y^* \in \partial g(y) \\ f(x) - \beta \geq \langle y^*, x - y \rangle, \quad \forall x \in S, \ h(x) \leq h(z). \end{array}\right\} \tag{6.2}$$

In order to transmute (E_0)–(6.2) in sufficient OC let us introduce two supplimentary assumptions (S is convex)

$$(G): \qquad \exists v \in S: \ F(v) < 0 \tag{6.3}$$

$$(H): \qquad \left.\begin{array}{c} \forall y \in S: \ g(y) = f(y) \ (i.e.\ F(y) = 0), \\ \exists y^* = y^* \in \partial g(y), \ \exists p = p(y, y^*) \in S, \\ f(p) - f(y) < \langle y^*, p - y \rangle . \end{array}\right\} \tag{6.4}$$

Now one can give the sufficient OC .

Theorem 6 *Let assumption (6.3) and (H)-(6.4) be satisfied, the function $h(\cdot)$ be continuous and $F(z) = 0$. Then, the condition*

$$(E): \qquad \left.\begin{array}{c} \forall (y,\beta): \ g(y) = \beta \geq f(y), \\ \inf(f,S) \leq \beta \leq \sup(f,S), \\ \exists y^* \in \partial g(y): \ \forall x \in S, \ h(x) \leq h(z), \\ f(x) - \beta \geq \langle y^*, x - y \rangle , \end{array}\right\} \tag{6.5}$$

becomes sufficient for z to be a global solution to $(P1)$.

#

Remarks . 1) Firstly, look at the difference between the conditions (E_0)–(6.2) and (E)–(6.5) due to the additional constraints on β

$$\beta \geq f(y), \ \inf(f,S) \leq \beta \leq \sup(f,S). \tag{6.6}$$

The two last inequality help us the immediate derivation from (E_0) of OC for the reverse-convex problem (1.4) that is the particular case of (1.2)–$(P1)$. Thus, if $f \equiv 0$, then (6.6) implies $\beta = 0$, and from (E)-(6.5) the well-known GOC for (1.4) immediately follows.

2) The condition (6.3) is rather natural, since it means that there exist points in S that are not feasible with respect to the constraint $F \geq 0$. Otherwise, the constraint would be senseless.

3) Clearly, the assumption (H)-(6.4) means that as soon as we remove the constraint $h(x) \leq h(z)$ from the condition (E)-(6.5), it becomes possible to violate the basic inequality

$$f(x) - \beta \geq \langle y^*, x - y \rangle$$

in the condition. So the assumption (H) reminds us of the normality conditions (e.g., the Slater and Mangasarian-Fromovitz conditions[3–5]). As well-known, these conditions ensure the nontriviality ($\lambda_0 \neq 0$) of Lagrange multiplier corresponding to the goal function in KKT-theorems. If $\lambda_0 = 0$ the Lagrange multiplier rule becomes inconsistent, since it does not characterize the solution to the problem of interest any more but rather describe a property of the constrains (for example, a linear dependence of the constraint's gradients, etc.)

So, one concludes that the objective function must be involved in OC in an influent form such that the necessary conditions could be transformed into sufficient ones. Remember the case, when the Slater conditions in convex problems make it possible to transform John's necessary conditions into KT sufficiency theorem (see [1–5]).

The situation with condition (E)–(6.5) and Assumption (H)–(6.4) is similar. Indeed, if the constraint $h(x) \leq h(z)$ is "not important" or, more precisely, (H) is violated at a point $y \in S : F(y) = 0$; i.e. the inequality

$$f(x(y)) - f(y) \geq \langle y^*, x(y) - y \rangle$$

holds, where $x(y)$ is a solution to Linearized Problem (LP)

$$(LP): \qquad \begin{array}{c} \Psi(x) := f(x) - \langle y^*, x \rangle \downarrow \min, \\ x \in S, \;\; h(x) \leq h(z) \end{array} \tag{6.7}$$

without the inequality $h(x) \leq h(z)$, then this means that OC (E) holds without the goal function $h(\cdot)$. In other words, it would be nonsense if OC (E) "did not pay attention" to the goal function $h(\cdot)$. Fortunately, this is not the case. #

So, the GOC for d.c. constraints optimization problems differ from those ones for d.c. minimization by several particular features. On the other hand, one can see certain unsolved questions for d.c. constraints problems.

For instance, they are

1) Optimality Conditions for minimizing sequences.

As a development of Theorem 6 we expect to prove a result similar to Theorem 2 and using the function

$$\phi(z) := \sup_{(x,y)} \{ \langle \nabla g(y), x - y \rangle + \beta - f(x) \mid$$

$$f(y) \leq \beta = g(y) \leq \sup(f, S), \; x \in S : \; h(x) \leq h(z) \}.$$

Note, that for reverse convex problem (1.4) the pertinant result is got and is going to appear. This fact yields a supplementary reason to get the desired assertion.

2) *Theoretical GS scheme.* Regardless the only sufficient (since necessity is not proven) character of the rules like (3.1), it would be curious to obtain some theoretical Global Search scheme, following from the preceding (supposed) GOC for minimizing sequences and having the same character as the rule (3.1)–(3.2).

Naturally, one supposes to get the corresponding result on the convergence of GS scheme.

3) Basing on Theorem from [19] and probable results from 1) and 2), one would be able to propose the principal Global Search Strategy having as a particular case $\mathcal{R}$-strategy for the reverse convex problem (PR). Notice, that the latter principal GS algorithm is not only advanced but has already been tested [13] and used for solving the well-known integer programming problem - so-called knapsack problem [14].

4) It will be very difficult to prove the convergence theorem for Global Search Strategy in (PR), but nevertheless we have to do it as an indispensable step for constructing the theory of solving the d.c. constraint problems.

Note that the appropriate investigations for reverse convex problems were undertaken in [6,13,14].

5) As an also indispensable step of researches on d.c. constraints problems, $\mathcal{R}$-strategy has to undergo the testings on some well-known examples.

It may happen, that it will be a rather difficult problem to find convenient tests examples. That's why we ask the assistance of all informed researchers.

As noted above, the similar testing in a reverse convex problem was carried out in [13,14].

6) As a final point, may-be, it would be interesting to find a d.c. constraint problem of real practice, say, quadratic or combinatorial, and to undertake an attempt to solve this problem developing each step of $\mathcal{R}$-strategy taking into account the nature and all the features of the concrete problem, as it was carried out in [17].

In the next section we give some observation on special local search methods for reverse convex problems and we do some prognostics about such methods for d.c. constraint problems.

7 Local Search methods for d.c. constraint's problems

After the reachable references one can conclude that, apparently, the first variants of Linearization Methods for reverse convex problems were proposed by Rosen J.B. [22] in 1966 for problems with several inequality constraints:

$$\left.\begin{array}{l} h(x) \downarrow \min, \quad x \in S, \\ g_i \geq 0, \quad i = 1, \ldots, m. \end{array}\right\} \tag{7.1}$$

where S is a compact convex set in $I\!R^n$ and $g_i \in C^2(\Omega)$, $S \subset \Omega$, $\Omega \subset I\!R^n$ is open set. $h_i, g_i, i = 1, ..., m$ are convex functions. The idea of the method consists in linearization w.r.t the basic nonconvexity about the current point and the consecutive solvings of Linearized Problems.

This means that for given $x^s \in S, g(x^s) \geq 0$, one take for the next iteration x^{s+1} an exact global solution to the following problem:

$$\left.\begin{aligned} h(x) \downarrow \min, \quad x \in S, \\ g_i(x^s) + \langle \nabla g_i(x^s), x - x^s \rangle \geq 0, \quad i = 1, ..., m; \end{aligned}\right\} \tag{7.2}$$

which is evidently convex.

Investigating the properties of the set-valued mapping $y \to \Gamma(y)$, where

$$\Gamma(y) = \{x \in S \mid g_i(y) + \langle \nabla g_i(y), x - y \rangle \geq 0, \quad i = 1, ..., m\} \tag{7.3}$$

J.B.Rosen proved the following result.

Theorem 7 (22) *i) Under the assumptions from above every vector of the sequence* $\{x^s\}$ *is in* D

$$D \triangleq \{x \in S \mid g_i(x) \geq 0, \quad i = 1, ..., m\}.$$

The corresponding numerical sequence $\{h(x^s)\}$ *is monotonously decreasing. The sequence* $\{x^s\}$ *contains a subsequence converging to a point* $x_* \in D$ *such that*

$$\inf(h, D) \leq h(x_*) \leq h(x^s), \quad s = 0, 1, ... \tag{7.4}$$

and

$$h(x_*) = \min(h, \Gamma(x_*)). \tag{7.5}$$

ii) If x_* *is a limit point of* $\{x^s\}$, *then* x_* *is the global minimum of Linearized problem about the point* x_*, *i.e. (7.2) with* $x^s = x_*$. *Furthermore, the optimality conditions (the KKT-necessary conditions) which must be satisfied at a global solution to (7.1) are, in fact, satisfied at* x_*.

Further, in 1970 R.Meyer studied in details the properties of the mapping $\Gamma(\cdot)$ and investigated the convergence of the Rosen's method under more general assumptions. But the result of Theorem 7 stay, generally speaking, without changes.

Later, in 1977, B.Bulatov also presented his version of Rosen's method and gave almost the same convergence theorem. Nevertheless, all the questions from Remark 3 in Section 5 stay without answers for modified Rosen's method (MRM), where the auxiliary problem (7.2) is solved approximatively.

We studied MRM and have proved a convergence theorem similar to Theorem 7. Besides we have got some concordance condition between the solving accuracies for problem 7.1 and 7.2.

On the other hand, we proposed another Local Search Method based on Theorem 5 and 6 and using the idea of the famous Tuy's theorem [1] for reverse convex problems. In this method one considers the following Linearized Problem ($m = 1$):

$$\left.\begin{aligned} \langle \nabla g(x^s), x \rangle \uparrow \max, \\ x \in S, h(x) \leq \eta_s := h(x^s) \end{aligned}\right\} \tag{7.6}$$

and for the next iteration x^{s+1} one takes an approximate solution to (7.6).

We have also investigated the features of the method and established a convergence theorem to a KKT-point. Moreover, in order to compare MRM and the latter one we have carried out a number of computational tests and got rather interesting conclusions.

References

[1] Horst R., and Tuy H. (1993), *Global Optimization Deterministic Approaches*, 2nd revised edition, Springer Verlag, Heidelberg.

[2] Horst R., Pardalos P. and Thoai N. (1995), *Introduction to Global Optimization*, Kluwer Academic Publishers, Dordrecht/Boston/London.

[3] *Handbook of Global Optimization* (1995), eds. Horst R. and Pardalos P., Kluwer Academic Publishers, Dordrecht/Boston/London.

[4] Polak E. (1971), *Computational methods in optimization. An unified approach.* Academic Press. New-York - London.

[5] Vasilyev F.P. (1988), *Numerical Methods of Extremum Problem's Solving.* Nauka, Moscow.

[6] Strekalovsky A. (1993), *Extremal Problems on complements of a convex sets* Translated from Kibernetika i Sistemnyi Analiz, N 1, 88–100, Plenum Publishing Corporation.

[7] Strekalovsky A.S. (1993), *The search for a global maximum of a convex functional on an admissible set*, Comput. Math. and Math. Physics, **33**, 315–328, Pergamon Press.

[8] Strekalovsky A.S. (1994), *On global search based on global optimality condition,* System Modelling and Optimization, Proceedings of the 16-th IFIP-TC7 Conference, Compiegne, France, July, 5-9, 1993, in Lectures Notes in Control and Information Sciences, Springer-Verlag, 261–267.

[9] Strekalovsky A.S. (1995), *On global maximum of a convex terminal functional in optimal control problems,* Journal of global optimization, **7**, 75–91.

[10] Strekalovsky A.S. and Vasiliev I.L. (1997) *On global search for non-convex optimal control problems,* Developments in global optimization, In: Im. Bomze, Tiibor Csendes, Reiner Horst and Panos M. Pardalos (eds) *Nonconvex Optimization and its Applications*, Kluwer Academic Publishers, 121–133.

[11] Strekalovsky A.S. (1997) *On Global Optimality Conditions for D.C. Programming Problems*, Irkutsk State University, seria "Optimization and Optimal Control", 62p.

[12] Strekalovsky A.S. (1998) *Global optimality conditions for nonconvex optimization,* Journal of Global Optimization, **12**, 415–434.

[13] Strekalovsky A.S. and Tsevendorj I. (1998), *Testing the $\Re$-strategy for a Reverse Convex Problem*, Journal of Global Optimization, **13**, 61–74.

[14] Kuznetsova A.A., Strekalovsky A.S. and Tsevendorj, I. (1999), *An approach to the solution of integer optimization problem*, Comput. Math. and Math. Physics, **39**, 6–13.

[15] Strekalovsky A.S., Kuznetsova A. . (1999), *On the convergence of a Global Search Algorithm for convex maximization over a feasible set.* Izvestia Vysshih Uchebnyh Zavedenii, seria "Matematika". 12.

[16] Strekalovsky A.S. (2000), *One way to Construct a Global Search Algorithm for d. c. Minimization problem*, In: Di Pillo, G., Giannessi, F. (eds.), *Nonlinear Optimization and Related Topics*, vol. 36, Applied Optimization Series, Kluwer Academic Publishers, 429–443.

[17] Kuznetsova A. ., Strekalovsky A.S. (2001), *On solving the Maximum Clique Problem.* Journal of Global Optimization, **21(3)**, 265–288.

[18] Strekalovsky A.S., Kuznetsova A., Yakovleva T.V. (2001), *On computional solving the problem of nonconvex optimization.* Siberian Journal of Computational Mathematics, **4(2)**, 185-199 (in Russian).

[19] Strekalovsky A.S. (2001), *Extremal problems with D.C. constraints*, Comput. Math. and Math. Physics, **41**, 12, 1742–1751.

[20] Strekalovsky A.S. (2002) *Minimization of difference of two convex functions*, Irkutsk State University, seria "Optimization and Optimal Control", 58 p.

[21] Strekalovsky A.S. (2002) *Minimization of the difference of two convex functions*, Comput. Math. and Math. Physics, **42(12)** (to appear).

[22] Rosen J.B. (1966) *Iterative solution of nonlinear optimal control problems*, SIAM J. on Control, **4**, 223–244.

[23] Meyer R. (1970) *The validity of a family of optimization methods*, SIAM J. on Control, **8**, 4–54.

[24] Bulatov V.P. (1977) *Embedding methods in optimization problems*, Nauka, Novosibirsk (in Russian).

[25] Tsevendorj I. (2001), *Piecewise-Convex Maximization Problems: Global Optimality Conditions.* Journal of Global Optimization, **21**, 1–14.

[26] Fortin D., Tsevendorj I. (2002), *Piecewise-Convex Maximization Problems: Algorithm and Computational Experiments.* Journal of Global Optimization, **42**, 61–77.

